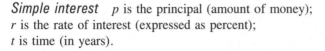

Simple interest p is the principal (amount of money); r is the rate of interest (expressed as percent); t is time (in years).

$$I = prt$$

Circle   The diameter is d; the radius is r.
Use the value $\pi = 3.14$.

Diameter	$d = 2r$
Circumference	$C = 2\pi r$ or $C = \pi d$
Area	$A = \pi r^2$

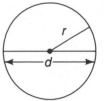

Right circular cylinder   The radius, r, is that of the top or bottom circle. The height is h.

Surface area	$S = 2\pi r^2 + 2\pi rh$
Volume	$V = \pi r^2 h$

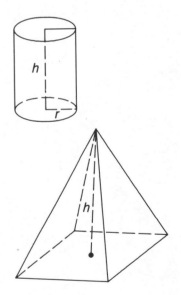

Right pyramid   The area of the base is B.
The height is h.

Volume	$V = \dfrac{1}{3}Bh$

Temperature C is the Celsius temperature.
F is the Fahrenheit temperature.

$$C = \frac{5}{9}(F - 32)$$

$$F = \frac{9}{5}C + 32$$

Beginning Algebra • 5th

Beginning Algebra • 5th

Margaret L. Lial
American River College

Charles D. Miller
American River College

HarperCollins*Publishers*

To the Student

If you need further help with algebra, you may want to get copies of both the *Study Guide* and the *Student's Solutions Manual* that accompany this textbook. The additional worked-out examples and practice problems that these books provide can help you study and understand the course material. Your college bookstore either has these books or can order them for you.

Cover: a detail from *Rhythm/Color: Morris Men,* copyright 1985 by Michael James, Somerset Village, Massachusetts. A quilt, pieced and quilted by machine in cottons, cotton satins, and silks, which first appeared in *The Art Quilt,* published by The Quilt Digest Press, San Francisco.

Lial, Margaret L.
   Beginning algebra / Margaret L. Lial, Charles D. Miller. — 5th ed.
     p.   cm.
   Includes index.
   ISBN 0-673-18808-6
   1. Algebra. I. Miller, Charles David,   II. Title.
QA152.2.L5 1988
512.9—dc19   87–17930
                                    CIP

ISBN 0-673-18808-6

   8-RRC-9291

Preface

The fifth edition of *Beginning Algebra* continues to address the needs of both the student and the instructor. Clear explanations keyed to objectives, numerous examples detailing the structure of each problem, carefully graded exercise sets keyed to examples, interactive tutorial software, a study guide, and a solutions manual are designed for the student. For the instructor, a complete instructional package is provided, including alternate forms of tests, additional test items keyed to objectives, computer-assisted testing, a complete answer key, complete solutions to all exercises, audiotapes, and videotapes.

Key Features **Keyed Objectives** Each section begins with a list of skills that the student should learn in that section. The objectives are keyed to appropriate discussions in the text by numbered symbols such as **1**.

Examples About 675 worked-out examples clearly illustrate concepts and techniques. Second color is used to identify pertinent steps within examples and to highlight explanatory side comments. For clarity, the end of each example is indicated with the symbol ✛.

Word Problems A problem-solving approach gives students early and repeated experience in solving applied problems. A list of steps for solving word problems is first presented in Chapter 2, yet even Chapter 1 has some simple word problems. Throughout the text students are given practice translating words into algebraic symbols. In this way, students see word problems early, work with them continuously, and, thus, gradually improve their problem solving. To enhance this feature, new word problems using roots and radicals and quadratic equations have been added. Other new word problems require the use of geometric formulas.

Pedagogical Use of Second Color Key definitions, formulas, and procedures are printed inside color boxes, helping students review easily. Color side comments within examples explain the structure of the problem. Cautionary comments and reminders are printed in color type where appropriate.

Chapter Summaries A new summary, giving important definitions, formulas, and procedures, has been added to each chapter.

Exercises **Graded Exercises** The range of difficulty in the exercise sets affords ample practice with drill exercises. Students are eased gradually through problems of increasing difficulty to those that will challenge outstanding students. More than 4100 drill exercises and 320 word problems, keyed to examples, are provided at the ends of sections. Including end-of-chapter exercises, the book gives a total of more than 5300 exercises, nearly 400 of them word problems. Answers to odd-numbered exercises are given at the back of the book.

Challenging Exercises   A group of challenging exercises is given at the end of most exercise sets. These are intended to give students an opportunity to go somewhat beyond the discussion in the text. Because of this intent, these exercises have no corresponding examples. These challenging exercises are not labeled in the text but are listed in the *Instructor's Guide* and the *Instructor's Answer Manual*.

Calculator Exercises   Calculator exercises have been included throughout the book. These optional exercises are identified by colored exercise numbers.

Review Exercises   Beginning in Chapter 2, most exercise sets end with a few problems reviewing earlier concepts that help students prepare for the next section.

Supplementary Exercises   A few sets of supplementary exercises, designed to integrate and clarify difficult or confusing topics, have been included. For example, a set in Chapter 4 requires students to distinguish among the different types of factoring problems.

Chapter Review Exercises   Extensive review exercises at the end of each chapter, approximately 650 in all, provide further opportunity for mastery of the material before taking an examination. These exercises are divided into two groups in most chapters. Exercises in group ❶ are keyed to appropriate sections of the chapter. Exercises in group ❷ consist of a mixed group of exercises that are not keyed to the sections. This helps students practice identifying problems by type.

Chapter Tests   Sample tests, of a length comparable to that of actual classroom tests, help students prepare for examinations. Answers to every test question are given at the back of the book.

New Content Highlights

— In Chapter 2 the section on word problems has been split into two sections, one on word problems and one on formulas and solving for a specified variable. Since both topics are especially difficult for students, this arrangement should make the material more accessible.
— On the other hand, most instructors prefer to teach the addition and multiplication properties of inequalities in one variable together, so this material in Chapter 2 has been rewritten as one section.
— The presentation of exponents in Chapter 3 has been rearranged so that the product rule and the power rules are presented for only positive exponents in Section 3.1. Then in Section 3.2, after students have had some practice with positive exponents, negative exponents and the quotient rule are introduced.
— The "trial and error" method of factoring trinomials is now presented before the grouping method in Section 4.2, since it generally is a more efficient method for students to use.

— In Section 6.6 the concepts of relation and function are now introduced in a more intuitive manner with many examples from real life. The ordered pair definition has been added to make the topic more understandable.
— Topics in Chapter 8 were rearranged to delay introducing rationalizing denominators until the basic operations of adding, subtracting, multiplying, and dividing radicals are presented.
— The exercises on using the quadratic formula in Chapter 9 have been separated into two groups, those in which the answers are already in simplified form and those in which the answers need to be simplified by the student.

Supplements   An extensive supplemental package is available for *Beginning Algebra,* fifth edition, including testing materials, solutions, and electronic media support.

The **Instructor's Guide** features six ready-to-duplicate tests for each chapter, including two multiple-choice tests. In addition, two forms of a final examination are provided, as well as a pretest that can be used for diagnostic purposes. Nearly 4000 additional test items, keyed to each objective in the text, are also included. Answers are provided for all tests and test items. A list of the challenging exercises in the textbook is given, to help in making assignments.

An **Instructor's Answer Manual** provides answers to all exercises in the text, for quick reference in the classroom if desired.

A **Computer-Assisted Testing System (CATS),** for both Apple II and IBM computers, provides more than 7500 questions, organized by objectives, and is available free to departments that adopt the book. CATS features an editing capability that allows instructors to add their own problems or edit existing problems. The system enables instructors to test by chapter, section, or objective, or by groups of chapters, sections, or objectives. Instructors can elect to generate "instant" tests, create multiple forms of tests, or design item-specific tests of their own choosing.

A set of **computer-assisted tutorials,** based on a mastery learning approach, has been developed for this text. Available for Apple II and IBM computers, the package not only gives additional worked-out examples and practice exercises but also requires students to demonstrate mastery of the skills associated with a particular concept or objective.

A set of nine **videotapes,** one per chapter, is available at low cost to users of the book. The tapes follow the outline of each chapter—section by section and objective by objective.

A set of nine **audiotapes,** covering the material in each section in the textbook, is available at no charge to users of the book. Students who need help with a particular topic or those who have missed class will find these tapes helpful.

A **Study Guide,** in semiprogrammed format, provides additional practice problems and reinforcement for the students. A self-test is given at the end of each chapter as well.

A **Student's Solutions Manual** contains solutions to every other odd-numbered exercise in the textbook plus solutions to all the exercises in the chapter tests.

An **Instructor's Solutions Manual** includes solutions to all even-numbered exercises and solutions to the odd-numbered exercises not included in the *Student's Solutions Manual*.

Acknowledgments

Thanks are due to the many users of the previous editions of this book who shared their ideas and made many good suggestions for improvements. The following people reviewed the manuscript for the fifth edition: John E. Alberghini, Manchester Community College; Thomas Alexander, University of Alabama in Birmingham; Donley A. Chandler, El Paso Community College; Teri Chiang, Mission College; Dale Ewen, Parkland College; Mary Jane Gates, University of Arkansas at Little Rock; Elayn Gay, University of New Orleans; Linda B. Holden, Indiana University at Bloomington; Donald R. Johnson, Scottsdale Community College; Bill E. Jordan, Seminole Community College; Keith Jorgensen, Orange Coast College; Laurence Maher, North Texas State University; Scott M. Newey, American River College; William Radulovich, Florida Junior College at Jacksonville; Thomas J. Ribley, Valencia Community College; John Snyder, Sinclair Community College; and James Thorpe, Saddleback Community College.

Special thanks and appreciation must go to those at Scott, Foresman who contributed so greatly to this book, in particular, Bill Poole, Linda Youngman, Carol Leon, and Kathy Cunningham.

Contents

Appendices

Answers to Selected Exercises 413

Beginning Algebra • 5th

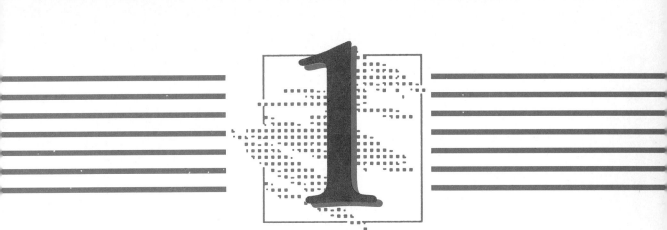

Number Systems

Objectives

1 Learn the definition of *factor.*
2 Write fractions in lowest terms.
3 Multiply and divide fractions.
4 Add and subtract fractions.

As preparation for the study of algebra, this section begins with a brief review of arithmetic. In everyday life, the numbers seen most often are the **whole numbers,**

$$0, \quad 1, \quad 2, \quad 3, \quad 4, \quad 5, \quad . \quad . \quad . \quad ,$$

and the **fractions,** such as

$$\frac{1}{2}, \quad \frac{2}{3}, \quad \frac{11}{12},$$

and so on. In a fraction, the top number is called the **numerator** and the bottom number is called the **denominator.**

1   In the statement $2 \times 9 = 18$, the numbers 2 and 9 are called **factors** of 18. Other factors of 18 include 1, 3, 6, and 18. The result of the multiplication, 18, is called the **product.**

The number 18 is **factored** by writing it as the product of two or more numbers. For example, 18 can be factored in several ways, as $6 \cdot 3$, or $18 \cdot 1$, or $9 \cdot 2$, or $3 \cdot 3 \cdot 2$. In algebra, raised dots are used instead of the $\times$ symbol to indicate multiplication.

A whole number (except 1) is **prime** if it has only itself and 1 as factors. (By agreement, the number 1 is not a prime number.) The first dozen primes are listed here.

$$2, \quad 3, \quad 5, \quad 7, \quad 11, \quad 13, \quad 17, \quad 19, \quad 23, \quad 29, \quad 31, \quad 37$$

It is often useful to find all the **prime factors** of a number—those factors that are prime numbers. For example, the only prime factors of 18 are 2 and 3.

Example 1   Write each number as the product of prime factors.

(a) 35

Write 35 as the product of the prime factors 5 and 7, or as

$$35 = 5 \cdot 7.$$

(b) 24

Divide by the smallest prime, 2, to get

$$24 = 2 \cdot 12.$$

Now divide 12 by 2 to find factors of 12.

$$24 = 2 \cdot 2 \cdot 6$$

Since 6 can be written as $2 \cdot 3$,

$$24 = 2 \cdot 2 \cdot 2 \cdot 3,$$

where all factors are prime.   ✚

Table 1, a table giving the prime factored form of the numbers 2 through 100, is given inside the back cover of this book.

2   Prime factors are used to write fractions in **lowest terms.** A fraction is in lowest terms when the numerator and the denominator have no factors in common (other than 1). Use the following steps to write a fraction in lowest terms.

Writing a Fraction in Lowest Terms	*Step 1* Write the numerator and the denominator as the product of prime factors. *Step 2* Divide the numerator and the denominator by the **greatest common factor,** the product of all factors common to both.

Example 2 Write each fraction in lowest terms.

(a) $\dfrac{10}{15} = \dfrac{2 \cdot 5}{3 \cdot 5} = \dfrac{2}{3}$

Since 5 is the only common factor of 10 and 15, dividing both numerator and denominator by 5 gives the fraction in lowest terms.

(b) $\dfrac{15}{45} = \dfrac{3 \cdot 5}{3 \cdot 3 \cdot 5} = \dfrac{1}{3}$

The factored form shows that 3 and 5 are the common factors of both 15 and 45. Dividing both 15 and 45 by $3 \cdot 5 = 15$ gives 15/45 in lowest terms as 1/3. ✚

3 The basic operations on whole numbers are addition, subtraction, multiplication, and division. These same operations apply to fractions. Multiply two fractions by first multiplying their numerators and then multiplying their denominators. This rule is written in symbols as follows.

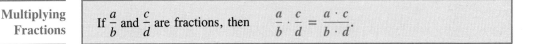

Multiplying Fractions

If $\dfrac{a}{b}$ and $\dfrac{c}{d}$ are fractions, then $\dfrac{a}{b} \cdot \dfrac{c}{d} = \dfrac{a \cdot c}{b \cdot d}$.

Example 3 Find the product of 3/8 and 4/9, and write it in lowest terms.

First, multiply 3/8 and 4/9.

$$\frac{3}{8} \cdot \frac{4}{9} = \frac{3 \cdot 4}{8 \cdot 9}$$

It is easiest to write a fraction in lowest terms while the product is in factored form. Factor 8 and 9 and then divide out common factors in the numerator and denominator.

$$\frac{3 \cdot 4}{8 \cdot 9} = \frac{3 \cdot 4}{2 \cdot 4 \cdot 3 \cdot 3}$$

$$= \frac{1}{2 \cdot 3}$$

$$= \frac{1}{6} \quad ✚$$

Two fractions are **reciprocals** of each other if their product is 1. For example, 3/4 and 4/3 are reciprocals since

$$\frac{3}{4} \cdot \frac{4}{3} = 1.$$

Also, 7/11 and 11/7 are reciprocals of each other. The reciprocal is used to divide fractions. To *divide* two fractions, multiply the first fraction and the reciprocal of the second one.

<table>
<tr><td>**Dividing Fractions**</td><td>For the fractions $\dfrac{a}{b}$ and $\dfrac{c}{d}$, $\qquad \dfrac{a}{b} \div \dfrac{c}{d} = \dfrac{a}{b} \cdot \dfrac{d}{c}.$</td></tr>
</table>

The reason this method works will be explained in a later chapter. The answer to a division problem is called a **quotient.** For example, the quotient of 20 and 10 is 2, since $20 \div 10 = 2$.

Example 4   Find the following quotients, and write them in lowest terms.

(a) $\dfrac{3}{4} \div \dfrac{8}{5} = \dfrac{3}{4} \cdot \dfrac{5}{8} = \dfrac{3 \cdot 5}{4 \cdot 8} = \dfrac{15}{32}$

(b) $\dfrac{3}{4} \div \dfrac{5}{8} = \dfrac{3}{4} \cdot \dfrac{8}{5} = \dfrac{3 \cdot 8}{4 \cdot 5} = \dfrac{3 \cdot 4 \cdot 2}{4 \cdot 5} = \dfrac{6}{5}$ ✦

4   The **sum** of two fractions having the same denominator is found by adding the numerators.

<table>
<tr><td>**Adding Fractions**</td><td>If $\dfrac{a}{b}$ and $\dfrac{c}{b}$ are fractions, then $\qquad \dfrac{a}{b} + \dfrac{c}{b} = \dfrac{a+c}{b}.$</td></tr>
</table>

Example 5   Add.

(a) $\dfrac{3}{7} + \dfrac{2}{7} = \dfrac{3+2}{7} = \dfrac{5}{7}$

(b) $\dfrac{2}{10} + \dfrac{5}{10} = \dfrac{2+5}{10} = \dfrac{7}{10}$ ✦

If two fractions to be added do not have the same denominators, the rule above can still be used, but only after the fractions are rewritten with a common denominator.

For example, to rewrite 3/4 as a fraction with a denominator of 32,

$$\frac{3}{4} = \frac{\blacksquare}{32},$$

find the number that can be multiplied by 4 to give 32. Since $4 \cdot 8 = 32$, use the number 8. Multiplying a number by 1 does not change its value. The value of 3/4 will be the same if it is multiplied by 8/8, which equals 1.

$$\frac{3}{4} = \frac{3}{4} \cdot \frac{8}{8} = \frac{3 \cdot 8}{4 \cdot 8} = \frac{24}{32}$$

Example 6 Write 5/8 as a fraction with a denominator of 72.

Since 8 must be multiplied by 9 to get 72, multiply both numerator and denominator by 9.

$$\frac{5}{8} = \frac{5}{8} \cdot \frac{9}{9} = \frac{5 \cdot 9}{8 \cdot 9} = \frac{45}{72} \quad \clubsuit$$

Example 7 Add the following fractions.

(a) $\dfrac{1}{2} + \dfrac{1}{3}$

These fractions cannot be added until both have the same denominator. Use 6 as a common denominator, since both 2 and 3 divide into 6. Write both 1/2 and 1/3 as fractions with a denominator of 6.

$$\frac{1}{2} = \frac{1}{2} \cdot \frac{3}{3} = \frac{1 \cdot 3}{2 \cdot 3} = \frac{3}{6} \quad \text{and} \quad \frac{1}{3} = \frac{1}{3} \cdot \frac{2}{2} = \frac{1 \cdot 2}{3 \cdot 2} = \frac{2}{6}$$

Now add.

$$\frac{1}{2} + \frac{1}{3} = \frac{3}{6} + \frac{2}{6} = \frac{3 + 2}{6} = \frac{5}{6}$$

(b) $3\dfrac{1}{2} + 2\dfrac{3}{4}$

Rewrite both numbers as follows.

$$3\frac{1}{2} = 3 + \frac{1}{2} = \frac{3}{1} + \frac{1}{2} = \frac{6}{2} + \frac{1}{2} = \frac{6 + 1}{2} = \frac{7}{2}$$

$$2\frac{3}{4} = 2 + \frac{3}{4} = \frac{8}{4} + \frac{3}{4} = \frac{8 + 3}{4} = \frac{11}{4}$$

Now add. The common denominator is 4.

$$3\frac{1}{2} + 2\frac{3}{4} = \frac{7}{2} + \frac{11}{4} = \frac{14}{4} + \frac{11}{4} = \frac{25}{4} \quad \clubsuit$$

The **difference** of two numbers is found by subtraction. For example, 9 − 5 = 4 so the difference of 9 and 5 is 4. Subtraction of fractions is similar to addition. Just subtract the numerators instead of adding them, according to the following definition.

Subtracting Fractions

If $\dfrac{a}{b}$ and $\dfrac{2}{b}$ are fractions, then $\quad \dfrac{a}{b} - \dfrac{c}{b} = \dfrac{a - c}{b}$

Example 8   Subtract.

(a) $\dfrac{5}{8} - \dfrac{3}{8} = \dfrac{5-3}{8} = \dfrac{2}{8} = \dfrac{1}{4}$

(b) $\dfrac{3}{4} - \dfrac{1}{3}$

These numbers must be written with a common denominator to use the rule for subtraction. Here a common denominator is 12.

$$\frac{3}{4} - \frac{1}{3} = \frac{9}{12} - \frac{4}{12} = \frac{9-4}{12} = \frac{5}{12} \quad \blacklozenge$$

1.1   Exercises

Write each number as the product of prime factors. See Example 1.

1. 30 **2.** 40 **3.** 50 **4.** 72

5. 65 **6.** 85 **7.** 100 **8.** 110

9. 17 **10.** 13 **11.** 124 **12.** 120

Write each fraction in lowest terms. See Example 2.

13. $\dfrac{7}{14}$ **14.** $\dfrac{3}{9}$ **15.** $\dfrac{10}{12}$ **16.** $\dfrac{8}{10}$

17. $\dfrac{16}{18}$ **18.** $\dfrac{14}{20}$ **19.** $\dfrac{50}{75}$ **20.** $\dfrac{32}{48}$

21. $\dfrac{72}{108}$ **22.** $\dfrac{96}{120}$ **23.** $\dfrac{120}{144}$ **24.** $\dfrac{77}{132}$

Find the products or quotients. Write answers in lowest terms. See Examples 3 and 4.

25. $\dfrac{3}{4} \cdot \dfrac{3}{5}$ **26.** $\dfrac{3}{8} \cdot \dfrac{5}{7}$ **27.** $\dfrac{1}{10} \cdot \dfrac{6}{5}$ **28.** $\dfrac{6}{7} \cdot \dfrac{1}{3}$

29. $\dfrac{9}{4} \cdot \dfrac{8}{15}$ **30.** $\dfrac{3}{5} \cdot \dfrac{20}{15}$ **31.** $\dfrac{3}{8} \div \dfrac{5}{4}$ **32.** $\dfrac{9}{16} \div \dfrac{3}{8}$

33. $\dfrac{5}{12} \div \dfrac{15}{4}$ **34.** $\dfrac{15}{16} \div \dfrac{30}{8}$ **35.** $\dfrac{15}{32} \div \dfrac{25}{8}$ **36.** $\dfrac{24}{25} \div \dfrac{3}{50}$

37. $\dfrac{5}{9} \cdot \dfrac{7}{10}$ **38.** $\dfrac{21}{30} \cdot \dfrac{5}{7}$ **39.** $1\dfrac{3}{10} \cdot 1\dfrac{2}{3}$ **40.** $1\dfrac{5}{16} \cdot 1\dfrac{1}{7}$

41. $9\dfrac{1}{3} \div 1\dfrac{1}{6}$ **42.** $13\dfrac{4}{9} \div \dfrac{11}{18}$ **43.** $\dfrac{28}{15} \div \dfrac{14}{5}$ **44.** $\dfrac{120}{7} \div \dfrac{45}{3}$

Add or subtract the following fractions. Write answers in lowest terms. Work from left to right in Exercises 63–68. See Examples 5–8.

45. $\dfrac{1}{12} + \dfrac{5}{12}$

46. $\dfrac{2}{5} + \dfrac{1}{5}$

47. $\dfrac{1}{10} + \dfrac{7}{10}$

48. $\dfrac{3}{8} + \dfrac{1}{8}$

49. $\dfrac{4}{9} + \dfrac{2}{3}$

50. $\dfrac{3}{5} + \dfrac{2}{15}$

51. $\dfrac{8}{11} + \dfrac{3}{22}$

52. $\dfrac{9}{10} + \dfrac{3}{5}$

53. $\dfrac{2}{3} - \dfrac{3}{5}$

54. $\dfrac{7}{12} - \dfrac{5}{9}$

55. $\dfrac{5}{6} - \dfrac{3}{10}$

56. $\dfrac{11}{4} - \dfrac{5}{8}$

57. $3\dfrac{1}{4} + \dfrac{1}{8}$

58. $5\dfrac{2}{3} + \dfrac{1}{4}$

59. $4\dfrac{1}{2} + 3\dfrac{2}{3}$

60. $7\dfrac{5}{8} + 3\dfrac{3}{4}$

61. $6\dfrac{2}{3} - 5\dfrac{1}{4}$

62. $8\dfrac{8}{9} - 7\dfrac{4}{5}$

63. $\dfrac{2}{5} + \dfrac{1}{3} + \dfrac{9}{10}$

64. $\dfrac{3}{8} + \dfrac{5}{6} + \dfrac{2}{3}$

65. $\dfrac{5}{14} + \dfrac{1}{6} - \dfrac{1}{9}$

66. $\dfrac{2}{15} + \dfrac{1}{6} - \dfrac{1}{10}$

67. $\dfrac{3}{10} - \dfrac{2}{15} + \dfrac{1}{8}$

68. $\dfrac{7}{10} - \dfrac{5}{8} + \dfrac{3}{20}$

Work each word problem.

69. John Rizzo paid 1/8 of a debt in January, 1/3 in February, and 1/4 in March. What portion of the debt was paid in these three months?

70. A rectangle is 5/16 yard on each of two sides, and 7/12 yard on each of the other two sides. Find the total distance around the rectangle.

71. Last month the Eastside Wholesale Market sold 3 1/4 tons of broccoli, 2 3/8 tons of spinach, 7 1/2 tons of corn, and 1 5/16 tons of turnips. Find the total number of tons of vegetables sold by the firm during the month.

72. Sharkey's Resort decided to expand by buying a piece of property next to the resort. The property has an irregular shape, with five sides. The lengths of the five sides are 146 1/2 feet, 98 3/4 feet, 196 feet, 76 5/8 feet, and 100 7/8 feet. Find the total distance around the piece of property.

73. Joann Kaufmann worked 40 hours during a certain week. She worked 8 1/4 hours on Monday, 6 3/8 hours on Tuesday, 7 2/3 hours on Wednesday, and 8 3/4 hours on Thursday. How many hours did she work on Friday?

74. A concrete truck is loaded with 9 7/8 cubic yards of concrete. The driver delivers 1 1/2 cubic yards at the first stop and 2 3/4 cubic yards at the second stop. At the third stop, the customer receives 3 5/12 cubic yards. How much concrete is left in the truck?

75. Rosario wants to make 16 dresses to sell at the company bazaar. Each dress needs 2 1/4 yards of material. How many yards should be bought?

76. Lindsay allows 1 3/5 bottles of beverage for each guest at a party. If he expects 35 guests, how many bottles of beverage will he need?

77. If 2 1/4 yards of fabric are needed to upholster a chair, how many chairs can be upholstered with 23 2/3 yards of fabric?

78. A cake recipe calls for 1 3/4 cups of sugar. A caterer has 15 1/2 cups of sugar on hand. How many cakes can he make?

1.2 Symbols

Objectives

1. Know the meaning of $<$ and $>$.
2. Translate word phrases to symbols.
3. Know the meaning of $\leq$ and $\geq$.
4. Write statements that change the direction of inequality symbols.

So far only the symbols of arithmetic, such as $+$, $-$, $\times$ (or $\cdot$), and $\div$, have been used. Another common symbol is the one for equality, $=$, which shows that two numbers are equal. This symbol with a slash through it, $\neq$, means "is not equal to." For example,

$$7 \neq 8$$

indicates that 7 is not equal to 8.

1   If two numbers are not equal, then one of the numbers must be smaller than the other. The symbol $<$ represents "is less than," so that "7 is less than 8" is written

$$7 < 8.$$

Also, write "6 is less than 9" as $6 < 9$.

The symbol $>$ means "is greater than." Write "8 is greater than 2" as

$$8 > 2.$$

The statement "17 is greater than 11" becomes $17 > 11$.

Keep the symbols $<$ and $>$ straight by remembering that the symbol always points to the smaller number. For example, write "8 is less than 15" by pointing the symbol toward the 8:

$$8 < 15.$$

2   In algebra, it is often necessary to convert word phrases to symbols. The next example shows this.

Example 1   Write each word statement in symbols.

(a) Twelve equals ten plus two.

$$12 = 10 + 2$$

(b) Nine is less than ten.

$$9 < 10$$

(c) Fifteen is not equal to eighteen.

$$15 \neq 18$$

(d) Seven is greater than four.

$$7 > 4 \quad \clubsuit$$

3 Two other symbols, $\leq$ and $\geq$, also represent the idea of inequality. The symbol $\leq$ means "is less than or equal to," so that

$$5 \leq 9$$

means "5 is less than or equal to 9." This statement is true, since $5 < 9$ is true. If either the $<$ part or the $=$ part is true, then the inequality $\leq$ is true.
    The symbol $\geq$ means "is greater than or equal to." Again,

$$9 \geq 5$$

is true because $9 > 5$ is true. Also, $8 \leq 8$ is true since $8 = 8$ is true. But it is not true that $13 \leq 9$ because neither $13 < 9$ nor $13 = 9$ is true.

Example 2 Tell whether or not each statement is true.

(a) $15 \leq 20$
The statement $15 \leq 20$ is true, since $15 < 20$.

(b) $25 \geq 30$
Both $25 > 30$ and $25 = 30$ are false. Because of this, $25 \geq 30$ is false.

(c) $12 \geq 12$
Since $12 = 12$, this statement is true. ✚

4 Any statement with $<$ can be converted to one with $>$, and any statement with $>$ can be converted to one with $<$. Do this by reversing the order of the numbers and the direction of the symbol. For example, the statement $6 < 10$ can be written with $>$ as $10 > 6$. Similarly, the statement $4 \leq 10$ can be changed to $10 \geq 4$.

Example 3 The following list shows the same statement written in two equally correct ways.

(a) $9 < 16$ $16 > 9$ **(b)** $5 > 2$ $2 < 5$

(c) $3 \leq 8$ $8 \geq 3$ **(d)** $12 \geq 5$ $5 \leq 12$ ✚

Here is a summary of the symbols discussed in this section.

Symbols of Equality and Inequality		
$=$ is equal to	$\neq$ is not equal to	
$<$ is less than	$>$ is greater than	
$\leq$ is less than or equal to	$\geq$ is greater than or equal to	

1.2 Exercises

Insert $<$ or $>$ to make the following statements true.

1. 6 _____ 9   **2.** 5 _____ 3   **3.** 12 _____ 15   **4.** 8 _____ 10

5. 25 _____ 12   **6.** 17 _____ 9   **7.** 32 _____ 50   **8.** 41 _____ 72

9. $\dfrac{3}{4}$ ____ 1   **10.** $\dfrac{2}{3}$ ____ 0   **11.** $1\dfrac{5}{8}$ ____ 1   **12.** $3\dfrac{7}{9}$ ____ 2

Insert ≤ or ≥ to make the following statements true.

13. 12 ____ 17   **14.** 28 ____ 42   **15.** 16 ____ 14   **16.** 39 ____ 17

17. 8 ____ 28   **18.** 10 ____ 15   **19.** 35 ____ 42   **20.** 51 ____ 62

Which of the symbols <, >, ≤, and ≥ make the following statements true?
Give all possible correct answers.

21. 6 ____ 9   **22.** 18 ____ 12   **23.** 51 ____ 50   **24.** 0 ____ 12

25. 5 ____ 5   **26.** 10 ____ 10   **27.** 48 ____ 0   **28.** 100 ____ 1000

29. 16 ____ 10   **30.** 5 ____ 3   **31.** $\dfrac{1}{4}$ ____ $\dfrac{2}{5}$ **32.** $\dfrac{2}{3}$ ____ $\dfrac{5}{8}$

33. .609 ____ .61   **34.** .5 ____ .499   **35.** $3\dfrac{1}{2}$ ____ 4   **36.** $5\dfrac{7}{8}$ ____ 6

Write the following word statements in symbols. See Example 1.

37. Seven equals five plus two.

38. Nine is greater than the product of four and two.

39. Three is less than the quotient of fifty and five.

40. Five equals ten minus five.

41. Twelve is not equal to five.

42. Fifteen does not equal sixteen.

43. Zero is greater than or equal to zero.

44. Six is less than or equal to six.

Tell whether each statement is true or false. See Example 2.

45. $8 + 2 = 10$ **46.** $8 \neq 9 - 1$ **47.** $12 \geq 10$ **48.** $45 < 45$

49. $0 < 15$ **50.** $16 \geq 10$ **51.** $1\dfrac{2}{3} + 2\dfrac{3}{4} = \dfrac{53}{12}$ **52.** $3\dfrac{2}{5} < 6\dfrac{1}{4}$

53. $\dfrac{25}{3} \geq \dfrac{19}{2}$ **54.** $\dfrac{18}{5} < \dfrac{5}{4}$ **55.** $9 < 0$ **56.** $15 \leq 32$

57. $6 \neq 5 + 1$ **58.** $15 < 21$ **59.** $2.13 < 2.13$ **60.** $1.95 \geq 1.96$

61. $8 \leq 0$ **62.** $26 \geq 50$ **63.** $12 \geq 12$ **64.** $5 \leq 5$

Rewrite the following true statements so the inequality symbol points in the opposite direction. See Example 3.

65. $6 < 14$ **66.** $8 \leq 9$ **67.** $15 \geq 3$ **68.** $29 > 4$

69. $9 > 8$ **70.** $12 < 17$ **71.** $0 \leq 6$ **72.** $7 \leq 12$

73. $\dfrac{18}{5} \geq \dfrac{15}{7}$ **74.** $\dfrac{25}{3} \geq \dfrac{1}{2}$ **75.** $.481 \geq .439$ **76.** $.762 < .763$

1.3 Exponents and Order of Operations

Objectives

1 Use exponents.

2 Use the order of operations.

3 Use more than one grouping symbol.

4 Insert parentheses to make a statement true.

1 It is common for a multiplication problem to have the same factor appearing several times. For example, in the product

$$3 \cdot 3 \cdot 3 \cdot 3 = 81$$

the factor 3 appears four times. In algebra, repeated factors are written with an *exponent*. For example, in $3 \cdot 3 \cdot 3 \cdot 3$, the number 3 appears as a factor four times, so the product is written as 3^4.

$$3 \cdot 3 \cdot 3 \cdot 3 = 3^4$$

The number 4 is the **exponent** and 3 is the **base.** An exponent, then, tells how many times the base is used as a factor.

Example 1 Find the values of the following.

(a) 5^2

$$\underbrace{5 \cdot 5}_{} = 25$$

5 is used as a factor 2 times

Read 5^2 as "5 squared."

(b) 6^3

$$\underbrace{6 \cdot 6 \cdot 6}_{} = 216$$

6 is used as a factor 3 times

Read 6^3 as "6 cubed."

(c) $2^5 = 2 \cdot 2 \cdot 2 \cdot 2 \cdot 2 = 32$ 2 is used as a factor 5 times

Read 2^5 as "2 to the fifth power."

(d) $\left(\dfrac{2}{3}\right)^3 = \dfrac{2}{3} \cdot \dfrac{2}{3} \cdot \dfrac{2}{3} = \dfrac{8}{27}$ $\dfrac{2}{3}$ is used as a factor 3 times   ✦

2 Many problems involve more than one operation of arithmetic. For example, to simplify the expression

$$5 + 2 \cdot 3$$

to a single number, should we add first, or multiply first? One way to make the order of operations clear is to use **grouping symbols.** For example, to show that the multiplication should be performed before the addition, parentheses can be used to write

$$5 + (2 \cdot 3) = 5 + 6 = 11.$$

If addition is to be performed first, the parentheses should group $5 + 2$ as follows.

$$(5 + 2) \cdot 3 = 7 \cdot 3 = 21$$

Other grouping symbols used in more complicated expressions are

brackets, [   ], braces, {   },

and fraction bars.

The most useful way to work problems with more than one operation is to use the following **order of operations.** This is the order used by calculators and computers.

Order of Operations	*If grouping symbols are present*, simplify within them, innermost first (and above and below fraction bars separately), in the following order. *Step 1* Apply all exponents. *Step 2* Do any multiplications or divisions in the order in which they occur, working from left to right. *Step 3* Do any additions or subtractions in the order in which they occur, working from left to right. *If no grouping symbols are present*, start with Step 1.

A dot has been used to show multiplication; another way to show multiplication is with parentheses. For example, 3(7), (3)7, and (3)(7) each mean $3 \cdot 7$ or 21. The next example shows the use of parentheses for multiplication.

Example 2   Find the values of the following.

(a) $9(6 + 11)$

Using the order of operations given above, work first inside the parentheses.

$$9(6 + 11) = 9(17) \qquad \text{Work inside parentheses}$$
$$= 153 \qquad \text{Multiply}$$

(b) $6 \cdot 8 + 5 \cdot 2$

Do any multiplications, working from left to right, and then add.

$$6 \cdot 8 + 5 \cdot 2 = 48 + 10 \qquad \text{Multiply}$$
$$= 58 \qquad \text{Add}$$

(c) $2(5 + 6) + 7 \cdot 3 = 2(11) + 7 \cdot 3 \qquad \text{Work inside parentheses}$
$$= 22 + 21 \qquad \text{Multiply}$$
$$= 43 \qquad \text{Add}$$

(d) $9 + 2^3 - 5$

Find 2^3 first.

$$9 + 2^3 - 5 = 9 + 2 \cdot 2 \cdot 2 - 5 \qquad \text{Use the exponent}$$
$$= 9 + 8 - 5$$
$$= 17 - 5 \qquad \text{Add}$$
$$= 12 \qquad \text{Subtract}$$

(e) $16 - 3^2 + 4^2 = 16 - 3 \cdot 3 + 4 \cdot 4 \qquad \text{Use the exponents}$
$$= 16 - 9 + 16$$
$$= 7 + 16 \qquad \text{Subtract}$$
$$= 23 \qquad \text{Add}$$

Notice that 3^2 must be evaluated before subtracting. This is always the case.   ❖

3   An expression with double parentheses, such as $2(8 + 3(6 + 5))$, can be confusing. To eliminate this, square brackets, [   ], often are used instead of one of the pairs of parentheses.

Example 3   Simplify $2[8 + 3(6 + 5)]$.

Work first within the parentheses, and then simplify until a single number is found inside the brackets.

$$2[8 + 3(6 + 5)] = 2[8 + 3(11)]$$
$$= 2[8 + 33]$$
$$= 2[41]$$
$$= 82 \quad ❖$$

Sometimes fraction bars are grouping symbols, as the next example shows.

Example 4   Simplify $\dfrac{4(5 + 3) + 3}{2(3) - 1}$.

The expression can be written as the quotient

$$[4(5 + 3) + 3] \div [2(3) - 1],$$

which shows that the fraction bar serves to group the numerator and denominator separately. Simplify both numerator and denominator, then divide, if possible.

$$\frac{4(5 + 3) + 3}{2(3) - 1} = \frac{4(8) + 3}{2(3) - 1} \qquad \text{Work inside parentheses}$$
$$= \frac{32 + 3}{6 - 1} \qquad \text{Multiply}$$
$$= \frac{35}{5} \qquad \text{Add or subtract}$$
$$= 7 \qquad \text{Simplify} \quad ❖$$

4   The final example shows how to decide where to insert grouping symbols so that an expression equals a particular number.

Example 5   Insert parentheses so that the following are true.

(a) $9 - 3 - 2 = 8$

This statement would be true if parentheses were inserted around $3 - 2$.

$$9 - (3 - 2) = 8$$

It is not true that $(9 - 3) - 2 = 8$, since $6 - 2 \ne 8$.

(b) $9 \cdot 2 - 4 \cdot 3 = 6$

Since $9 \cdot 2 - 4 \cdot 3 = 18 - 12 = 6$, no parentheses are needed here. If desired, parentheses may be placed as follows.

$$(9 \cdot 2) - (4 \cdot 3) = 6 \quad \clubsuit$$

Example 6   In her pay envelope an employee received three $20 bills. She bought a book for $16 and a $1.50 pen using one of the bills. Write an expression using numerical and grouping symbols to describe these transactions.

The three $20 bills are represented by 3(20). To represent the subtraction of 16 and 1.50 from this product, write either

$$3(20) - 16 - 1.50 \quad \text{or} \quad 3(20) - (16 + 1.50). \quad \clubsuit$$

1.3   Exercises

Find the values of the following. In Exercises 21–24, round to the nearest thousandth. See Example 1.

1. 6^2	**2.** 9^2	**3.** 8^2	**4.** 10^2
5. 17^2	**6.** 22^2	**7.** 5^3	**8.** 7^3
9. 6^4	**10.** 3^4	**11.** 2^5	**12.** 4^5
13. 3^6	**14.** 2^6	**15.** $\left(\dfrac{1}{2}\right)^2$	**16.** $\left(\dfrac{3}{4}\right)^2$
17. $\left(\dfrac{2}{5}\right)^3$	**18.** $\left(\dfrac{3}{7}\right)^3$	**19.** $\left(\dfrac{4}{5}\right)^3$	**20.** $\left(\dfrac{2}{3}\right)^5$
*21. $(.83)^4$	**22.** $(.712)^2$	**23.** $(1.46)^3$	**24.** $(2.85)^4$

Find the values of the following expressions. See Examples 2–4.

25. $4 + 6 \cdot 2$	**26.** $9 + 3 \cdot 4$
27. $12 - 5 \cdot 2$	**28.** $16 - 3 \cdot 5$
29. $3 \cdot 8 - 4 \cdot 6$	**30.** $2 \cdot 20 - 8 \cdot 5$
31. $6 \cdot 5 + 3 \cdot 10$	**32.** $5 \cdot 8 + 10 \cdot 4$

---

*Calculator exercises have been included throughout the book and are identified with an exercise number printed in color.

33. $5[8 + (2 + 3)]$

34. $9[(14 + 5) - 10]$

35. $(6 - 3)[8 - (2 + 1)]$

36. $(7 - 1)[9 + (6 - 3)]$

37. $\dfrac{2(5 + 3) + 2 \cdot 2}{2(4 - 1)}$

38. $\dfrac{9(7 - 1) - 8 \cdot 2}{4(6 - 1)}$

39. $\dfrac{8^2 + 2}{5 - 2^2}$

40. $\dfrac{4^2 - 8}{15 - 3^2}$

First simplify both sides of each inequality. Then tell whether the given statement is true or false.

41. $9 \cdot 3 - 11 \le 16$

42. $6 \cdot 5 - 12 \le 18$

43. $5 \cdot 11 + 2 \cdot 3 \le 60$

44. $9 \cdot 3 + 4 \cdot 5 \ge 48$

45. $0 \ge 12 \cdot 3 - 6 \cdot 6$

46. $10 \le 13 \cdot 2 - 15 \cdot 1$

47. $45 \ge 2[2 + 3(2 + 5)]$

48. $55 \ge 3[4 + 3(4 + 1)]$

49. $[3 \cdot 4 + 5(2)] \cdot 3 > 72$

50. $2 \cdot [7 \cdot 5 - 3(2)] \le 58$

51. $\dfrac{3 + 5(4 - 1)}{2 \cdot 4 + 1} \ge 3$

52. $\dfrac{7(3 + 1) - 2}{3 + 5 \cdot 2} \le 2$

53. $3 \ge \dfrac{2(5 + 1) - 3(1 + 1)}{5(8 - 6) - 4 \cdot 2}$

54. $7 \le \dfrac{3(8 - 3) + 2(4 - 1)}{9(6 - 2) - 11(5 - 2)}$

55. $7.43^2 - 5.77^2 \ge 21.92$

56. $(.841)^3 - (.58)^4 < .479$

Insert parentheses in each expression so that the resulting statement is true. Some problems require no parentheses. See Example 5.

57. $10 - 7 - 3 = 6$

58. $16 - 4 - 3 = 15$

59. $3 \cdot 5 + 7 = 22$

60. $3 \cdot 5 + 7 = 36$

61. $3 \cdot 5 - 4 = 3$

62. $3 \cdot 5 - 4 = 11$

63. $3 \cdot 5 + 2 \cdot 4 = 23$

64. $3 \cdot 5 + 2 \cdot 4 = 84$

65. $3 \cdot 5 + 2 \cdot 4 = 68$

66. $100 \div 20 \div 5 = 1$

67. $360 \div 18 \div 4 = 5$

68. $100 \div 20 \div 5 = 25$

69. $4096 \div 256 \div 4 = 4$

70. $2^2 + 4 \cdot 2 = 16$

71. $6 + 5 \cdot 3^2 = 99$

72. $3^3 - 2 \cdot 4 = 100$

73. $8 - 2^2 \cdot 2 = 8$

74. $6^2 - 2 \cdot 5 = 170$

Write an expression using numerical and grouping symbols to describe the situation in each of the following problems. See Example 6.

75. Marjorie Jensen invested $600. After one year, her investment had tripled. She then took $150 and made a car payment.

76. John Wilson had 5 decks of cards, each containing 52 cards. He removed all 4 aces from one deck.

77. A bus has 63 passengers. At one stop, 23 people get off and 17 new people get on.

78. An elevator has 5 passengers. At the first stop, 6 get on and 1 gets off. At the next stop, 3 get on and 2 get off.

1.4 Variables and Equations

Objectives

☐1 Define *variable*.
☐2 Find the value of algebraic expressions, given values for the variables.
☐3 Convert statements from words to algebraic expressions.
☐4 Identify solutions of equations.
☐5 Define and use the domain of a set of numbers.

☐1 A **variable** is a symbol, usually a letter, such as x, y, or z, used to represent any unknown number. An **algebraic expression** is a collection of numbers, variables, symbols for operations, and symbols for grouping (such as parentheses). For example,

$$6(x + 5), \qquad 2m - 9, \qquad \text{and} \qquad 8p^2 + 6p + 2$$

are all algebraic expressions. In the algebraic expression $2m - 9$, the expression $2m$ indicates the product of 2 and m, just as $8p^2$ shows the product of 8 and p^2. Also, $6(x + 5)$ means the product of 6 and $x + 5$.

☐2 An algebraic expression takes on different numerical values as the variables take on different values.

Example 1 Find the numerical values of the following algebraic expressions when $m = 5$.

(a) $8m$

Replace m with 5, to get

$$8m = 8 \cdot 5 = 40.$$

(b) $3m^2$

For $m = 5$,

$$3m^2 = 3 \cdot 5^2 = 3 \cdot 25 = 75. \quad \clubsuit$$

In Example 1(b), it is important to notice that $3m^2$ means $3 \cdot m^2$; it *does not* mean $3m \cdot 3m$. The product $3m \cdot 3m$ is indicated by $(3m)^2$.

Example 2 Find the value of each expression when $x = 5$ and $y = 3$.

(a) $2x + 7y$

Replace x with 5 and y with 3. Do the multiplication first, and then add.

$$
\begin{aligned}
2x + 7y &= 2 \cdot 5 + 7 \cdot 3 &&\text{Let } x = 5 \text{ and } y = 3 \\
&= 10 + 21 &&\text{Multiply} \\
&= 31 &&\text{Add}
\end{aligned}
$$

(b) $\dfrac{9x - 8y}{2x - y}$

Replace x with 5 and y with 3.

$$\dfrac{9x - 8y}{2x - y} = \dfrac{9 \cdot 5 - 8 \cdot 3}{2 \cdot 5 - 3} \qquad \text{Let } x = 5 \text{ and } y = 3$$

$$= \dfrac{45 - 24}{10 - 3} \qquad \text{Multiply}$$

$$= \dfrac{21}{7} \qquad \text{Subtract}$$

$$= 3 \qquad \text{Divide}$$

(c) $x^2 - 2y^2 = 5^2 - 2 \cdot 3^2 \qquad$ Let $x = 5$ and $y = 3$

$\qquad = 25 - 2 \cdot 9 \qquad$ Use the exponents

$\qquad = 25 - 18 \qquad$ Multiply

$\qquad = 7 \qquad$ Subtract ✦

3   Variables are used in changing word phrases into algebraic expressions. The next example shows how to do this.

Example 3   Change the following word phrases to algebraic expressions. Use x as the variable.

(a) The sum of a number and 9

"Sum" is the answer to an addition problem. This phrase translates as

$$x + 9 \qquad \text{or} \qquad 9 + x.$$

(b) 7 minus a number

"Minus" indicates subtraction, so the answer is

$$7 - x.$$

Here $x - 7$ would *not* be correct; this statement translates as "a number minus 7," not "7 minus a number." The expressions $7 - x$ and $x - 7$ are not always equal. For example, if $x = 10$, $10 - 7 \neq 7 - 10$.

(c) 7 taken from a number

Since 7 is taken *from* a number, write

$$x - 7.$$

In this case $7 - x$ would not be correct, because "taken from" means "subtracted from."

(d) The product of 11 and a number

$$11 \cdot x \qquad \text{or} \qquad 11x$$

As mentioned above, $11x$ means 11 times x. No symbol is needed to indicate the product of a number and a variable.

(e) 5 divided by a number

$$\frac{5}{x}$$

(f) The product of 2, and the sum of a number and 8

$$2(x + 8) \quad \clubsuit$$

4   An **equation** is a statement that shows that two algebraic expressions are equal. Examples of equations are

$$x + 4 = 11, \qquad 2y = 16, \qquad \text{and} \qquad 4p + 1 = 25 - p.$$

Solving an Equation

> To **solve** an equation means to find the values of the variable that make the equation true. The values of the variable that make the equation true are called the **solutions** of the equation.

Example 4   Decide whether the given number is a solution of the equation.

(a) $5p + 1 = 36;$ 7

Replace p with 7.

$$5p + 1 = 36$$
$$5 \cdot 7 + 1 = 36 \qquad \text{Let } p = 7$$
$$35 + 1 = 36$$
$$36 = 36 \qquad \text{True}$$

The number 7 is a solution of the equation.

(b) $9m - 6 = 32;$ 4

$$9m - 6 = 32$$
$$9 \cdot 4 - 6 = 32 \qquad \text{Let } m = 4$$
$$36 - 6 = 32$$
$$30 = 32 \qquad \text{False}$$

The number 4 is not a solution of the equation. $\clubsuit$

5   Sometimes the solutions of an equation must come from a certain list of numbers. This list of numbers is often written as a *set;* for example, the set containing the numbers 1, 2, 3, 4, and 5 is written with **set braces** as

$$\{1, 2, 3, 4, 5\}.$$

A **set** is a collection of objects. For more information on sets, see Appendix B at the back of this book.

Domain	The set of numbers from which the solutions of an equation must be chosen is called the **domain** of the equation.

In an application, the domain is often determined by the natural restrictions of the problem. For example, if the answer to a problem is a number of people, only whole numbers would make sense, so the domain would be the set of whole numbers. In other situations the domain is frequently an arbitrary choice.

Example 5   Change each word statement to an equation. Use x as the variable. Then find all solutions for the equation from the domain

$$\{0, 2, 4, 6, 8, 10\}.$$

(a) The sum of a number and four is six.
   The word "is" suggests "equals." If x represents the unknown number, then translate as follows.

$$\begin{array}{ccc}
\text{The sum of} & & \\
\text{a number and four} & \text{is} & \text{six.} \\
\downarrow & \downarrow & \downarrow \\
x + 4 & = & 6
\end{array}$$

Try each number from the given domain $\{0, 2, 4, 6, 8, 10\}$, in turn, to see that 2 is the only solution of $x + 4 = 6$.

(b) 9 more than five times a number is 49.
   Use x to represent the unknown number.

$$\begin{array}{cccccc}
9 & \text{more than} & \text{five times a number} & \text{is} & 49. \\
\downarrow & \downarrow & \downarrow & \downarrow & \downarrow \\
9 & + & 5x & = & 49
\end{array}$$

Try each number from the domain $\{0, 2, 4, 6, 8, 10\}$. The solution is 8, since $9 + 5 \cdot 8 = 49$.   ✢

1.4   Exercises

Find the numerical values of the following when (a) $x = 3$ and (b) $x = 15$. See Example 1.

1. $x + 9$

2. $x - 1$

3. $5x$

4. $7x$

5. $2x + 8$

6. $9x - 5$

7. $\dfrac{x + 1}{3}$

8. $\dfrac{x - 2}{5}$

9. $\dfrac{3x - 5}{2x}$

10. $\dfrac{x + 2}{x - 1}$

11. $3x^2 + x$

12. $2x + x^2$

13. $6.459x$

14. $.74x^2$

15. $.0745(x^2 + 2)$

16. $.204(3 + x)$

Find the numerical values of the following when (a) $x = 4$ and $y = 2$ and (b) $x = 1$ and $y = 5$. Round to the nearest thousandth in Exercises 31 and 32. See Example 2.

17. $3(x + 2y)$ 　　　　**18.** $2(2x + y)$ 　　　　**19.** $x + \dfrac{4}{y}$ 　　　　**20.** $y + \dfrac{8}{x}$

21. $\dfrac{x}{3} + \dfrac{5}{y}$ 　　　**22.** $\dfrac{x}{5} + \dfrac{y}{4}$ 　　　**23.** $5(4x + 7y)$ 　　　**24.** $8(5x + 9y)$

25. $\dfrac{2x + 3y}{x + y + 1}$ 　　**26.** $\dfrac{5x + 3y + 1}{2x}$ 　　**27.** $\dfrac{2x + 4y - 6}{5y + 2}$ 　　**28.** $\dfrac{4x + 3y - 1}{x}$

29. $\dfrac{x^2 + y^2}{x + y}$ 　　**30.** $\dfrac{9x^2 + 4y^2}{3x^2 + 2y}$ 　　**31.** $.841x^2 + .32y^2$ 　　**32.** $\dfrac{3.4x + 2.59y}{0.8x + 0.3y^2}$

Change the word phrases to algebraic expressions. Use x to represent the variable. See Example 3.

33. Eight times a number

34. Fifteen divided by a number

35. The quotient of five and a number

36. Six added to a number

37. A number subtracted from eight

38. Nine subtracted from a number

39. Eight added to the product of a number and three

40. The difference of five times a number and six

41. Eight times a number, added to fifty-two

42. Six added to two-thirds of a number

Decide whether the given number is a solution of the equation. See Example 4.

43. $p - 5 = 12;$ 　17 　　　　**44.** $x + 6 = 15;$ 　9 　　　　**45.** $5m + 2 = 7;$ 　2

46. $3r + 5 = 8;$ 　2 　　　　**47.** $2y + 3(y - 2) = 14;$ 　4 　　**48.** $6a + 2(a + 3) = 14;$ 　1

49. $6p + 4p - 9 = 11;$ 　2 　　**50.** $2x + 3x + 8 = 38;$ 　6 　　**51.** $3r^2 - 2 = 46;$ 　4

52. $2x^2 + 1 = 19;$ 　3 　　　**53.** $\dfrac{z + 4}{z - 2} = 2;$ 　8 　　　**54.** $\dfrac{x + 6}{x - 2} = 9;$ 　3

55. $9.54x + 3.811 = 0.4273x + 16.57718;$ 　1.4

56. $0.935(y + 6.1) + 0.0142 = 7.83y + .2017;$ 　.8

Change the word statements to equations. Use x as the variable. Find the solutions from the domain $\{0, 1, 2, 3, 4, 5, 6\}$. See Example 5.

57. The sum of a number and 8 is 12.

58. A number minus three equals two.

59. Twice a number plus two is ten.

60. The sum of twice a number and 6 is 18.

61. Five more than twice a number is 13.

62. The product of a number and 8 is 24.

63. Three times a number is equal to two more than twice the number.

64. Twelve divided by a number equals three times that number.

65. The quotient of twenty and five times a number is 2.

66. A number divided by 2 is 0.

Find the value of the following when x = 4.

67. $\dfrac{(2x - 3)(5x + 2)}{x - 1}$

68. $\dfrac{7x - 3}{(x + 2)(x - 1)}$

69. $\dfrac{2[4(x + 3) - x]}{2(x + 1)}$

70. $\dfrac{3[x(x - 1) + 2]}{5(2x - 5)}$

1.5   *Real Numbers and the Number Line*

Objectives

1 Set up number lines.

2 Identify whole numbers, integers, rational numbers, irrational numbers, and real numbers.

3 Tell which of two real numbers is smaller.

4 Find additive inverses.

5 Find absolute values of real numbers.

1 Graphs can be a helpful way to picture sets of numbers. Numbers are graphed on **number lines** like the one in Figure 1.1.

Figure 1.1

Draw a number line by locating any point on the line and calling it 0. Choose any point to the right of 0 and call it 1. The distance between 0 and 1 gives a unit of measure used to locate other points, as shown in Figure 1.1. The points labeled in Figure 1.1 and those continuing in the same way to the right correspond to the set of whole numbers.

Whole Numbers

$$\{0, 1, 2, 3, 4, 5, 6, \cdots\}$$

The three dots show that the list of numbers continues in the same way indefinitely.

All the whole numbers starting with 1 are located to the right of 0 on the number line. But numbers may also be placed to the left of 0. These numbers, written -1, -2, -3, and so on, are shown in Figure 1.2. (The minus sign is used to show that the numbers are located to the *left* of 0.)

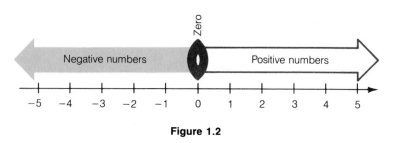

Figure 1.2

The numbers to the *left* of 0 are **negative numbers.** The numbers to the *right* of 0 are **positive numbers.** The number 0 itself is neither positive nor negative. Positive numbers and negative numbers are called **signed numbers.**

There are many practical applications of negative numbers. For example, a temperature on a cold January day might be $-10°$, or 10 degrees below zero. A business that spends more than it takes in has a negative "profit." An altitude of 30 feet below sea level corresponds to a height of -30 feet.

2   The set of numbers marked on the number line in Figure 1.2, including positive and negative numbers and zero, is part of the set of integers.

Integers	$\{\cdots, -3, -2, -1, 0, 1, 2, 3, \cdots\}$

Not all numbers are integers. For example, 1/2 is not; it is a number halfway between the integers 0 and 1. Also, 3 1/4 is not an integer. Several numbers that are not integers are *graphed* in Figure 1.3. The **graph** of a number is a point on the number line. Think of the graph of a set of numbers as a picture of the set.

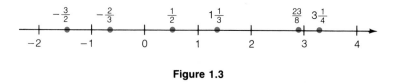

Figure 1.3

All the numbers in Figure 1.3 can be written as quotients of integers. These numbers are examples of **rational numbers.**

<table>
<tr><td>Rational
Numbers</td><td>$\{x|x$ is a quotient of two integers, with denominator not $0\}$

(Read this "the set of all numbers x such that x is a quotient of two integers, with denominator not 0.")</td></tr>
</table>

Since any integer can be written as the quotient of itself and 1, all integers also are rational numbers.

All numbers that can be represented by points on the number line are called **real numbers.**

<table>
<tr><td>Real Numbers</td><td>$\{x|x$ is a number that can be represented by a point
on the number line$\}$</td></tr>
</table>

Although a great many numbers are rational, not all are. For example, a floor tile one foot on a side has a diagonal whose length is the square root of 2 (written $\sqrt{2}$). It can be shown that $\sqrt{2}$ cannot be written as a quotient of integers. Because of this, $\sqrt{2}$ is not rational; it is **irrational.**

<table>
<tr><td>Irrational
Numbers</td><td>$\{x|x$ is a real number that is not rational$\}$</td></tr>
</table>

Examples of irrational numbers include $\sqrt{3}$, $\sqrt{7}$, $-\sqrt{10}$, and π, which is the ratio of the distance around a circle to the distance across it.

An example of a number that is not a real number is the square root of a negative number. These numbers are discussed in the last chapter of this book.

Two ways to represent the relationships among the various types of numbers are shown in Figure 1.4.

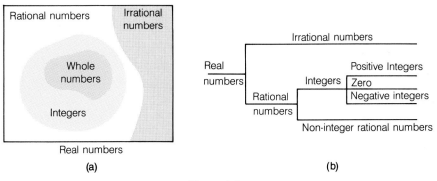

Figure 1.4

Example 1   List the numbers in the set

$$\left\{ -5, \quad -\frac{2}{3}, \quad 0, \quad \sqrt{2}, \quad 3\frac{1}{4}, \quad 5, \quad 5.8 \right\}$$

that belong to each of the following sets of numbers.

(a) Integers

The integers in the set are -5, 0, and 5.

(b) Rational numbers

The rational numbers are -5, $-2/3$, 0, 3 1/4, 5, 5.8, since each of these numbers *can* be written as the quotient of two integers. For example, 5.8 = 58/10.

(c) Real numbers

All the numbers in the set are real numbers.   ✛

3   Given any two whole numbers, you probably can tell which number is smaller. But what happens with negative numbers, as in the set of integers? Positive numbers decrease as the corresponding points on the number line go to the left. For example, $8 < 12$, and 8 is to the left of 12 on the number line. This ordering is extended to all real numbers by definition.

Ordering Real Numbers

For any two real numbers a and b, **a is less than b** if a is to the left of b on the number line.

This means that any negative number is smaller than 0, and any negative number is smaller than any positive number. Also, 0 is smaller than any positive number.

Example 2   Is it true that $-3 < -1$?

To decide whether the statement $-3 < -1$ is true, locate both numbers, -3 and -1, on a number line, as shown in Figure 1.5. Since -3 is to the left of -1 on the number line, -3 is smaller than -1. The statement $-3 < -1$ is true.   ✛

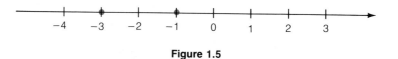

Figure 1.5

4   By a property of the real numbers, for any real number x (except 0), there is exactly one number on the number line the same distance from 0 as x but on the opposite side of 0.

For example, Figure 1.6 shows that the numbers 3 and -3 are each the same distance from 0 but are on opposite sides of 0. The numbers 3 and -3 are called **additive inverses,** or **opposites,** of each other.

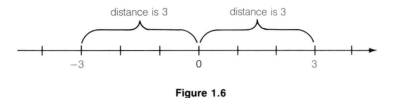

distance is 3   distance is 3

-3 0   3

Figure 1.6

Additive Inverse

> The **additive inverse** of a number x is the number that is the same distance from 0 on the number line as x, but on the opposite side of 0.

The additive inverse of the number 0 is 0 itself. This makes 0 the only real number that is its own additive inverse. Other additive inverses occur in pairs. For example, 4 and -4, and 5 and -5, are additive inverses of each other. Several pairs of additive inverses are shown in Figure 1.7.

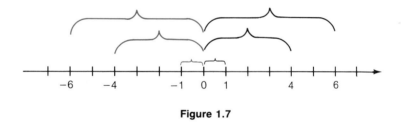

-6 -4 -1 0 1   4   6

Figure 1.7

The additive inverse of a number can be indicated by writing the symbol $-$ in front of the number. With this symbol, the additive inverse of 7 is written -7. The additive inverse of -4 can be written $-(-4)$. Figure 1.7 suggests that 4 is an additive inverse of -4. Since a number can have only one additive inverse, the symbols 4 and $-(-4)$ must represent the same number, which means that

$$-(-4) = 4.$$

A generalization of this idea is given below.

Double Negative Rule

> For any real number x, $-(-x) = x.$

Example 3 Give the additive inverse of each number.

Number	Additive Inverse
-4	$-(-4)$, or 4
0	0
-2	2
19	-19
3	-3 ✦

Example 3 suggests the following rule.

Find the additive inverse of a number by changing the sign of the number.

5 As mentioned above, additive inverses are numbers that are the same distance from 0 on the number line. This idea can also be expressed by saying that a number and its additive inverse have the same absolute value. The **absolute value** of a real number can be defined as the distance between 0 and the number on the number line. The symbol for the absolute value of the number x is $|x|$, read ''the absolute value of x.'' For example, the distance between 2 and 0 on the number line is 2 units, so that

$$|2| = 2.$$

Because the distance between -2 and 0 on the number line is also 2 units,

$$|-2| = 2.$$

Since distance is a physical measurement, which is never negative,

the absolute value of a number is never negative.

For example, $|12| = 12$ and $|-12| = 12$, since both 12 and -12 lie at a distance of 12 units from 0 on the number line. Also, since 0 is a distance 0 units from 0, $|0| = 0$.

In symbols, the absolute value of x is defined as

$$|x| = \begin{cases} x & \text{if } x \geq 0 \\ -x & \text{if } x < 0. \end{cases}$$

By this definition, if x is a positive number or 0, then its absolute value is x itself. For example, since 8 is a positive number, $|8| = 8$. However, if x is a negative number, then its absolute value is the additive inverse of x. This means that if $x = -9$, then $|-9| = -(-9) = 9$, since the additive inverse of -9 is 9.

Example 4 Simplify by removing absolute value symbols.

    (a) $|5| = 5$ **(b)** $|-5| = -(-5) = 5$

(c) $-|5| = -(5) = -5$ **(d)** $-|-14| = -(14) = -14$

(e) $|8 - 2| = |6| = 6$ **(f)** $|3 - 15| = |-12| = -(-12) = 12$ ✚

Parts (e) and (f) of Example 4 show that absolute value bars are also grouping symbols.

1.5 Exercises

Find the additive inverse of each number. For the exercises with absolute value, simplify first before deciding on the additive inverse. See Examples 3 and 4.

1. 8 **2.** 12 **3.** -9 **4.** -11

5. -2 **6.** -3 **7.** $|15|$ **8.** $|5|$

9. $|8|$ **10.** $|0|$ **11.** $|-12|$ **12.** $|-3|$

Select the smaller of the two given numbers. See Examples 2 and 4.

13. $-12, -4$ **14.** $-9, -14$ **15.** $-8, -1$ **16.** $-15, -16$

17. $3, |-4|$ **18.** $5, |-2|$ **19.** $|-3|, |-4|$ **20.** $|-8|, |-9|$

21. $-|-6|, -|-4|$ **22.** $-|-2|, -|-3|$ **23.** $|5 - 3|, |6 - 2|$ **24.** $|7 - 2|, |8 - 1|$

Write true *or* false *for each statement in Exercises 25–46. See Examples 2 and 4.*

25. $-2 < -1$ **26.** $-8 < -4$ **27.** $-3 \geq -7$

28. $-9 \geq -12$ **29.** $-15 \leq -20$ **30.** $-21 \leq -27$

31. $-8 \leq -(-4)$ **32.** $-9 \leq -(-6)$ **33.** $0 \leq -(-4)$

34. $0 \geq -(-6)$ **35.** $6 > -(-2)$ **36.** $-8 > -(-2)$

37. $-4 < -(-5)$ **38.** $-6 \leq -0$ **39.** $|-6| < |-9|$

40. $|-12| < |-20|$ **41.** $-|8| > |-9|$ **42.** $-|12| > |-15|$

43. $-|-5| \geq -|-9|$ **44.** $-|-12| \leq -|-15|$ **45.** $|6 - 5| \geq |6 - 2|$

46. $|13 - 8| \leq |7 - 4|$

47. List all numbers from the set

$$\left\{-9, -\sqrt{7}, -1\frac{1}{4}, -\frac{3}{5}, 0, \sqrt{5}, 3, 5.9, 7\right\}$$

that are

(a) whole numbers; (b) integers; (c) rational numbers;

(d) irrational numbers; (e) real numbers.

48. List all numbers from the set

$$\left\{-5.3, -5, -\sqrt{3}, -1, -\frac{1}{9}, 0, 1.2, 1.8, 3, \sqrt{11}\right\}$$

that are

(a) whole numbers; (b) integers; (c) rational numbers;

(d) irrational numbers; (e) real numbers.

Graph each group of numbers on a number line.

49. 0, 3, −5, −6

50. 2, 6, −2, −1

51. −2, −6, |−4|, 3, −|4|

52. −5, −3, −|−2|, −0, |−4|

53. $\dfrac{1}{4}$, $2\dfrac{1}{2}$, $-3\dfrac{4}{5}$, −4, $-1\dfrac{5}{8}$

54. $5\dfrac{1}{4}$, $4\dfrac{5}{9}$, $-2\dfrac{1}{3}$, 0, $-3\dfrac{2}{5}$

55. |3|, −|3|, −|−4|, −|−2|

56. |6|, −|6|, −|−8|, −|−3|

In Exercises 57–62, give three examples of numbers that satisfy the given condition.

57. Positive real numbers but not integers

58. Real numbers but not positive numbers

59. Real numbers but not whole numbers

60. Rational numbers but not integers

61. Real numbers but not rational numbers

62. Rational numbers but not negative numbers

In Exercises 63–72, answer true or false for each statement.

63. Every rational number is a real number.

64. Every integer is a rational number.

65. Some integers are not real numbers.

66. Every integer is positive.

67. Every whole number is positive.

68. Some irrational numbers are negative.

69. Some real numbers are not rational.

70. Not every rational number is positive.

71. Some whole numbers are not integers.

72. The number 0 is irrational.

For each of the following, give values of a and b that make the following statements true. Then give values of a and b that make the statement false.

73. $|a + b| = |a - b|$

74. $|a - b| = |b - a|$

75. $|a + b| = -|a + b|$

76. $|-(a + b)| = -(a + b)$

1.6   Addition of Real Numbers

Objectives

1   Add two numbers with the same sign on a number line.

2   Add positive and negative numbers.

3   Use the order of operations with real numbers.

1   The number line can be used to show the addition of real numbers, as in the following examples.

Example 1   Use the number line to find the sum 2 + 3.

Add the positive numbers 2 and 3 on the number line by starting at 0 and drawing an arrow two units to the *right*, as shown in Figure 1.8. This arrow represents the number 2 in the sum $2 + 3$. Then, from the right end of this arrow draw another arrow three units to the right. The number below the end of this second arrow is 5, so $2 + 3 = 5$. ✚

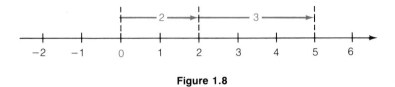

Figure 1.8

Example 2   Use the number line to find the sum $-2 + (-4)$. (Parentheses are placed around the -4 to avoid the confusing use of $+$ and $-$ next to each other.)

Add the negative numbers -2 and -4 on the number line by starting at 0 and drawing an arrow two units to the *left,* as shown in Figure 1.9. The arrow is drawn to the left to represent the addition of a *negative* number. From the left end of this first arrow, draw a second arrow four units to the left. The number below the end of this second arrow is -6, so $-2 + (-4) = -6$. ✚

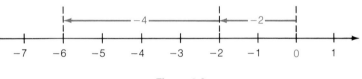

Figure 1.9

In Example 2, the sum of the two negative numbers -2 and -4 is a negative number whose distance from 0 is the sum of the distance of -2 from 0 and the distance of -4 from 0. That is, *the sum of two negative numbers is the negative of the sum of their absolute values.*

$$-2 + (-4) = -(|-2| + |-4|) = -(2 + 4) = -6$$

Add two numbers having the *same* signs by adding the absolute values of the numbers. The result has the same sign as the numbers being added.

Example 3   Find the sums.

(a) $-2 + (-9) = -11$

(b) $-8 + (-12) = -20$

(c) $-15 + (-3) = -18$ ✚

2 Use the number line again to give meaning to the sum of a positive number and a negative number.

Example 4   Use the number line to find the sum $-2 + 5$.

Find the sum $-2 + 5$ on the number line by starting at 0 and drawing an arrow two units to the left. From the left end of this arrow, draw a second arrow five units to the right, as shown in Figure 1.10. The number below the end of the second arrow is 3, so $-2 + 5 = 3$. ✚

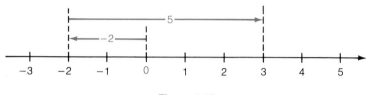

Figure 1.10

Addition of numbers with different signs can also be defined using absolute value.

> Add numbers with *different* signs by finding the difference of the absolute values of the numbers. The answer is given the sign of the number with the larger absolute value.

For example, to add -12 and 5, find their absolute values: $|-12| = 12$ and $|5| = 5$. Then find the difference of these absolute values: $12 - 5 = 7$. Since $|-12| > |5|$, the sum will be negative, so that the final answer is $-12 + 5 = -7$.

While a number line is useful in showing the rules for addition, it is important to be able to do the problems mentally.

Example 5   Check each answer, trying to work the addition mentally. If you get stuck, use a number line.

(a) $7 + (-4) = 3$

(b) $-8 + 12 = 4$

(c) $-\dfrac{1}{2} + \dfrac{1}{8} = -\dfrac{4}{8} + \dfrac{1}{8} = -\dfrac{3}{8}$ Remember to get a common denominator first

(d) $\dfrac{5}{6} + \left(-\dfrac{4}{3}\right) = -\dfrac{1}{2}$

(e) $-4.6 + 8.1 = 3.5$

(f) $-16 + 16 = 0$

(g) $42 + (-42) = 0$ ✚

Parts (f) and (g) of Example 5 suggest the following rule for additive inverses.

The sum of a number and its additive inverse is 0.

The rules for adding signed numbers are summarized below.

Adding Signed Numbers	*Like signs* Add the absolute values of the numbers. The sum has the same sign as the given numbers.
	Unlike signs Find the difference of the larger absolute value and the smaller. Give the answer the sign of the number having the larger absolute value.

3 Sometimes an addition problem involves adding more than two numbers. As mentioned earlier, do the calculations inside the brackets or parentheses until a single number is obtained. Remember to use the order of operations given in Section 1.3 when adding more than two numbers.

Example 6 Find the sums.

(a) $-3 + [4 + (-8)]$

First work inside the brackets. Follow the rules for the order of operations given in Section 1.3.

$$-3 + [4 + (-8)] = -3 + (-4) = -7$$

(b) $8 + [(-2 + 6) + (-3)] = 8 + [4 + (-3)] = 8 + 1 = 9$ ✚

1.6 Exercises

Find the sums. See Examples 1–6.

1. $5 + (-3)$

2. $11 + (-8)$

3. $6 + (-8)$

4. $3 + (-7)$

5. $-6 + (-2)$

6. $-8 + (-3)$

7. $-3 + (-9)$

8. $-11 + (-5)$

9. $12 + (-8)$

10. $10 + (-2)$

11. $4 + [13 + (-5)]$

12. $6 + [2 + (-13)]$

13. $8 + [-2 + (-1)]$

14. $12 + [-3 + (-4)]$

15. $-2 + [5 + (-1)]$

16. $-8 + [9 + (-2)]$

17. $-6 + [6 + (-9)]$

18. $-3 + [4 + (-8)]$

19. $[(-9) + (-14)] + 12$

20. $[(-8) + (-6)] + 10$

21. $-\dfrac{1}{6} + \dfrac{2}{3}$

22. $\dfrac{9}{10} + \left(-\dfrac{3}{5}\right)$

23. $\dfrac{5}{8} + \left(-\dfrac{17}{12}\right)$

24. $-\dfrac{6}{25} + \dfrac{19}{20}$

25. $2\dfrac{1}{2} + \left(-3\dfrac{1}{4}\right)$

26. $-4\dfrac{3}{8} + 6\dfrac{1}{2}$

27. $-6.1 + [3.2 + (-4.8)]$

28. $-9.4 + [-5.8 + (-1.4)]$

29. $[-3 + (-4)] + [5 + (-6)]$

30. $[-8 + (-3)] + [-7 + (-6)]$

31. $[-4 + (-3)] + [8 + (-1)]$

32. $[-5 + (-9)] + [16 + (-21)]$

33. $[-4 + (-6)] + [(-3) + (-8)] + [12 + (-11)]$

34. $[-2 + (-11)] + [12 + (-2)] + [18 + (-6)]$

35. $(-9.648 + 11.237) + [(-4.9123 + 1.8769) + 3.1589]$

36. $[-3.851 + (-2.4691)] + [11.809 + (-1.735)] + (-1.409)$

Write true *or* false *for each statement.*

37. $-9 + 5 + 6 = -2$

38. $-6 + (8 - 5) = -3$

39. $-3 + 5 = 5 + (-3)$

40. $11 + (-6) = -6 + 11$

41. $|-8 + 3| = 8 + 3$

42. $|-4 + 2| = 4 + 2$

43. $|12 - 3| = 12 - 3$

44. $|-6 + 10| = 6 + 10$

45. $[4 + (-6)] + 6 = 4 + (-6 + 6)$

46. $[(-2) + (-3)] + (-6) = 12 + (-1)$

47. $-7 + [-5 + (-3)] = [(-7) + (-5)] + 3$

48. $6 + [-2 + (-5)] = [(-4) + (-2)] + 5$

49. $-5 + (-|-5|) = -10$

50. $|-3| + (-5) = -2$

51. $-2 + |-5| + 3 = 8 + (-2)$

Find all solutions for the following equations from the domain
$\{-3, -2, -1, 0, 1, 2, 3\}$.

52. $x + 3 = 0$

53. $x + 1 = -2$

54. $x + 2 = -1$

55. $14 + x = 12$

56. $x + 8 = 7$

57. $x + (-4) = -6$

58. $x + (-2) = -5$

59. $-8 + x = -6$

60. $-2 + x = -1$

*The word "sum" indicates addition. Write a numerical expression for each
statement and simplify.*

61. The sum of -9 and 2 and 6

62. The sum of 4 and -7 and -3

63. 12 added to the sum of -17 and -6

64. -3 added to the sum of 15 and -1

65. The sum of -11 and -4 increased by -5

66. The sum of -8 and -15 increased by -3

*Solve each word problem in Exercises 67–76 by writing a sum of real numbers
and adding. (No variables are needed.)*

67. Joann has $15. She then spends $6. How much is left?

68. An airplane is flying at an altitude of 6000 feet. It then descends 4000 feet.
What is its final altitude?

69. Chuck is standing 15 feet below sea level in Death Valley. He then goes down another 120 feet. Find his final altitude.

70. Donna has $11 and spends $19. What is her final balance? (Write the answer with a negative number.)

71. One number of Nancy's blood pressure was 120, but then it changed by -30. Find her present blood pressure.

72. The temperature was $-14°$, but then it went down $12°$. Find the new temperature.

73. The temperature at 4 A.M. was $-22°$, but it went up $35°$ by noon. What was the temperature at noon?

74. A man owes $94 to a credit card company. He makes a payment of $60. What amount does he still owe?

75. Sarah Post owes $983.72 on her Visa credit card. She returns items costing $74.18 and $12.53. She makes two purchases of $11.79 each and further purchases of $106.58, $29.81, and $73.24. She makes a payment of $186.50. Find the amount that she then owes.

76. A welder working with stainless steel must use precise measurements. Suppose a welder attaches two pieces of steel that are each 3.589 inches in length, then attaches an additional three pieces that are each 9.089 inches long, and finally cuts off a piece that is 7.612 inches long. Find the length of the welded piece of steel.

77. What number must be added to -9 to get 8?

78. What number must be added to -15 to get 3?

79. The sum of what number and 5 is -11?

80. The sum of what number and -6 is -9?

1.7 Subtraction of Real Numbers

Objectives

1 Find a difference on the number line.

2 Use the definition of subtraction.

3 Work subtraction problems that involve grouping symbols.

1 Recall that the answer to a subtraction problem is a **difference.** Differences of signed numbers can be found by using a number line. Since *addition* of a positive number on the number line is shown by drawing an arrow to the *right*, *subtraction* of a positive number is shown by drawing an arrow to the *left*.

Example 1   Use the number line to find the difference $7 - 4$.

To find the difference $7 - 4$ on the number line, begin at 0 and draw an arrow seven units to the right. From the right end of this arrow, draw an arrow four units to the left, as shown in Figure 1.11. The number at the end of the second arrow shows that $7 - 4 = 3$.   ✛

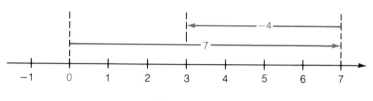

Figure 1.11

The procedure used in Example 1 to find $7 - 4$ is exactly the same procedure that would be used to find $7 + (-4)$, so that

$$7 - 4 = 7 + (-4).$$

2   Example 1 suggests that *subtraction* of a positive number from a larger positive number is the same as *adding* the additive inverse of the smaller number to the larger. This result is extended as the definition of subtraction for all real numbers.

Definition of Subtraction	For any real numbers x and y, $x - y = x + (-y)$.

That is, to **subtract** y from x, *add the additive inverse* (or opposite) of y to x. This definition gives the following procedure for subtracting signed numbers.

Subtracting Signed Numbers	*Step 1* Change the subtraction symbol to addition. *Step 2* Change the sign of the number being subtracted. *Step 3* Add, as in the previous section.

Example 2   Subtract.

┌Change − to +
No change─┐ ↓   ┌Additive inverse of 3
(a) $12 - 3 = 12 + (-3) = 9$

(b) $5 - 7 = 5 + (-7) = -2$

┌Change − to +
No change─┐ ↓   ┌Additive inverse of -5
(c) $-3 - (-5) = -3 + (5) = 2$

(d) $-6 - (-9) = -6 + (9) = 3$ ✛

Subtraction can be used to reverse the result of an addition problem. For example, if 4 is added to a number and then subtracted from the sum, the original number is the result. For example,

$$12 + 4 = 16 \quad \text{and} \quad 16 - 4 = 12.$$

The symbol $-$ has now been used for three purposes:

1. to represent subtraction, as in $9 - 5 = 4$;
2. to represent negative numbers, such as -10, -2, and -3;
3. to represent the additive inverse of a number, as in "the additive inverse of 8 is -8."

More than one use may appear in the same problem, such as $-6 - (-9)$, where -9 is subtracted from -6. The meaning of the symbol depends on its position in the algebraic expression.

3   As before, with problems that have grouping symbols, first do any operations inside the parentheses and brackets. Work from the inside out.

Example 3   Work each problem.

(a)
$$
\begin{aligned}
-6 - [2 - (8 + 3)] &= -6 - [2 - 11] \\
&= -6 - [2 + (-11)] \\
&= -6 - (-9) \\
&= -6 + (9) \\
&= 3
\end{aligned}
$$

(b)
$$
\begin{aligned}
5 - [(-3 - 2) - (4 - 1)] &= 5 - [(-3 + (-2)) - 3] \\
&= 5 - [(-5) - 3] \\
&= 5 - [(-5) + (-3)] \\
&= 5 - (-8) \\
&= 5 + 8 \\
&= 13 \quad \clubsuit
\end{aligned}
$$

1.7   Exercises

Subtract. See Examples 1–3.

1. $3 - 6$
2. $7 - 12$
3. $5 - 9$
4. $8 - 13$

5. $-6 - 2$
6. $-11 - 4$
7. $-9 - 5$
8. $-12 - 15$

9. $6 - (-3)$
10. $12 - (-2)$
11. $-6 - (-2)$
12. $-7 - (-5)$

13. $2 - (3 - 5)$
14. $-3 - (4 - 11)$
15. $\dfrac{1}{2} - \left(-\dfrac{1}{4}\right)$
16. $\dfrac{1}{3} - \left(-\dfrac{4}{3}\right)$

17. $-\dfrac{3}{4} - \dfrac{5}{8}$
18. $-\dfrac{5}{6} - \dfrac{1}{2}$
19. $\dfrac{5}{8} - \left(-\dfrac{1}{2} - \dfrac{3}{4}\right)$
20. $\dfrac{9}{10} - \left(\dfrac{1}{8} - \dfrac{3}{10}\right)$

21. $3.4 - (-8.2)$

22. $5.7 - (-11.6)$

23. $-6.4 - 3.5$

24. $-4.4 - 8.6$

25. $-4.1128 - (7.418 - 9.80632)$

26. $(-1.8142 - 3.7256) - (-9.8025)$

27. $[-7.6892 - (-3.2512)] - (-8.1243)$

28. $[(-2.1463 - 1.8374) - .28174] - [-3.258 - (-1.0926)]$

Work each problem. See Example 3.

29. $(4 - 6) + 12$

30. $(3 - 7) + 4$

31. $(8 - 1) - 12$

32. $(9 - 3) - 15$

33. $6 - (-8 + 3)$

34. $8 - (-9 + 5)$

35. $2 + (-4 - 8)$

36. $6 + (-9 - 2)$

37. $(-5 - 6) - (9 + 2)$

38. $(-4 + 8) - (6 - 1)$

39. $(-8 - 2) - (-9 - 3)$

40. $(-4 - 2) - (-8 - 1)$

41. $-9 + [(3 - 2) - (-4 + 2)]$

42. $-8 - [(-4 - 1) + (9 - 2)]$

43. $-3 + [(-5 - 8) - (-6 + 2)]$

44. $-4 + [(-12 + 1) - (-1 - 9)]$

45. $-9.1237 + [(-4.8099 - 3.2516) + 11.27903]$

46. $-7.6247 - [(-3.9928 + 1.42773) - (-2.80981)]$

47. $[-12.1035 - (8.11725 + 3.83122)] - 17.40963$

48. $[-34.9122 + (6.45378 - 12.14273)] - 8.46922$

Write the given problem in symbols (no variables are needed). Then solve.

49. Subtract -6 from 12.

50. Subtract -8 from 15.

51. From -25, subtract -4.

52. What number is 6 less than -9?

53. -24 is how much greater than -27?

54. How much greater is 8 than -5?

55. How much greater is -7.3 than -8.4?

56. -12.4 is how much greater than -14.3?

The word ''difference'' indicates subtraction. Write a numerical expression for each statement and simplify.

57. Find the difference of 8 and -2.

58. Find the difference of 3 and -7.

59. Add -11 to the difference of -4 and 2.

60. Add 8 to the difference of 1 and -3.

61. From the sum of -12 and -3, subtract 4.

62. From the sum of 8 and -13, subtract -2.

Solve each word problem by writing it as a difference of real numbers and subtracting.

63. The temperature dropped $10°$ below the previous temperature of $-5°$. Find the new temperature.

64. Bill owed his brother \$10. He repaid \$6 and later borrowed \$7. What positive or negative number represents his present financial status?

65. The bottom of Death Valley is 282 feet below sea level. The top of Mt. Whitney, visible from Death Valley, has an altitude of 14,494 feet above sea level. Using zero as sea level, find the difference between these two elevations.

66. Harriet has $15, and Joann is $12 in debt. Find the difference between these amounts.

67. A chemist is running an experiment at a temperature of $-174.6°$. She then raises the temperature by $2.3°$. Find the new temperature.

68. One year a company had a "profit" of $-\$25,000$. The next year, the profit decreased by 7200. Find the profit the next year.

69. One company made a profit of $76,000, and another company lost $29,000. Find the difference between these.

70. One reading of a dial was 7.904. A second reading was -3.291. By how much had the reading declined?

Let $x = -5$, $y = -4$, and $z = 8$. Find the value of each expression.

71. $x + y$ 72. $y - x$

73. $x + y - z$ 74. $z - y - x$

1.8 Multiplication of Real Numbers

Objectives

1. Find the product of a positive and a negative number.
2. Find the product of two negative numbers.
3. Use the order of operations in multiplication with signed numbers.
4. Evaluate expressions involving variables.

You already know the rules for multiplying positive numbers. For example,

the product of two positive numbers is positive.

But what about multiplying other real numbers? Any rules for multiplication of real numbers ought to be consistent with the rules for multiplication from arithmetic. For example, the product of 0 and any real number (positive or negative) should be 0.

Multiplication by Zero	For any real number x, $x \cdot 0 = 0$.

1 In order to define the product of a positive and a negative number so that the result is consistent with the multiplication of two positive numbers, look at the following pattern.

$$
\begin{array}{l}
3 \cdot 5 = 15 \\
3 \cdot 4 = 12 \\
3 \cdot 3 = 9 \\
3 \cdot 2 = 6 \\
3 \cdot 1 = 3 \\
3 \cdot 0 = 0 \\
3 \cdot (-1) = ?
\end{array}
\quad
\begin{array}{l}
\text{Numbers} \\
\text{decrease} \\
\text{by 3}
\end{array}
$$

What should $3(-1)$ equal? The product $3(-1)$ represents the sum

$$-1 + (-1) + (-1) = -3,$$

so the product should be -3. Also,

$$3(-2) = -2 + (-2) + (-2) = -6$$

and

$$3(-3) = -3 + (-3) + (-3) = -9.$$

These results maintain the pattern in the list above, which suggests the following rule.

For any positive real numbers x and y,

$$x(-y) = -(xy) \qquad \text{and} \qquad (-x)y = -(xy).$$

That is,

the product of two numbers with opposite signs is negative.

Example 1 Find the products using the multiplication rule given above.

(a) $8(-5) = -(8 \cdot 5) = -40$ **(b)** $5(-4) = -(5 \cdot 4) = -20$

(c) $(-7)(2) = -(7 \cdot 2) = -14$ **(d)** $(-9)(3) = -(9 \cdot 3) = -27$ ✠

2 The product of two positive numbers is positive, and the product of a positive and a negative number is negative. What about the product of two negative numbers? Look at another pattern.

$$
\begin{array}{l}
(-5)(4) = -20 \\
(-5)(3) = -15 \\
(-5)(2) = -10 \\
(-5)(1) = -5 \\
(-5)(0) = 0 \\
(-5)(-1) = ?
\end{array}
\quad
\begin{array}{l}
\text{Numbers} \\
\text{increase} \\
\text{by 5}
\end{array}
$$

The numbers on the left of the equals sign shown in color decrease by 1 for each step down the list. The products on the right increase by 5 for each step down the list. To maintain this pattern, $(-5)(-1)$ should be 5 more than $(-5)(0)$, or 5 more than 0, so

$$(-5)(-1) = 5.$$

The pattern continues with

$$(-5)(-2) = 10$$
$$(-5)(-3) = 15$$
$$(-5)(-4) = 20$$
$$(-5)(-5) = 25,$$

and so on. This pattern suggests the following rule.

For any positive real numbers x and y, $(-x)(-y) = xy$.

In words,

the product of two negative numbers is positive.

Example 2   Find the products using the multiplication rule given above.

(a) $(-9)(-2) = 9 \cdot 2 = 18$ **(b)** $(-6)(-12) = 6 \cdot 12 = 72$

(c) $(-8)(-1) = 8 \cdot 1 = 8$ **(d)** $(-15)(-2) = 15 \cdot 2 = 30$ ✖

A summary of the results for multiplying signed numbers is given here.

Multiplying Signed Numbers | The product of two numbers having the *same* sign is *positive*, and the product of two numbers having *different* signs is *negative*.

3   The next example shows the order of operations discussed earlier, used with the multiplication of positive and negative numbers.

Example 3   Perform the indicated operations.

(a) $(-9)(2) - (-3)(2)$

First find all products, working from left to right.

$$(-9)(2) - (-3)(2) = -18 - (-6)$$

Now perform the subtraction.

$$-18 - (-6) = -18 + 6 = -12$$

(b) $(-6)(-2) - (3)(-4) = 12 - (-12) = 12 + 12 = 24$

(c) $-5(-2 - 3) = -5(-5) = 25$ ✖

4   The last examples show numbers substituted for variables in multiplication problems.

Example 4   Evaluate the expression

$$(3x + 4y)(-2m)$$

given the following values.

(a) $x = -1$, $y = -2$, $m = -3$

First substitute the given values for the variables. Then find the value of the expression.

$(3x + 4y)(-2m)$	$= [3(-1) + 4(-2)][-2(-3)]$	Put parentheses around the number for each variable
	$= [-3 + (-8)][6]$	Find the products
	$= (-11)(6)$	Use order of operations
	$= -66$	

(b) $x = 7$, $y = -9$, $m = 5$

Substitute.

$(3x + 4y)(-2m)$	$= [3 \cdot 7 + 4(-9)](-2 \cdot 5)$	Use parentheses around -9
	$= [21 + (-36)](-10)$	Find the products
	$= (-15)(-10)$	
	$= 150$ ✚	

Example 5   Evaluate $2x^2 - 3y^2$ for $x = -3$ and $y = -4$.
Use parentheses as shown.

$$2(-3)^2 - 3(-4)^2 = 2(9) - 3(16)$$
$$= 18 - 48$$
$$= -30 \quad ✚$$

1.8   Exercises

Find the products. In Exercises 23–26, round to the nearest thousandth. See Examples 1 and 2.

1. $(-3)(-4)$ **2.** $(-3)(4)$ **3.** $3(-4)$ **4.** $-2(-8)$

5. $(-10)(-12)$ **6.** $9(-5)$ **7.** $0(-11)$ **8.** $3(-15)$

9. $(15)(-11)$ **10.** $(-9)(-4)$ **11.** $-\dfrac{3}{8} \cdot \left(-\dfrac{10}{9}\right)$ **12.** $-\dfrac{5}{4} \cdot \left(-\dfrac{5}{8}\right)$

13. $\left(-1\dfrac{1}{4}\right)\left(\dfrac{2}{15}\right)$ **14.** $\left(\dfrac{3}{7}\right)\left(-1\dfrac{5}{9}\right)$ **15.** $(-8)\left(-\dfrac{3}{4}\right)$ **16.** $(-6)\left(-\dfrac{5}{3}\right)$

17. $(-5.1)(.02)$ **18.** $(-3.7)(-2.1)$ **19.** $(1.8)(-7.2)$ **20.** $(-4.7)(-6.8)$

21. $(3.4)(-3.5)$ **22.** $(-5.2)(-7.4)$ **23.** $(-12.804)(4.12)$ **24.** $(3.871)(-5.463)$

25. $(-8.91)(-4.725)$ **26.** $(-6.4972)(-13.8015)$

Perform the indicated operations. See Example 3.

27. $6 - 4 \cdot 5$ **28.** $3 - 2 \cdot 9$ **29.** $-9 - (-2) \cdot 3$

30. $-11 - (-7) \cdot 4$ **31.** $9(6 - 10)$ **32.** $5(12 - 15)$

33. $-6(2 - 4)$ **34.** $-9(5 - 8)$ **35.** $(4 - 9)(2 - 3)$

36. $(6 - 11)(3 - 6)$ **37.** $(2 - 5)(3 - 7)$ **38.** $(5 - 12)(2 - 6)$

39. $(-4 - 3)(-2) + 4$ **40.** $(-5 - 2)(-3) + 6$ **41.** $5(-2) - 4$

42. $9(-6) - 8$ **43.** $3(-4) - (-2)$ **44.** $5(-2) - (-9)$

Evaluate the following expressions, given $x = -2$, $y = 3$, and $a = -4$. See Examples 4 and 5.

45. $5x - 2y + 3a$ **46.** $6x - 5y + 4a$ **47.** $(2x + y)(3a)$

48. $(5x - 2y)(-2a)$ **49.** $(3x - 4y)(-5a)$ **50.** $(6x + 2y)(-3a)$

51. $(-5 + x)(-3 + y)(2 - a)$ **52.** $(6 - x)(5 + y)(3 + a)^2$ **53.** $-2y^2 + 3(a + 2)$

54. $5(x - 4) - 4a^2$ **55.** $3a^2 - (x - 3)^2$ **56.** $(y - 4)^2 - 2x^3$

Find the solution for the following equations from the domain $\{-3, -2, -1, 0, 1, 2, 3\}$.

57. $2x = -4$ **58.** $-4m = 0$ **59.** $-9y = 0$

60. $-8p = 16$ **61.** $-9r = 27$ **62.** $2x + 1 = -3$

63. $3w + 3 = -3$ **64.** $-4a + 2 = 10$ **65.** $-5t + 6 = 11$

The word "product" indicates multiplication. Write a numerical expression for each statement and simplify.

66. The product of -9 and 2 is added to 6.

67. The product of 4 and -7 is added to -9.

68. After the product of -1 and 6 is found, the result is subtracted from -9.

69. Twice the product of -8 and 2 is subtracted from -4.

70. Nine is subtracted from the product of 7 and -6.

71. Three is subtracted from the product of -2 and 3.

Perform the indicated operations.

72. $(-8 - 2)(-4)^2 - (-5)$ **73.** $(-9 - 1)(-2)^2 - (-6)$ **74.** $|-4(-2)| + |-4|$

75. $|8(-5)| + |-2|$ **76.** $|2|(-4)^2 + |6| \cdot |-4|$ **77.** $|-3|(-2) + |-8| \cdot |5|$

Replace a, b, and c with various integers to decide whether the following statements are true or false. For each false statement, give an example showing it is false.

78. $ab = ba$ **79.** $a(bc) = (ab)c$

80. $a^2 = 2a$ **81.** $a(b + c) = ab + ac$

1.9   Division of Real Numbers

Objectives

1 Find the reciprocal, or multiplicative inverse, of a number.
2 Divide with signed numbers.
3 Simplify numerical expressions involving quotients.
4 Factor integers.

1 The difference of two numbers is found by adding the additive inverse of the second number to the first. The *quotient* of two numbers is found by *multiplying* by the *multiplicative inverse*. By definition, since

$$8 \cdot \frac{1}{8} = \frac{8}{8} = 1 \quad \text{and} \quad \frac{5}{4} \cdot \frac{4}{5} = \frac{20}{20} = 1,$$

the multiplicative inverse of 8 is 1/8, and of 5/4 is 4/5.

Multiplicative Inverse	Pairs of numbers whose product is 1 are **multiplicative inverses,** or **reciprocals,** of each other.

Example 1   Give the multiplicative inverse of each number.

Number	*Multiplicative inverse (Reciprocal)*
4	$\dfrac{1}{4}$
-5	$\dfrac{1}{-5}$ or $-\dfrac{1}{5}$
$-\dfrac{5}{8}$	$-\dfrac{8}{5}$
0	None ✦

Why is there no multiplicative inverse for the number 0? Suppose that k is to be the multiplicative inverse of 0. Then $k \cdot 0$ should equal 1. But $k \cdot 0 = 0$ for any number k. Since there is no value of k that is a solution of the equation $k \cdot 0 = 1$,

0 has no multiplicative inverse.

2 By definition, the quotient of x and y is the product of x and the multiplicative inverse of y.

Definition of Division	For any real numbers x and y, with $y \neq 0$, $\dfrac{x}{y} = x \cdot \dfrac{1}{y}.$

The definition of division indicates that y, the number to divide by, cannot be 0. The reason is that 0 has no multiplicative inverse, so that $1/0$ is not a number. For this reason

division by 0 is not defined

and is never permitted. If a division problem turns out to involve division by 0, write "not defined."

Since division is defined in terms of multiplication, all the rules for multiplication of signed numbers also apply to division.

Example 2   Find the quotients using the definition of division.

(a) $\dfrac{12}{3} = 12 \cdot \dfrac{1}{3} = 4$

(b) $\dfrac{-10}{2} = -10 \cdot \dfrac{1}{2} = -5$

(c) $\dfrac{-14}{-7} = -14\left(\dfrac{1}{-7}\right) = 2$

(d) $\dfrac{-10}{0}$ not defined   ✦

The following rule for division with signed numbers follows from the definition of division and the rules for multiplication with signed numbers.

Dividing Signed Numbers	The quotient of two numbers having the *same* sign is *positive;* the quotient of two numbers having *different* signs is *negative.*

Example 3   Find the quotients.

(a) $\dfrac{8}{-2} = -4$

(b) $\dfrac{-45}{-9} = 5$

(c) $-\dfrac{1}{8} \div \left(-\dfrac{3}{4}\right) = -\dfrac{1}{8} \cdot \left(-\dfrac{4}{3}\right) = \dfrac{1}{6}$ ✦

From the definitions of multiplication and division of real numbers,

$$\frac{-40}{8} = -40 \cdot \frac{1}{8} = -5,$$

and

$$\frac{40}{-8} = 40\left(\frac{1}{-8}\right) = -5,$$

so that

$$\frac{-40}{8} = \frac{40}{-8}.$$

Based on this example, the quotient of a positive and a negative number can be expressed in any of the following three forms.

For any positive real numbers x and y, $\dfrac{-x}{y} = \dfrac{x}{-y} = -\dfrac{x}{y}$.

The form $x/-y$ is seldom used.

The quotient of two negative numbers can be expressed as a quotient of two positive numbers.

For any positive real numbers x and y, $\dfrac{-x}{-y} = \dfrac{x}{y}$.

3 The next example shows how to simplify numerical expressions involving quotients.

Example 4   Simplify $\dfrac{5(-2) - (3)(4)}{2(1 - 6)}$.

Simplify the numerator and denominator separately. Then divide.

$$\frac{5(-2) - (3)(4)}{2(1 - 6)} = \frac{-10 - 12}{2(-5)} = \frac{-22}{-10} = \frac{11}{5} \quad \clubsuit$$

The rules for operations with signed numbers are summarized here.

Operations with Signed Numbers

> **Addition**
> *Like signs*   Add the absolute values of the numbers. The result has the same sign as the numbers.
> *Unlike signs*   Subtract the number with the smaller absolute value from the one with the larger. Give the result the sign of the number having the larger absolute value.
>
> **Subtraction**
> Add the additive inverse, or negative of the second number.
>
> **Multiplication and Division**
> *Like signs*   The product or quotient of two numbers with like signs is positive.
> *Unlike signs*   The product or quotient of two numbers with unlike signs is negative.
> Division by 0 is not defined.

4 In Section 1.1 the definition of a *factor* was given for whole numbers. (For example, since $9 \cdot 5 = 45$, both 9 and 5 are factors of 45.) The definition can now be extended to integers.

If the product of two integers is a third integer, then each of the two integers is a **factor** of the third. For example, $(-3)(-4) = 12$, so -3 and -4 are both factors of 12. The factors of 12 are the numbers -12, -6, -4, -3, -2, -1, 1, 2, 3, 4, 6, and 12.

Example 5   The following chart shows several integers and the factors of those integers.

Integer	Factors
18	-18, -9, -6, -3, -2, -1, 1, 2, 3, 6, 9, 18
20	-20, -10, -5, -4, -2, -1, 1, 2, 4, 5, 10, 20
15	-15, -5, -3, -1, 1, 3, 5, 15
7	-7, -1, 1, 7
1	-1, 1 ✚

1.9   Exercises

Find the multiplicative inverse where one exists. Round to the nearest thousandth in Exercises 9 and 10. See Example 1.

1. 9

2. 8

3. -4

4. -10

5. 0

6. $\dfrac{3}{4}$

7. $-\dfrac{9}{10}$

8. $-\dfrac{4}{5}$

9. .8697

10. 1.4385

Find the quotients. See Examples 2 and 3.

11. $\dfrac{-10}{5}$

12. $\dfrac{-12}{3}$

13. $\dfrac{18}{-3}$

14. $\dfrac{-280}{-20}$

15. $\dfrac{-180}{0}$

16. $\dfrac{-350}{0}$

17. $\dfrac{0}{-2}$

18. $\dfrac{0}{12}$

19. $-\dfrac{1}{2} \div \left(-\dfrac{3}{4}\right)$

20. $-\dfrac{5}{8} \div \left(-\dfrac{3}{16}\right)$

21. $(-4.2) \div (-2)$

22. $(-9.8) \div (-7)$

23. $\dfrac{4}{-.8}$

24. $\dfrac{-6}{.3}$

25. $\dfrac{12}{2-5}$

26. $\dfrac{15}{3-8}$

27. $\dfrac{50}{2-7}$

28. $\dfrac{30}{5-5}$

29. $\dfrac{-40}{8-(-8)}$

30. $\dfrac{-72}{6-(-2)}$

31. $\dfrac{-120}{-3-(-5)}$

32. $\dfrac{-200}{-6-(-4)}$

33. $\dfrac{-30-(-8)}{-11}$

34. $\dfrac{-17-(-12)}{5}$

Simplify the numerators and denominators separately. Then find the quotients. See Example 4.

35. $\dfrac{-8(-2)+4}{3-(-1)}$

36. $\dfrac{-12(-3)-6}{-15-(-3)}$

37. $\dfrac{-15(2)-10}{-7-3}$

38. $\dfrac{-20(6)+6}{-5-1}$

39. $\dfrac{-2(6)^2+3}{2-(-1)}$

40. $\dfrac{3(-8)^2+3}{-6+1}$

41. $\dfrac{6^2+4^2}{5(2+13)}$

42. $\dfrac{4^2-5^2}{3(6-9+2)}$

43. $\dfrac{3^2+5^2+1}{4^2-1^2}$

Find all integer factors of each number. See Example 5.

44. 36   **45.** 32   **46.** 25

47. 40   **48.** 17   **49.** 29

Find the solution of each equation from the domain
$\{-8, -6, -4, -2, 0, 2, 4, 6, 8\}.$

50. $\dfrac{x}{4} = -2$ **51.** $\dfrac{x}{2} = -1$ **52.** $\dfrac{n}{-2} = 3$ **53.** $\dfrac{t}{-2} = -2$

54. $\dfrac{q}{-3} = 0$ **55.** $\dfrac{p}{5} = 0$ **56.** $\dfrac{m}{-2} = -4$ **57.** $\dfrac{y}{-1} = 2$

Write each statement in symbols and find the solution. The domain is the set of integers from -12 to 12, inclusive.

58. Six times a number is -42.

59. Four times a number is -32.

60. When a number is divided by 3, the result is -3.

61. When a number is divided by -3, the answer is -4.

62. The quotient of a number and 2 is -6.
$\left(\text{Write the quotient as } \dfrac{x}{2}.\right)$

63. The quotient of a number and 5 is -2.

64. The quotient of 6 and one more than a number is 3.

65. When the square of a number is divided by 3, the result is 12.

Write a numerical expression for each statement and simplify.

66. Add -9 to the quotient of 15 and -3.

67. Add 4 to the quotient of -28 and 4.

68. Subtract 3 from the quotient of 8 and -2.

69. Subtract 3 from the quotient of -16 and 4.

70. Find the product of -10 and 2. Then find the quotient of this product and 5.

Find each quotient. Round to the nearest thousandth in Exercises 74, 75, 77, and 78.

71. $\dfrac{-5(2) + [3(-2) - 4]}{-3 - (-1)}$

72. $\dfrac{[4(-1) + 3](-2)}{-2 - 3}$

73. $\dfrac{-9(-2) - [(-4)(-2) + 3]}{-2(3) - 2(2)}$

74. $\dfrac{11.096 - [-8.1151^2 - 6.2431]}{.8 - .05(.3)}$

75. $\dfrac{[-6.42 - (-3.891)](2.471)}{-.05(6.31 - 2.43)}$

76. $\dfrac{10^2 - [5^2 - 15]}{2(8^2 + 3^2) + 2}$

77. $\dfrac{[1 - (.86)^2] + (2.5)^2}{(-1.43)^3 - [-3.76 + 6.45]}$

78. $\dfrac{(-.49)^2 - (.21)^2(.32)}{(4.11)(3.58) - (-1.12)^3}$

1.10   Properties of Addition and Multiplication

Objectives
Identify the use of the following properties:

1 commutative;

2 associative;

3 identity;

4 inverse;

5 distributive.

The basic properties of addition and multiplication of real numbers are discussed in this section. In the following statements, x, y, and z represent real numbers.

1   Commutative properties   By the commutative properties, two numbers can be added or multiplied in any order.

$$x + y = y + x$$
$$xy = yx$$

Example 1   Use a commutative property to complete each statement.

(a) $-8 + 5 = 5 +$ _____

By the commutative property for addition, the missing number is -8, since $-8 + 5 = 5 + (-8)$.

(b) $(-2)(7) =$ _____(-2)

By the commutative property for multiplication, the missing number is 7, since $(-2)(7) = (7)(-2)$.   ✚

2   Associative properties   By the associative properties, when three numbers are added or multiplied, the first two may be grouped together, or the last two may be grouped together, without affecting the answer.

$$(x + y) + z = x + (y + z)$$
$$(xy)z = x(yz)$$

Example 2   Use an associative property to complete each statement.

(a) $8 + (-1 + 4) = (8 +$ _____$) + 4$

The missing number is -1.

(b) $[2 \cdot (-7)] \cdot 6 = 2 \cdot$ _____

The completed expression on the right should be $2 \cdot [(-7) \cdot 6]$.   ✚

By the associative property of addition, the sum of three numbers will be the same however the numbers are "associated" in groups. For this reason, parentheses can be left out in many addition problems. For example, either

$$(-1 + 2) + 3 \qquad \text{or} \qquad -1 + (2 + 3)$$

can be written as just

$$-1 + 2 + 3.$$

In the same way, parentheses can also be left out of many multiplication problems.

Example 3   **(a)** Is $(2 + 4) + 5 = 2 + (4 + 5)$ an example of the associative property?

The order of the three numbers is the same on both sides of the equals sign. The only change is in the grouping of the numbers. Therefore, this is an example of the associative property.

(b) Is $6(3 \cdot 10) = 6(10 \cdot 3)$ an example of the associative property or the commutative property?

The same numbers, 3 and 10, are grouped on both sides of the equals sign. On the left, however, the 3 appears first in $(3 \cdot 10)$. On the right, the 10 appears first. Since the change involves the order of the numbers, this is an example of the commutative property.

(c) Is $(8 + 1) + 7 = 8 + (7 + 1)$ an example of the associative property or the commutative property?

In this statement, both the order and the grouping are changed. On the left the order of the three numbers is 8, 1, and 7. On the right it is 8, 7, and 1. On the left the 8 and 1 are grouped, and on the right the 7 and 1 are grouped. Therefore, both the associative and the commutative properties were used.   ✜

The commutative and associative properties are useful in reducing the work involved when adding several numbers.

Example 4   Find the sum: $23 + 41 + 2 + 9 + 25$.

The commutative and associative properties make it possible to choose pairs of numbers whose sums are easy to add.

$$23 + 41 + 2 + 9 + 25 = (41 + 9) + (23 + 2) + 25$$
$$= 50 + 25 + 25$$
$$= 100 \quad ✜$$

3   **Identity properties**   The identity properties say that the sum of 0 and any number equals that number, and the product of 1 and any number equals that number.

$$x + 0 = x \qquad \text{and} \qquad 0 + x = x$$
$$x \cdot 1 = x \qquad \text{and} \qquad 1 \cdot x = x$$

The number 0 leaves the identity, or value, of any real number unchanged by addition. For this reason, 0 is called the **identity element for addition.** Since multiplication by 1 leaves any real number unchanged, 1 is the **identity element for multiplication.**

Example 5   These statements are examples of the identity properties.

(a) $-3 + 0 = -3$

(b) $0 + \dfrac{1}{2} = \dfrac{1}{2}$

(c) $-\dfrac{3}{4} \cdot 1 = -\dfrac{3}{4}$

(d) $1 \cdot 25 = 25$ ✜

4   **Inverse properties**   By the inverse properties, the sum of the numbers x and $-x$ is 0, and the product of the nonzero numbers x and $1/x$ is 1.

$$x + (-x) = 0 \qquad \text{and} \qquad -x + x = 0$$

$$x \cdot \dfrac{1}{x} = 1 \qquad \text{and} \qquad \dfrac{1}{x} \cdot x = 1 \qquad (x \neq 0)$$

Recall that $-x$ is the **additive inverse** of x and $1/x$ is the **multiplicative inverse** of the nonzero number x.

Example 6   These statements are examples of the inverse properties.

(a) $\dfrac{2}{3} \cdot \dfrac{3}{2} = 1$

(b) $(-5)\left(-\dfrac{1}{5}\right) = 1$

(c) $-\dfrac{1}{2} + \dfrac{1}{2} = 0$

(d) $4 + (-4) = 0$ ✜

5   Look at the following statements.

$$2(5 + 8) = 2(13) = 26$$
$$2(5) + 2(8) = 10 + 16 = 26$$

Since both expressions equal 26,

$$2(5 + 8) = 2(5) + 2(8).$$

This result is an example of the *distributive property,* the only property involving *both* addition and multiplication. With this property, a product can be changed to a sum or difference.

Distributive property   By the distributive property, multiplying a number x by a sum of numbers $y + z$ gives the same result as multiplying x by y and x by z and then adding the two products.

$$x(y + z) = xy + xz \qquad \text{and} \qquad (y + z)x = yx + zx$$

As the arrows show, the x outside the parentheses is "distributed" over the y and z inside. The second statement follows from the first by the commutative property. Using the definition of subtraction gives another form of the distributive property for subtraction.

$$x(y - z) = xy - xz \quad \text{and} \quad (y - z)x = yx - zx$$

The distributive property also can be extended to more than two numbers.

$$x(y + z + w) = xy + xz + xw$$

Example 7   Simplify using the distributive property.

(a) $5(9 + 6) = 5 \cdot 9 + 5 \cdot 6$ Multiply both terms by 5

$$= 45 + 30$$
$$= 75$$

(b) $4(x + 5 + y) = 4x + 4 \cdot 5 + 4y = 4x + 20 + 4y$

(c) $-2(x + 3) = -2x + (-2)(3) = -2x - 6$

(d) $3(k - 9) = 3k - 3 \cdot 9 = 3k - 27$

(e) $6 \cdot 8 + 6 \cdot 2 = 6(8 + 2) = 6(10) = 60$

In this example, the distributive property was used backwards.

(f) $20\left(\dfrac{3}{5} + \dfrac{5}{4} - \dfrac{7}{10}\right) = 20\left(\dfrac{3}{5}\right) + 20\left(\dfrac{5}{4}\right) - 20\left(\dfrac{7}{10}\right)$

$$= 12 + 25 - 14$$
$$= 23$$

(g) $8(3r + 11t + 5z) = 8(3r) + 8(11t) + 8(5z)$

$$= (8 \cdot 3)r + (8 \cdot 11)t + (8 \cdot 5)z \qquad \text{Associative property}$$
$$= 24r + 88t + 40z \quad \clubsuit$$

The distributive property is used to remove the parentheses from expressions such as $-(2y + 3)$. Do this by writing $-(2y + 3)$ as $-1 \cdot (2y + 3)$.

$$-(2y + 3) = -1 \cdot (2y + 3)$$
$$= -1 \cdot (2y) + (-1) \cdot (3)$$
$$= -2y - 3$$

Example 8   Simplify each expression.

(a) $-(7r - 8) = -1(7r) + (-1)(-8) = -7r + 8$

(b) $-(-9w + 2) = 9w - 2 \quad \clubsuit$

The properties of addition and multiplication are summarized below.

Properties of Addition and Multiplication	For all real numbers x, y, and z:		
		Addition	*Multiplication*
	Commutative	$x + y = y + x$	$xy = yx$
	Associative	$(x + y) + z = x + (y + z)$	$(xy)z = x(yz)$
	Identity	$x + 0 = x$ and $0 + x = x$	$x \cdot 1 = x$ and $1 \cdot x = x$
	Inverse	$x + (-x) = 0$ and $-x + x = 0$	$x \cdot \dfrac{1}{x} = 1$ and $\dfrac{1}{x} \cdot x = 1$ $(x \neq 0)$
	Distributive	$x(y + z) = xy + xz$ and $(y + z)x = yx + zx$	

1.10   Exercises

Label each statement as an example of the commutative, associative, identity, inverse, or distributive property. See Examples 1–7.

1. $5(15 \cdot 8) = (5 \cdot 15)8$

2. $(23)(9) = (9)(23)$

3. $12(-8) = (-8)(12)$

4. $(-9)[6(-2)] = [-9(6)](-2)$

5. $2 + (p + r) = (p + r) + 2$

6. $(4m)n = 4(mn)$

7. $-6 + 12 = 12 + (-6)$

8. $(-9)(-11) = (-11)(-9)$

9. $6 + (-6) = 0$

10. $9 + (11 + 4) = (9 + 11) + 4$

11. $-4 + 0 = -4$

12. $0 + (-9) = -9$

13. $3\left(\dfrac{1}{3}\right) = 1$

14. $-7\left(-\dfrac{1}{7}\right) = 1$

15. $\dfrac{2}{3} \cdot 1 = \dfrac{2}{3}$

16. $-\dfrac{9}{4} \cdot 1 = -\dfrac{9}{4}$

17. $6(5 - 2x) = 6 \cdot 5 - 6(2x)$

18. $5(2m) + 5(7n) = 5(2m + 7n)$

Use the indicated property to write a new expression that is equal to the given expression. Simplify the new expression if possible. See Examples 1, 2, and 5–7.

19. $9k$;   commutative

20. $z + 5$;   commutative

21. $m + 0$;   identity

22. $(-9) + 0$;   identity

23. $3(r + m)$;   distributive

24. $11(k + z)$;   distributive

25. $8 \cdot \dfrac{1}{8}$;   inverse

26. $\dfrac{1}{6} \cdot 6$;   inverse

27. $12 + (-12)$;   inverse

28. $-8 + 8$;   inverse

29. $5 + (-5)$;   commutative

30. $-9\left(-\dfrac{1}{9}\right)$;   commutative

31. $-3(r + 2)$;   distributive

32. $4(k - 5)$;   distributive

33. $9 \cdot 1$;   identity

34. $1(-4)$;   identity

35. $5k(-6)$;   associative

36. $(m + 4) + (-2)$;   associative

Use the distributive property to rewrite each expression. See Example 7.

37. $5(m + 2)$

38. $6(k + 5)$

39. $-4(r + 2)$

40. $-3(m + 5)$

41. $-8(k - 2)$

42. $-4(z - 5)$

43. $-9(a + 3)$

44. $-3(p + 5)$

45. $(r + 8)4$

46. $(m + 12)6$

47. $(8 - k)(-2)$

48. $(9 - r)(-3)$

49. $2(5r + 6m)$

50. $5(2a + 4b)$

51. $-4(3x - 4y)$

52. $-9(5k - 12m)$

53. $5 \cdot 8 + 5 \cdot 9$

54. $4 \cdot 3 + 4 \cdot 9$

55. $7 \cdot 2 + 7 \cdot 8$

56. $6x + 6m$

57. $9p + 9q$

58. $8(2x) + 8(3p)$

59. $5(7z) + 5(8w)$

60. $11(2r) + 11(3s)$

Use the distributive property to rewrite each expression. See Example 8.

61. $-(3k + 5)$

62. $-(2z + 12)$

63. $-(4y - 8)$

64. $-(3r - 15)$

65. $-(-4 + p)$

66. $-(-12 + 3a)$

67. $-(-1 - 15r)$

68. $-(-14 - 6y)$

Decide whether or not the events in Exercises 69–72 are commutative.

69. Getting out of bed and taking a shower

70. Putting on your right shoe or your left shoe first

71. Taking English or taking history

72. Putting on your shoe or putting on your sock

73. Evaluate $25 - (6 - 2)$ and evaluate $(25 - 6) - 2$. Do you think subtraction is associative?

74. Evaluate $180 \div (15 \div 3)$ and evaluate $(180 \div 15) \div 3$. Do you think division is associative?

Replace a, b, and c with various integers to decide whether the following statements are true or false. For each false statement, give an example showing it is false.

75. $(-a) + (-b) = -(a + b)$

76. $(-a) + b = b - a$

77. $a + (b \cdot c) = (a + b)(a + c)$

78. $(a + b) + (-b) = a$

Chapter 1 *Summary*

Symbols of Equality and Inequality	$=$ is equal to	$\neq$ is not equal to
	$<$ is less than	$>$ is greater than
	$\leq$ is less than or equal to	$\geq$ is greater than or equal to

Order of Operations

If grouping symbols are present, simplify within them, innermost first (and above and below fraction bars separately), in the following order.

Step 1   Apply all exponents.

Step 2   Do any multiplications or divisions in the order in which they occur, working from left to right.

Step 3   Do any additions or subtractions in the order in which they occur, working from left to right.

If no grouping symbols are present, start with Step 1.

Operations with Signed Numbers

Addition

Like signs   Add the absolute values of the numbers. The result has the same sign as the numbers.

Unlike signs   Subtract the number with the smaller absolute value from the one with the larger. Give the result the sign of the number having the larger absolute value.

Subtraction

Add the additive inverse, or negative of the second number.

Multiplication and Division

Like signs   The product or quotient of two numbers with like signs is positive.

Unlike signs   The product or quotient of two numbers with unlike signs is negative.

Division by 0 is not defined.

Properties of the Real Numbers

For all real numbers a, b, and c:

Commutative properties $a + b = b + a$
$ab = ba$

Associative properties $a + (b + c) = (a + b) + c$
$a(bc) = (ab)c$

Identity properties   There is a real number 0 such that

$$a + 0 = a \quad \text{and} \quad 0 + a = a.$$

There is a real number 1 such that

$$a \cdot 1 = a \quad \text{and} \quad 1 \cdot a = a.$$

Inverse property of addition   For each real number a, there is a single real number $-a$ such that

$$a + (-a) = 0 \quad \text{and} \quad (-a) + a = 0.$$

Inverse property of multiplication	For each nonzero real number a, there is a single real number $1/a$ such that

$$a \cdot \frac{1}{a} = 1 \quad \text{and} \quad \frac{1}{a} \cdot a = 1.$$

Distributive property	$a(b + c) = ab + ac$

Chapter 1 *Review Exercises*

If you have trouble with any of these exercises, look in the section given in brackets.

① **[1.1]** *Write each fraction in lowest terms.*

1. $\dfrac{18}{54}$ **2.** $\dfrac{60}{72}$ **3.** $\dfrac{22}{110}$ **4.** $\dfrac{120}{150}$

Perform each operation. Write the answers in lowest terms.

5. $\dfrac{7}{10} \cdot \dfrac{1}{5}$ **6.** $\dfrac{32}{15} \div \dfrac{8}{5}$

7. $\dfrac{3}{8} + \dfrac{1}{2}$ **8.** $6\dfrac{1}{4} - 5\dfrac{3}{4}$

Work the following word problems.

9. John painted 1/4 of a room on Monday and 1/3 of the room on Tuesday. What portion was then painted?

10. A contractor installs toolsheds. Each requires 1 1/4 cubic yards of concrete. How much concrete would be needed for 25 sheds?

[1.2] *Which of the symbols $<$, $\leq$, $>$, and $\geq$ make the following statements true? Give all possible correct answers.*

11. .87 _____ .865 **12.** .94 _____ .904 **13.** $\dfrac{2}{3}$ _____ .7 **14.** $\dfrac{3}{4}$ _____ $\dfrac{4}{5}$

[1.3] *Find the following.*

15. 4^3 **16.** $\left(\dfrac{7}{3}\right)^3$

Simplify each statement and then decide whether the statement is true or false.

17. $70 < 4 - 9(6 + 2)$ **18.** $2 \cdot 3 + 5(4 + 8) \leq 68$ **19.** $\dfrac{2(1 + 3)}{3(2 + 1)} < 1$

20. $6^2 - 4^2 \geq 5$ **21.** $5^2 + 4^2 \leq 40$

[1.4] *Find the numerical value of the given expression when* $x = 2$ *and* $y = 5$.

22. $\dfrac{3x - y}{2x}$ **23.** $\dfrac{2x}{y + 1}$ **24.** $x^2 + y^2$ **25.** $\dfrac{3x^2 + 5y^2}{7x^2 - y^2}$

Write each word phrase using algebraic expressions. Use x as the variable.

26. Six subtracted from three times a number   **27.** The difference of 4 and a number

[1.5] *Select the smaller of the two given numbers.*

28. $5, |-8|$ **29.** $9, |-7|$ **30.** $-|-2|, -|-9|$ **31.** $-|-7|, -|-4|$

Decide whether each statement is true or false.

32. $3 \le -(-5)$ **33.** $9 \ge -(-10)$ **34.** $-3 > -2$ **35.** $-5 < -4$

36. $-|7| \le -|-2|$ **37.** $-|4| > -|-3|$ **38.** $|-(-2)| > -|-2|$ **39.** $|-5| < |5|$

Graph each group of numbers on a number line.

40. $-1, -3, |-4|, |-1|$ **41.** $3\dfrac{1}{4}, -2\dfrac{4}{5}, -1\dfrac{1}{8}, \dfrac{2}{3}$ **42.** $|-2|, -|-5|, -|3|, -|0|$

[1.6] *Find the sums.*

43. $(-9 + 6) + (-10)$ **44.** $[-3 + (-5)] + (-8)$ **45.** $\dfrac{7}{8} + \left(-\dfrac{3}{10}\right)$

46. $\dfrac{7}{12} + \left(-\dfrac{2}{9}\right)$ **47.** $14.2 + (-8.6)$ **48.** $-11.3 + (-2.9)$

Solve the word problems.

49. Tom has $9. He spends $11. Find his new balance.

50. The temperature is $15°$. It goes down $22°$. Find the new temperature.

[1.7] *Find any sums or differences.*

51. $15 - (-3)$ **52.** $-8 - 9$ **53.** $\dfrac{3}{4} - \left(-\dfrac{2}{3}\right)$

54. $-\dfrac{1}{5} - \left(-\dfrac{7}{10}\right)$ **55.** $(-9 + 6) - (-3)$ **56.** $(-15 - 7) - (-9)$

[1.8] *Simplify each expression.*

57. $-\dfrac{4}{5}\left(-\dfrac{10}{7}\right)$ **58.** $-\dfrac{3}{8}\left(-\dfrac{16}{15}\right)$ **59.** $(-11.3)(2.5)$

60. $(-9.4)(-2.8)$ **61.** $8(-9) - (6)(-2)$ **62.** $-6(-2) - (-4)(3)$

Evaluate the following expressions, given $x = -5$, $y = 4$, *and* $z = -3$.

63. $y^2 - 2z^2$ **64.** $(2x - 8y)(z^2)$ **65.** $3z^2 - 4x^2$

[1.9] *Find the quotients.*

66. $\dfrac{-8 - (-9)}{10}$ **67.** $\dfrac{-7 - (-21)}{8}$ **68.** $\dfrac{8 - 4(-2)}{-5(3) - 1}$

[1.10] *Decide whether each statement is an example of the commutative, associative, identity, inverse, or distributive property.*

69. $8.974 \cdot 1 = 8.974$

70. $-\dfrac{2}{3} + \dfrac{2}{3} = 0$

71. $7 + 4m = 4m + 7$

72. $8(4 \cdot 3) = (8 \cdot 4)3$

73. $\dfrac{5}{8} \cdot \dfrac{8}{5} = 1$

74. $9p + 0 = 9p$

Simplify using the distributive property.

75. $6(3k + 5r)$

76. $-3(2m - 7k)$

❷ *Perform the indicated operations.*

77. $[(-4) + 6 + (-9)] + [-3 + (-5)]$

78. $\left(\dfrac{5}{8}\right)^2$

79. $-|(-9)(4)| - |-2|$

80. 6^3

81. $\dfrac{5(-3) - 8(3)}{(-5)(-4) + (-7)}$

82. $\dfrac{7}{10} - \dfrac{1}{4}$

83. $\dfrac{11^2 - 5^2 + 2}{5^2 + 5^2 - 1^2}$

84. $\dfrac{7^2 - 3^2}{2^2 + 4^2}$

85. $-11(-4) - (3)(-7)$

86. $2\dfrac{3}{4} + 5\dfrac{5}{12}$

87. $-3 + [(-12 + 15) - (-4 - 5)]$

88. $\dfrac{8}{5} \div \dfrac{32}{15}$

89. $(-8 - 2) - [(-1 - 4) - (-1)]$

90. $[(-6) + (-7) + 8] + [8 + (-15)]$

91. $\dfrac{3}{7} \cdot \dfrac{5}{6}$

92. $|6(-8)| - |-3|$

Write a numerical expression and simplify it. Use x as the variable, if one is needed.

93. One year a company spent $28,000 on advertising. The next year the amount spent on advertising was reduced by $7000. How much was spent on advertising the second year?

94. The product of a number and one more than the number

95. The sum of a number and 4, divided by 8

96. On a cold day, the temperature was $-17°$. It then increased by $19°$. Find the new temperature.

Chapter 1 *Test*

1. Write $\dfrac{84}{132}$ in lowest terms.

2. Add: $\dfrac{3}{8} + \dfrac{7}{12} + \dfrac{11}{15}$.

3. Divide: $\dfrac{6}{5} \div \dfrac{19}{15}$.

Answer true *or* false *for the following.*

4. $6 - |-4| \geq 10$

5. $4[-20 + 7(-2)] \leq 135$

6. $-2 \geq \dfrac{-36 - 3(-6)}{(-2)(-4) - (-1)}$

7. $-14 < \dfrac{32 + 4(1 + 5)}{6(3 + 5) - 2}$

8. $(-3)^2 + 2^2 = 5^2$

Find the numerical value of the given expression when m = 6 and p = 2.

9. $5m + 2p^3$

10. $\dfrac{7m^2 - p^2}{m + 4}$

Select the smaller number from each pair.

11. $6, \quad -|-8|$

12. .742,   .705

Write each word phrase as an algebraic expression. Use x as the variable.

13. Twice a number subtracted from 11

14. The quotient of 9 and the difference of a number and 8

Decide whether the given number is a solution for the equation.

15. $6m + m + 2 = 37; \quad 5$

16. $8(y - 3) + 2y = 18; \quad 4$

Perform the indicated operations wherever possible.

17. $-9 - (4 - 11) + (-5)$

18. $-2\dfrac{1}{5} + 5\dfrac{1}{4}$

19. $-6 - [-5 + (8 - 9)]$

20. $3^2 + (-7) - (2^3 - 5)$

21. $|-6| \cdot (-5) + 2 \cdot |8|$

22. $\dfrac{-7 - (-5 + 1)}{-4 - (-3)}$

23. $\dfrac{-6[5 - (-1 + 4)]}{-9[2 - (-1)] - 6(-4)}$

24. $\dfrac{15(-4 - 2)}{16(-2) + (-7 - 1)(-3 - 1)}$

Find the solution for each equation. Choose solutions from the domain
$\{-9, -5, -4, -3, -2, -1, 3, 9\}.$

25. $\dfrac{p}{-3} = 3$

26. $2x + 1 = -7$

27. $-4x - 7 = 5$

Match the property in Column I with all examples of it from Column II.

Column I	Column II

28. Commutative

A. $12 + (-12) = -12 + 12$

29. Associative

B. $\dfrac{6}{5}\left(\dfrac{5}{6}\right) = 1$

30. Identity

31. Inverse

C. $1(-9) = -9$

32. Distributive

D. $2(5) + 2(9) = 2(5 + 9)$

E. $8x = x \cdot 8$

F. $-(2 - p) = -2 + p$

G. $-7 + 7 = 0$

H. $9 + (2 + y) = (9 + 2) + y$

I. $-12 + 0 = -12$

33. Use the distributive property to simplify $-(3 - 4m)$.

Solving Equations and Inequalities

Objectives

1. Simplify expressions.
2. Identify terms and numerical coefficients.
3. Identify like terms.
4. Combine like terms.
5. Simplify expressions from word problems.

1 It is often necessary to simplify the expressions on either side of an equation as the first step in solving it. This section shows how to simplify expressions using the properties of addition and multiplication introduced in Chapter 1.

Example 1 Simplify the following expressions.

(a) $4x + 8 + 9$

Since $8 + 9 = 17$,

$$4x + 8 + 9 = 4x + 17.$$

(b) $4(3m - 2n)$

Use the distributive property first.

$$4(3m - 2n) = 4(3m) - 4(2n)$$
$$= (4 \cdot 3)m - (4 \cdot 2)n \qquad \text{Associative property}$$
$$= 12m - 8n$$

(c) $6 + 3(4k + 5) = 6 + 3(4k) + 3(5) \qquad$ Distributive property
$$= 6 + (3 \cdot 4)k + 3(5) \qquad \text{Associative property}$$
$$= 6 + 12k + 15$$
$$= 6 + 15 + 12k \qquad\qquad \text{Commutative property}$$
$$= 21 + 12k$$

(d) $5 - (2y - 8) = 5 - 1 \cdot (2y - 8) \qquad -(2y - 8) = -1 \cdot (2y - 8)$
$$= 5 - 2y + 8 \qquad\qquad \text{Distributive property}$$
$$= 5 + 8 - 2y \qquad\qquad \text{Commutative property}$$
$$= 13 - 2y \quad \clubsuit$$

The steps using the commutative and associative properties will not be shown in the rest of the examples, but you should be aware that they usually are involved.

2   A **term** is a single number, or an indicated product or quotient, raised to powers. Examples of terms include

$$-9x^2, \quad 15y, \quad -3, \quad 8m^2n, \quad \frac{2}{p}, \quad \text{and} \quad k.$$

The **numerical coefficient** of the term $9m$ is 9, the numerical coefficient of $-15x^3y^2$ is -15, the numerical coefficient of x is 1, and the numerical coefficient of 8 is 8.

Example 2   Give the numerical coefficient of the following terms.

Term	Numerical Coefficient
$-7y$	-7
$8p$	8
$34r^3$	34
$-26x^5yz^4$	-26
$-k$	$-1 \quad \clubsuit$

3   Terms with exactly the same variables that have the same exponents are **like terms.** For example, $9m$ and $4m$ have the same variables and are like terms. Also, $6x^3$ and $-5x^3$ are like terms. The terms $-4y^3$ and $4y^2$ have different exponents and are **unlike terms.**

Recall the distributive property:

$$x(y + z) = xy + xz.$$

This statement can also be written "backward" as

$$xy + xz = x(y + z).$$

The sum or difference of like terms may be expressed using this form of the distributive property. For example,

$$3x + 5x = (3 + 5)x = 8x.$$

4   This process is called **combining like terms.** It is important to notice that **only like terms may be combined.**

Example 3   Combine like terms in the following expressions.

(a) $9m + 5m$

Use the distributive property as given above to combine the like terms.

$$9m + 5m = (9 + 5)m = 14m$$

(b) $6r + 3r + 2r = (6 + 3 + 2)r = 11r$

(c) $4x + x = 4x + 1x = (4 + 1)x = 5x$ (*Note:* $x = 1x.$)

(d) $16y^2 - 9y^2 = (16 - 9)y^2 = 7y^2$

(e) $32y + 10y^2$ cannot be simplified because $32y$ and $10y^2$ are unlike terms. The distributive property cannot be used here to combine coefficients.   ✚

When an expression involves parentheses, the distributive property is used both "forward" and "backward" to simplify the expression by combining like terms as shown in the following examples.

Example 4   Combine like terms in the following expressions.

(a) $14y + 2(6 + 3y) = 14y + 2(6) + 2(3y)$ Distributive property

$\qquad\qquad\qquad = 14y + 12 + 6y$ Multiply

$\qquad\qquad\qquad = 20y + 12$ Combine like terms

(b) $9k - 6 - 3(2 - 5k) = 9k - 6 - 3(2) - 3(-5k)$ Distributive property

$\qquad\qquad\qquad\qquad = 9k - 6 - 6 + 15k$ Multiply

$\qquad\qquad\qquad\qquad = 24k - 12$ Combine like terms

(c) $-(2 - r) + 10r = -1(2 - r) + 10r$ $-(2 - r) = -1(2 - r)$

$\qquad\qquad\qquad = -1(2) - 1(-r) + 10r$ Distributive property

$\qquad\qquad\qquad = -2 + r + 10r$ Multiply

$\qquad\qquad\qquad = -2 + 11r$ Combine like terms

(d) $5(2a - 6) - 3(4a - 9)$

$$= 10a - 30 - 12a + 27 \qquad \text{Distributive property}$$
$$= 10a + (-30) + (-12a) + 27 \quad \text{Definition of subtraction}$$
$$= 10a + (-12a) + (-30) + 27 \quad \text{Commutative property}$$
$$= -2a - 3 \qquad\qquad\qquad \text{Combine like terms} \quad \clubsuit$$

Example 4(d) shows that the commutative property can be used with subtraction by treating the subtracted terms as adding a negative term.

5   The next example shows how to simplify the result of converting a word phrase into a mathematical expression.

Example 5   Five times a number, four times a number, and six times a number are added to 9.

$$9 + 5x + 4x + 6x \qquad \text{Write as a mathematical expression}$$
$$= 9 + 15x \qquad \text{Combine like terms} \quad \clubsuit$$

2.1   Exercises

Give the numerical coefficient of the following terms. See Example 2.

1. $15y$

2. $7z$

3. $-22m^4$

4. $-2k^7$

5. $35a^4b^2$

6. $12m^5n^4$

7. -9

8. 21

9. y^2

10. x^4

11. $-r$

12. $-z$

Write like *or* unlike *for the following groups of terms.*

13. $6m, \ -14m$

14. $-2a, \ 5a$

15. $7z^3, \ 7z^2$

16. $10m^5, \ 10m^6$

17. $25y, \ -14y, \ 8y$

18. $-11x, \ 5x, \ 7x$

19. $2, \ 5, \ -2$

20. $-8, \ 3, \ 9$

Simplify the following expressions by combining terms. See Examples 1 and 3.

21. $9y + 8y$

22. $15m + 12m$

23. $-4a - 2a$

24. $-3z - 9z$

25. $12b + b$

26. $30x + x$

27. $2k + 9 + 5k + 6$

28. $2 + 17z + 1 + 2z$

29. $-5y + 3 - 1 + 5 + y - 7$

30. $2k - 7 - 5k + 7k - 3 - k$

31. $-2x + 3 + 4x - 17 + 20$

32. $r - 6 - 12r - 4 + 6r$

33. $16 - 5m - 4m - 2 + 2m$

34. $6 - 3z - 2z - 5 + z - 3z$

35. $-10 + x + 4x - 7 - 4x$

36. $-p + 10p - 3p - 4 - 5p$

37. $1 + 7x + 11x - 1 + 5x$

38. $-r + 2 - 5r + 3 + 4r$

39. $6y^2 + 11y^2 - 8y^2$

40. $-9m^3 + 3m^3 - 7m^3$

41. $2p^2 + 3p^2 - 8p^3 - 6p^3$

42. $5y^3 + 6y^3 - 3y^2 - 4y^2$

43. $-7.913q^2 + 2.804q - 11.723 + 5.069q^2 - 8.124q - 6.977$

44. $8.271m^3 - 3.722m^2 + 5.006 - 3.994m^3 + 4.129m^2 - 1.728$

Use the distributive property to simplify the following expressions. See Example 4.

45. $6(5t + 11)$

46. $2(3x + 4)$

47. $-3(n + 5)$

48. $-4(y - 8)$

49. $-3(2r - 3) + 2(5r + 3)$

50. $-4(5y - 7) + 3(2y - 5)$

51. $8(2k - 1) - (4k - 3)$

52. $6(3p - 2) - (5p + 1)$

53. $-2(-3k + 2) - (5k - 6) - 3k - 5$

54. $-2(3r - 4) - (6 - r) + 2r - 5$

55. $-7.916(3y - 2.8) - 4.72(9.1 - 5y)$

56. $4.873(8.2q - 7.3) + 1.29(3.9 - .42q)$

Convert the following word phrases into mathematical expressions. Use x as the variable. Combine terms whenever possible. See Example 5.

57. Two times a number is subtracted from the sum of a number and 2.

58. Four times a number is added to the sum of the number and -15.

59. Three times a number is subtracted from twice the number. This result is subtracted from 9 times the number.

60. A number is subtracted from 4 times the number, with this result subtracted from the sum of 6 and five times the number.

61. Nine is multiplied times the sum of five times a number and 4, with the result subtracted from the difference of 4 and twice the number.

62. Seven times a number is added to -9. This result is subtracted from four times the sum of three times the number and 5.

Combine terms.

63. $5 - 3x + 2y + x - 7 + 4y$

64. $3x^2 + 17 - 4x + 3x^2 + 9x + 5x$

65. $-2p + p^2 - p + 4p^2 + 6p^2 + 8p + 1$

66. $-3a + 2 + 5b - a - b - 3a$

67. $6(r - 5) + 3r - (4 - r)$

68. $-(m + 6) - 4(m + 1) + 3$

69. $z + 2(5 - z) - (2z + 3) - z^2$

70. $2k^2 + 4(k - 1) - 3(k + 2) + k$

Review Exercises *Most of the exercise sets in the rest of the book end with brief sets of review exercises. These exercises are designed to help you review ideas needed for the next few sections in the chapter. If you need help with these review exercises, look in the sections indicated each time.*

Find the additive inverse of each number. For help see Section 1.5.

71. 6

72. 15

73. -4

74. -8

Add a number to each expression so that the result is just x.

75. $x - 7$

76. $x + 10$

77. $x + 6$

78. $x - 3$

2.2   *The Addition Property of Equality*

Objectives

1   Identify linear equations.

2   Use the addition property of equality.

3   Simplify equations, and then use the addition property of equality.

1   Methods of solving linear equations are introduced in this section.

Linear Equation

> A linear equation can be written in the form
> $$ax + b = 0,$$
> for real numbers a and b, with $a \neq 0$.

Linear equations are solved by using a series of steps to produce a simpler equation of the form

$$x = \text{a number.}$$

2   According to the equation

$$x - 5 = 2,$$

both $x - 5$ and 2 represent the same number, since this is the meaning of the equals sign. Solve the equation by changing the left side from $x - 5$ to just x. This is done by adding 5 to $x - 5$. Keep the two sides equal by also adding 5 on the right side.

$$x - 5 = 2 \qquad \text{Given equation}$$
$$x - 5 + 5 = 2 + 5 \qquad \text{Add 5 on both sides}$$

Here 5 was added on both sides of the equation. Now simplify each side separately to get

$$x = 7.$$

The solution of the given equation is 7. Check by replacing x with 7 in the given equation.

$$x - 5 = 2 \qquad \text{Given equation}$$
$$7 - 5 = 2 \qquad \text{Let } x = 7$$
$$2 = 2 \qquad \text{True}$$

Since this final result is true, 7 checks as the solution.

The equation above was solved by adding the same number to both sides, as justified by the **addition property of equality.**

Addition Property of Equality	If A, B, and C are algebraic expressions, then the equations
	$$A = B \quad \text{and} \quad A + C = B + C$$
	have exactly the same solution. In words, the same number may be added to both sides of an equation.

In the addition property, C represents a real number. That means that numbers or terms with variables, or even sums of terms that represent real numbers, can be added to both sides of an equation.

Example 1   Solve the equation $x - 16 = 7$.

If the left side of this equation were just x, the solution would be found. Get x alone by using the addition property of equality and adding 16 on both sides.

$$x - 16 = 7$$
$$(x - 16) + 16 = 7 + 16 \qquad \text{Add 16 on both sides}$$
$$x = 23$$

Check by substituting 23 for x in the original equation.

$$x - 16 = 7 \qquad \text{Given equation}$$
$$23 - 16 = 7 \qquad \text{Let } x = 23$$
$$7 = 7 \qquad \text{True}$$

Since the check results in a true statement, 23 is the solution.   ❖

In this example, why was 16 added to both sides of the equation $x - 16 = 7$? The equation would be solved if it could be rewritten so that one side contained only the variable and the other side contained only a number. Since $x - 16 + 16 = x + 0 = x$, adding 16 on the left side simplifies that side to just x, the variable, as desired.

The addition property of equality says that the same number may be *added* to both sides of an equation. As was shown in Chapter 1, subtraction is defined in terms of addition. Because of the way subtraction is defined, the addition property also permits *subtracting* the same number on both sides of an equation.

Example 2   Solve the equation $3k + 17 = 4k$.

First, get all terms that contain variables on the same side of the equation. One way to do this is to subtract $3k$ from each side.

$$3k + 17 = 4k$$
$$3k + 17 - 3k = 4k - 3k \qquad \text{Subtract } 3k \text{ from both sides}$$
$$17 = k$$

The solution is 17.

The equation $3k + 17 = 4k$ could also be solved by first subtracting $4k$ from each side, as follows.

$$3k + 17 = 4k$$

$$3k + 17 - 4k = 4k - 4k \qquad \text{Subtract } 4k$$

$$17 - k = 0$$

Now subtract 17 from both sides.

$$17 - k - 17 = 0 - 17 \qquad \text{Subtract } 17$$

$$-k = -17$$

This result gives the value of $-k$, but not of k itself. However, this result does say that the additive inverse of k is -17, which means that k must be 17.

$$-k = -17$$

$$k = 17$$

This solution agrees with the first one; however, the first method required fewer steps. Check the solution in the original equation. ✚

3 Sometimes an equation must be simplified as a first step in its solution.

Example 3 Solve the equation $4r + 5r - \dfrac{3}{2} + 8 - 3r - 5r = 12 + \dfrac{5}{2}$.

First, simplify on each side of the equation by combining like terms.

$$4r + 5r - \frac{3}{2} + 8 - 3r - 5r = 12 + \frac{5}{2}$$

$$4r + 5r - 3r - 5r - \frac{3}{2} + \frac{16}{2} = \frac{24}{2} + \frac{5}{2} \qquad \text{Get a common denominator}$$

$$r + \frac{13}{2} = \frac{29}{2} \qquad \text{Combine like terms}$$

Subtract 13/2 from both sides of this equation.

$$r + \frac{13}{2} - \frac{13}{2} = \frac{29}{2} - \frac{13}{2}$$

$$r = \frac{16}{2} = 8$$

The solution of the given equation is 8. (Check this in the original equation.) ✚

Example 4   Solve the equation $3(2 + 5x) - (1 + 14x) = 6$.

Use the distributive property first to simplify the equation.

$$3(2 + 5x) - (1 + 14x) = 6$$
$$6 + 15x - 1 - 14x = 6 \quad \text{Distributive property}$$
$$x + 5 = 6 \quad \text{Combine like terms}$$

Subtract 5 on both sides of the equation to get the variable term alone on one side.

$$x = 1$$

Check this answer by substituting 1 for x in the original equation.   ✦

2.2 Exercises

Solve each equation by using the addition property of equality. Check each solution. See Examples 1 and 2.

1. $x - 3 = 7$

2. $x + 5 = 13$

3. $7 + k = 5$

4. $9 + m = 4$

5. $3r - 10 = 2r$

6. $2p = p + 3$

7. $7z = -8 + 6z$

8. $4y = 3y - 5$

9. $m + 5 = 0$

10. $k - 7 = 0$

11. $2 + 3x = 2x$

12. $10 + r = 2r$

13. $2p + 6 = 10 + p$

14. $5r + 2 = -1 + 4r$

15. $2k + 2 = -3 + k$

16. $6 + 7x = 6x + 3$

17. $x - 5 = 2x + 6$

18. $-3r + 7 = -4r - 19$

19. $6z + 3 = 5z - 3$

20. $6t + 5 = 5t + 7$

21. $2p = p + \dfrac{1}{2}$

Solve the following equations. First simplify each side of the equation as much as possible. Check each solution. See Examples 2 and 3.

22. $4x + 3 + 2x - 5x = 2 + 8$

23. $3x + 2x - 6 + x - 5x = 9 + 4$

24. $9r + 4r + 6 - 8 = 10r + 6 + 2r$

25. $-3t + 5t - 6t + 4 - 3 = -3t + 2$

26. $11z + 2 + 4z - 3z = 5z - 8 + 6z$

27. $2k + 8k + 6k - 4k - 8 + 2 = 3k + 2 + 10k$

28. $4m + 8m - 9m + 2 - 5 = 4m + 6$

29. $15y - 4y + 8 - 2 + 7 - 4 = 4y + 2 + 8y$

30. $-9p + 4p - 3p + 2p - 6 = -5p - 6$

31. $5x - 2x + 3x - 4x + 8 - 2 + 4 = x + 10$

Solve the following equations. Check each solution. See Example 4.

32. $5(m - 1) - 6m = -8$

33. $2(k + 5) - 3k = 8$

34. $-4(3y + 2) = -13y - 10$

35. $-5(2q - 3) = -11q + 19$

36. $(11y + 10) - 3(3 + 4y) = 2$

37. $(15p - 3) - 2(7p + 1) = -3$

38. $2(r + 5) - (9 + r) = -1$

39. $4(y - 6) - (3y + 2) = 8$

40. $-6(2a + 1) + (13a - 7) = 4$

41. $-5(3k - 3) + (1 + 16k) = 2$

42. $4(7x - 1) + 3(2 - 5x) = 4(3x + 5)$

43. $9(2m - 3) - 4(5 + 3m) = 5(4 + m)$

44. $-2(8p + 7) - 3(4 - 7p) = 2(3 + 2p) - 6$

45. $-5(8 - 2z) + 4(7 - z) = 7(8 + z) - 3$

46. $5m = 4m + \dfrac{2}{3}$

47. $\dfrac{4}{3}z = \dfrac{1}{3}z - 5$

48. $\dfrac{9}{5}m = \dfrac{4}{5}m + 6$

49. $-2 - \dfrac{3}{4}y = \dfrac{1}{4}y$

50. $-11 - \dfrac{7}{9}p = \dfrac{2}{9}p$

51. $\dfrac{11}{4}r - \dfrac{1}{2} = \dfrac{7}{4}r + \dfrac{2}{3}$

52. $\dfrac{9}{2}y - \dfrac{3}{4} = \dfrac{7}{2}y + \dfrac{5}{8}$

53. $2.7a + 5 = 1.7a$

54. $4.7p - 3 = 3.7p$

Write an equation using the following information. For example, for the sentence "Twice the sum of a number and six is 18," write $2(x + 6) = 18$. Then find the number.

55. Three times a number is 17 more than twice the number. Find the number.

56. If six times a number is subtracted from seven times a number, the result is -9. Find the number.

57. If five times a number is added to three times the number, the result is the sum of seven times the number and 9. Find the number.

58. If nine times a number is subtracted from eleven times the number, the result is -4 more than three times the number. Find the number.

59. The sum of twice a number and 5 is multiplied by 6. The result is 8 less than 13 times the number. Find the number.

60. Four times a number is subtracted from 7. The result is multiplied by 5, giving 3 more than -19 times the number. Find the number.

Review Exercises   Use the associative and inverse properties to simplify. See Section 1.10.

61. $3\left(\dfrac{1}{3}m\right)$

62. $5\left(\dfrac{1}{5}q\right)$

63. $\dfrac{1}{7}(7y)$

64. $\dfrac{1}{12}(12z)$

65. $-\dfrac{1}{9}(-9q)$

66. $-\dfrac{1}{4}(-4m)$

67. $-\dfrac{5}{8}\left(-\dfrac{8}{5}r\right)$

68. $-\dfrac{9}{10}\left(-\dfrac{10}{9}x\right)$

2.3 The Multiplication Property of Equality

Objectives

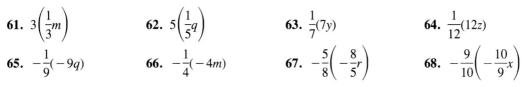

1 Use the multiplication property of equality.

2 Simplify equations, and then use the multiplication property of equality.

3 Solve equations such as $-r = a$.

The addition property of equality by itself is not enough to solve an equation like $3x + 2 = 17$.

$$3x + 2 = 17$$
$$3x + 2 - 2 = 17 - 2 \qquad \text{Subtract 2 from both sides}$$
$$3x = 15 \qquad \text{Simplify}$$

The variable x is not alone on one side of the equation: the equation has $3x$ instead. Another property is needed to change $3x = 15$ to $x = $ a number.

1 If $3x = 15$, then $3x$ and 15 both represent the same number. Multiplying both $3x$ and 15 by the same number will also result in an equality. The **multiplication property of equality** states that both sides of an equation can be multiplied by the same number.

<table>
<tr><td>Multiplication
Property of
Equality</td><td>If A, B, and C are real numbers, then the equations

$$A = B \qquad \text{and} \qquad AC = BC$$

have exactly the same solution. (Assume that $C \neq 0$.) In words, both sides of an equation may be multiplied by the same nonzero number.</td></tr>
</table>

This property can be used to solve $3x = 15$. The $3x$ on the left must be changed to $1x$, or x, instead of $3x$. To get x, multiply both sides of the equation by 1/3. Use 1/3 since it is the reciprocal of the coefficient of x. This works because $3 \cdot 1/3 = 3/3 = 1$.

$$3x = 15$$
$$\frac{1}{3}(3x) = \frac{1}{3} \cdot 15 \qquad \text{Multiply both sides by } \frac{1}{3}$$
$$\left(\frac{1}{3} \cdot 3\right)x = \frac{1}{3} \cdot 15 \qquad \text{Associative property}$$
$$1x = 5 \qquad \text{Multiply}$$
$$x = 5 \qquad \text{Inverse property}$$

The solution of the equation is 5. Check this by substituting 5 for x in the given equation.

Just as the addition property of equality permits subtracting the same number from both sides of an equation, the multiplication property of equality permits dividing both sides of an equation by the same nonzero number. For example, the equation $3x = 15$, solved above by multiplication, could also be solved by dividing both sides by 3, as follows.

$$3x = 15$$
$$\frac{3x}{3} = \frac{15}{3} \qquad \text{Divide by 3}$$
$$x = 5 \qquad \text{Simplify}$$

Example 1   Solve the equation $25p = 30$.

Get p (instead of $25p$) by using the multiplication property of equality and dividing both sides of the equation by 25, the coefficient of p.

$$25p = 30$$

$$\frac{25p}{25} = \frac{30}{25} \qquad \text{Divide by 25}$$

$$p = \frac{30}{25} = \frac{6}{5}$$

The solution is $\frac{6}{5}$. Check by substituting $\frac{6}{5}$ for p in the given equation.

$$25p = 30$$

$$\frac{25}{1}\left(\frac{6}{5}\right) = 30$$

$$30 \stackrel{.}{=} 30$$

The solution is $\frac{6}{5}$. ✦

In the next two examples, multiplication produces the solution more quickly than division would.

Example 2   Solve the equation $\frac{a}{4} = 3$.

Replace $\frac{a}{4}$ by $\frac{1}{4}a$ since division by 4 is the same as multiplication by $\frac{1}{4}$. Get a alone by multiplying both sides by 4, the reciprocal of the coefficient of a.

$$\frac{a}{4} = 3$$

$$\frac{1}{4}a = 3 \qquad \text{Change } \frac{a}{4} \text{ to } \frac{1}{4}a$$

$$4 \cdot \frac{1}{4}a = 4 \cdot 3 \qquad \text{Multiply by 4}$$

$$1a = 12$$

$$a = 12$$

Check the answer.

$$\frac{a}{4} = 3 \qquad \text{Given equation}$$

$$\frac{12}{4} = 3 \qquad \text{Let } a = 12$$

$$3 = 3 \qquad \text{True}$$

The solution 12 is correct. ✦

Example 3 Solve the equation $\frac{3}{4}h = 6$.

To get h alone, multiply both sides of the equation by $\frac{4}{3}$, the reciprocal of $\frac{3}{4}$.

Use $\frac{4}{3}$ because $\frac{4}{3} \cdot \frac{3}{4}h = 1 \cdot h = h$.

$$\frac{3}{4}h = 6$$

$$\frac{4}{3}\left(\frac{3}{4}h\right) = \frac{4}{3} \cdot 6 \qquad \text{Multiply by } \frac{4}{3}$$

$$1 \cdot h = \frac{4}{3} \cdot \frac{6}{1}$$

$$h = 8$$

The solution is 8. Check the answer by substitution in the given equation. ✤

2 In the next example, it is necessary to simplify the equation first before finding its solution.

Example 4 Solve the equation $5.2m + 6.4m = 34.8$.
Simplify the equation by using the distributive property to combine terms.

$$5.2m + 6.4m = 34.8$$

$$11.6m = 34.8 \qquad \text{Distributive property}$$

Now divide both sides by 11.6.

$$\frac{11.6m}{11.6} = \frac{34.8}{11.6} \qquad \text{Divide by 11.6}$$

$$m = 3 \qquad \text{Simplify}$$

The solution is 3. Check this solution. ✤

3 In the previous section, an equation like $-r = 4$ was solved by reasoning that since the additive inverse of r is 4, r must be -4. The following example shows how to use the multiplication property of equality to solve equations such as $-r = 4$.

Example 5 Solve the equation $-r = 4$.
Change $-r$ to r by first writing $-r$ as $-1 \cdot r$.

$$-r = 4$$

$$-1 \cdot r = 4 \qquad -r = -1 \cdot r$$

Now multiply both sides of this equation by -1.

$$-1(-1 \cdot r) = -1 \cdot 4 \qquad \text{Multiply by } -1$$
$$(-1)(-1) \cdot r = -4 \qquad \text{Associative property}$$
$$1 \cdot r = -4 \qquad \text{Simplify}$$
$$r = -4$$

Check this solution.

It is not necessary to show all these steps. If $-x = a$, just multiply both sides by -1 to get $x = -a$. ✤

2.3 Exercises

Solve each equation. See Examples 1–5.

1. $5x = 25$ **2.** $7x = 28$ **3.** $3a = -24$ **4.** $5k = -60$

5. $8s = -56$ **6.** $10t = -36$ **7.** $-6x = 16$ **8.** $-6x = 24$

9. $-18z = 108$ **10.** $-11p = 77$ **11.** $5r = 0$ **12.** $2x = 0$

13. $-y = 6$ **14.** $-m = 2$ **15.** $-n = -4$ **16.** $-p = -8$

17. $2x + 3x = 20$ **18.** $3k + 4k = 14$

19. $7r - 13r = -24$ **20.** $12a - 18a = -36$

21. $5m + 6m - 2m = 72$ **22.** $11r - 5r + 6r = 84$

23. $k + k + 2k = 80$ **24.** $4z + z + 2z = 28$

25. $3.5r - 5.8r = 6.9$ **26.** $.09p - .13p = 1.2$

27. $6y + 8y - 17y = 9$ **28.** $14a - 19a + 2a = 15$

29. $-7y + 8y - 9y = -56$ **30.** $-11b + 7b + 2b = -100$

31. $\dfrac{p}{5} = 3$ **32.** $\dfrac{x}{7} = 7$ **33.** $\dfrac{2}{3}t = 6$ **34.** $\dfrac{4}{3}m = 18$

35. $\dfrac{15}{2}z = 20$ **36.** $\dfrac{12}{5}r = 18$ **37.** $\dfrac{3}{4}p = -60$ **38.** $\dfrac{5}{8}z = -40$

39. $\dfrac{2}{3}k = 5$ **40.** $\dfrac{5}{3}m = 6$ **41.** $\dfrac{-2}{7}p = -7$ **42.** $-\dfrac{3}{11}y = -2$

43. $1.7p = 5.1$ **44.** $2.3k = 11.04$ **45.** $-4.2m = 25.62$ **46.** $-3.9a = -15.6$

47. $8.974z = 6.2818$ **48.** $11.506y = 3.4518$ **49.** $-9.273k = 10.2003$ **50.** $-1.565r = 3.9125$

Solve each equation.

51. $3m + 5 - 2m - 1 = 4m + 2 + 7$ **52.** $6a + 2a - 4 = 1 + 8a - 6 + 2a$

53. $-p + 3p - 6 = 7p + 3 - 5 + p$ **54.** $2z - z + 4z + 5 = -7 + 3 + 8z$

55. $4k - 1 + 3 + 2k = 6 - 5 - 3k + 9k$ **56.** $12x - 2x - 4 - x + 1 = 4x + 7 + 2x + 2$

Write an equation for each problem. Then solve the equation.

57. When a number is multiplied by 4, the result is 6. Find the number.

58. When a number is divided by -5, the result is 2. Find the number.

59. Chuck decided to divide a sum of money equally among four relatives, Dennis, Mike, Ed, and Joyce. Each relative received $62. Find the sum that was originally divided.

60. If twice a number is divided by 5, the result is 4. Find the number.

61. Twice a number is divided by 1.74, producing -8.38 as a quotient. Find the number.

62. A number is multiplied by 12 and then divided by 54.96. The result is -3.1. Find the number.

Review Exercises   *Simplify each expression. See Section 2.1.*

63. $9(2q + 7)$ **64.** $4(3m - 5)$ **65.** $-4(5p - 1) + 6$ **66.** $-3(2y + 7) - 9$

67. $-(2 - 5r) + 6r$ **68.** $-(12 - 3y) - 2$ **69.** $6 - 3(4a + 3)$ **70.** $9 - 5(7 - 8p)$

2.4   Solving Linear Equations

Objectives

1   Learn the four steps for solving a linear equation and how to use them.

2   Write word phrases as mathematical phrases.

1   This section shows how to use the properties to solve more complicated equations.

Solving a Linear Equation	*Step 1* Simplify each side of the equation by combining like terms. Use the commutative, associative, and distributive properties as needed.
	Step 2 If necessary, use the addition property of equality to simplify further, so that the variable is on one side of the equals sign and the number is on the other.
	Step 3 If necessary, use the multiplication property of equality to simplify further. This gives an equation of the form $x =$ a number.
	Step 4 Check the solution by substituting into the original equation. (Do *not* substitute into an intermediate step.)

Example 1   Solve the equation $2x + 3x + 3 = 38$.

Follow the four steps summarized above.

Step 1   Combine like terms.

$$2x + 3x + 3 = 38$$
$$5x + 3 = 38 \qquad \text{Combine like terms}$$

Step 2   Use the addition property of equality. Subtract 3 from both sides.

$$5x + 3 - 3 = 38 - 3 \qquad \text{Subtract 3}$$
$$5x = 35$$

Step 3   Use the multiplication property of equality. Divide both sides by 5.

$$\frac{5x}{5} = \frac{35}{5} \qquad \text{Divide by 5}$$
$$x = 7$$

Step 4   Check the solution. Substitute 7 for x in the given equation.

$$2x + 3x + 3 = 38$$
$$2(7) + 3(7) + 3 = 38 \qquad \text{Let } x = 7$$
$$14 + 21 + 3 = 38$$
$$38 = 38 \qquad \text{True}$$

Since the final statement is true, 7 is the solution.   ✚

Example 2   Solve the equation $3r + 4 - 2r - 7 = 4r + 3$.

Step 1 $\qquad 3r + 4 - 2r - 7 = 4r + 3$

$\qquad\qquad\qquad r - 3 = 4r + 3 \qquad\qquad$ Combine like terms

Step 2 $\qquad\quad r - 3 - r = 4r + 3 - r \qquad$ Subtract r

$\qquad\qquad\qquad\quad -3 = 3r + 3$

$\qquad\qquad -3 - 3 = -3 + 3r + 3 \qquad$ Subtract 3

$\qquad\qquad\qquad\quad -6 = 3r$

Step 3 $\qquad\qquad\quad \dfrac{-6}{3} = \dfrac{3r}{3} \qquad\qquad$ Divide by 3

$\qquad\qquad\qquad\quad -2 = r \quad \text{or} \quad r = -2$

Step 4   Substitute -2 for r in the original equation.

$$3r + 4 - 2r - 7 = 4r + 3$$
$$3(-2) + 4 - 2(-2) - 7 = 4(-2) + 3$$
$$-6 + 4 + 4 - 7 = -8 + 3$$
$$-5 = -5 \qquad \text{True}$$

The solution of the equation is -2.   ✚

In Step 2 of Example 2, the terms were added and subtracted in such a way that the variable term ended up on the right. Choosing differently would lead to the variable term being on the left side of the equation. Usually there is no advantage either way.

Example 3   Solve the equation $4(k - 3) - k = k - 6$.

Step 1   Before combining like terms, use the distributive property to simplify $4(k - 3)$.

$$4(k - 3) - k = k - 6$$
$$4 \cdot k - 4 \cdot 3 - k = k - 6 \qquad \text{Distributive property}$$
$$4k - 12 - k = k - 6$$
$$3k - 12 = k - 6 \qquad \text{Combine like terms}$$

Step 2
$$3k - 12 + 12 = k - 6 + 12 \qquad \text{Add 12}$$
$$3k = k + 6$$
$$3k - k = k + 6 - k \qquad \text{Subtract } k$$
$$2k = 6$$

Step 3
$$\frac{2k}{2} = \frac{6}{2} \qquad \text{Divide by 2}$$
$$k = 3$$

Step 4   Check your answer by substituting 3 for k in the given equation. Remember to do the work inside the parentheses first.

$$4(k - 3) - k = k - 6$$
$$4(3 - 3) - 3 = 3 - 6$$
$$4(0) - 3 = 3 - 6$$
$$0 - 3 = 3 - 6$$
$$-3 = -3 \qquad \text{True}$$

The solution of the equation is 3.   ✛

Example 4   Solve the equation $8a - (3 + 2a) = 3a + 1$.

Step 1   Simplify.

$$8a - (3 + 2a) = 3a + 1$$
$$8a - 3 - 2a = 3a + 1 \qquad \text{Distributive property}$$
$$6a - 3 = 3a + 1$$

Step 2
$$6a - 3 + 3 = 3a + 1 + 3 \qquad \text{Add 3}$$
$$6a = 3a + 4$$
$$6a - 3a = 3a + 4 - 3a \qquad \text{Subtract } 3a$$
$$3a = 4$$

Step 3
$$\frac{3a}{3} = \frac{4}{3} \qquad \text{Divide by 3}$$
$$a = \frac{4}{3}$$

Step 4   Check that the solution is $\dfrac{4}{3}$.

$$8a - (3 + 2a) = 3a + 1$$

$$8\left(\frac{4}{3}\right) - \left[3 + 2\left(\frac{4}{3}\right)\right] = 3\left(\frac{4}{3}\right) + 1 \qquad \text{Let } a = \frac{4}{3}$$

$$\frac{32}{3} - \left[3 + \frac{8}{3}\right] = 4 + 1$$

$$\frac{32}{3} - \left[\frac{9}{3} + \frac{8}{3}\right] = 5$$

$$\frac{32}{3} - \frac{17}{3} = 5$$

$$5 = 5 \qquad\qquad \text{True}$$

The check shows that $\dfrac{4}{3}$ is the solution.   ✛

Example 5   Solve the equation $4(8 - 3t) = 32 - 8(t + 2)$.

Step 1   Simplify.

$$4(8 - 3t) = 32 - 8(t + 2)$$

$$32 - 12t = 32 - 8t - 16$$

$$32 - 12t = 16 - 8t$$

Step 2

$$32 - 12t + 12t = 16 - 8t + 12t \qquad \text{Add } 12t$$

$$32 = 16 + 4t$$

$$32 - 16 = 16 + 4t - 16 \qquad \text{Subtract } 16$$

$$16 = 4t$$

Step 3

$$\frac{16}{4} = \frac{4t}{4} \qquad\qquad \text{Divide by } 4$$

$$4 = t \quad \text{or} \quad t = 4$$

Step 4   Check this solution in the given equation.

$$4(8 - 3t) = 32 - 8(t + 2)$$

$$4(8 - 3 \cdot 4) = 32 - 8(4 + 2) \qquad \text{Let } t = 4$$

$$4(8 - 12) = 32 - 8(6)$$

$$4(-4) = 32 - 48$$

$$-16 = -16 \qquad\qquad \text{True}$$

The solution, 4, checks.   ✛

2   The next section includes a detailed discussion of methods of solving word problems. One of the main steps in solving word problems is converting or translating the phrases in the word problems into mathematical expressions. The next few examples show how to translate phrases that occur frequently in word problems.

Example 6   Write the following phrases as mathematical expressions. Use x to represent the unknown. (Other letters could be used to represent the unknown.)

(a) 5 plus a number

The word *plus* indicates addition. If x represents the unknown number, then "5 plus a number" can be written as either

$$5 + x \quad \text{or} \quad x + 5.$$

(b) Add 20 to a number.

If x represents the unknown number, then "add 20 to a number" becomes

$$x + 20.$$

(c) "The sum of a number and 12" is $x + 12$ or $12 + x$.

(d) "7 more than a number" is $7 + x$ or $x + 7$.   ✛

Notice the difference between the wording in part (d) and "7 *is* more than a number," which translates as $7 > x$.

Example 7   Write each phrase as a mathematical expression. Use x as the variable.

(a) 3 less than a number

The words *less than* indicate subtraction, so "3 less than a number" is written $x - 3$. ($3 - x$ *would not* be correct.) Compare this example with "3 *is* less than a number," written $3 < x$.

(b) "A number decreased by 14" is $x - 14$.

(c) "Ten fewer than x" is $x - 10$.   ✛

Example 8   Write the following as mathematical expressions. Use x as the variable.

(a) "The product of a number and 3" is written $3 \cdot x$ or just $3x$, since *product* indicates multiplication.

(b) "Three times a number" is also $3x$.

(c) "Two thirds of a number" is $\frac{2}{3}x$.

(d) "The quotient of a number and 2" is $\frac{x}{2}$. (The word *quotient* indicates division; use a fraction bar instead of $\div$.)

(e) "The reciprocal of a nonzero number" is $\frac{1}{x}$.   ✛

Some word problems involve a combination of symbols. The next example shows how to translate such phrases.

Example 9   Write the following phrases as mathematical expressions. Use x as the variable.

(a) "Seven less than four times a number" is written $4x - 7$.

(b) "A nonzero number plus its reciprocal" is $x + \dfrac{1}{x}$.

(c) "Five times the sum of a number and two" is written as $5(x + 2)$.

(d) "A number divided by the sum of 4 and the number" is $\dfrac{x}{4 + x}$.

(e) "The quotient of a number and eight, plus the number."

The comma indicates that the quotient is to be $\dfrac{n}{8}$ not $\dfrac{n}{8 + n}$. Thus the correct translation is $\dfrac{n}{8} + n$.   ✛

2.4 Exercises

Solve each equation. Check your solutions. See Examples 1–5.

1. $4h + 8 = 16$　　　　　　**2.** $3x - 15 = 9$　　　　　　**3.** $6k + 12 = -12 + 7k$

4. $2m - 6 = 6 + 3m$　　　　**5.** $12p + 18 = 14p$　　　　　**6.** $10m - 15 = 7m$

7. $3x + 9 = -3(2x + 3)$　　**8.** $4z + 2 = -2(z + 2)$　　　**9.** $2(2r - 1) = -3(r + 3)$

10. $3(3k + 5) = 2(5k + 5)$　　**11.** $2(3x + 4) = 8(2 + x)$　　　**12.** $4(3p + 3) = 3(3p - 1)$

13. $3(5 + 1.4x) = 3x$　　　　　　　　　**14.** $2(-3 + 2.1x) = 2x + x$

15. $7.492y - 3.86 = 5.562y$　　　　　　**16.** $4.813q + 1.769 = 5.25525q$

17. $.291z + 3.715 = -.874z + 1.9675$　　**18.** $1.043k - 2.816 = .359k - 4.2524$

Combine terms as necessary. Then solve the equations. See Examples 3–5.

19. $-4 - 3(2x + 1) = 11$　　　　　　**20.** $8 - 2(3x - 4) = 2x$

21. $-5k - 8 = 2(k + 6) + 1$　　　　　**22.** $4a - 7 = 3(2a + 5) - 2$

23. $5(2m - 1) = 4(2m + 1) + 7$　　　**24.** $3(3k - 5) = 4(3k - 1) - 17$

25. $5(4t + 3) = 6(3t + 2) - 1$　　　　**26.** $7(2y + 6) = 9(y + 3) + 5$

27. $5(x - 3) + 2 = 5(2x - 8) - 3$　　**28.** $6(2v - 1) - 5 = 7(3v - 2) - 24$

29. $-2(3s + 9) - 6 = -3(3s + 11) - 6$　　**30.** $-3(5z + 24) + 2 = 2(3 - 2z) - 10$

31. $6(2p - 8) + 24 = 3(5p - 6) - 6$　　**32.** $2(5x + 3) - 3 = 6(2x - 3) + 15$

33. $3(m - 4) - (4m - 11) = -5$　　　**34.** $4(2a + 6) - (a + 1) = 2$

35. $-(4m + 2) - (-3m - 5) = 3$　　　**36.** $-(6k - 5) - (-5k + 8) = -4$

37. $2x + 2(3x + 2) - 9 = 3x - 9 + 3$　　**38.** $3(z - 2) + 4z = 8 + z + 1 - z$

39. $2(r - 3) + 5(r + 4) = 9$　　　　　**40.** $-4(m - 8) + 3(2m + 1) = 6$

41. $1.2(x + 5) = 3(2x - 8) + 23.28$ **42.** $4.7(m - 1) = 2(3m + 5) - 17.69$

43. $5.1(p + 6) - 3.8p = 4.9(p + 3) + 34.62$ **44.** $2.7(k - 3) + 1.1k = 3.4(k + 1) - 10.54$

45. $-3.2(1.4k + .8) - (2.97k - .3) = 1.32(2k - 1.8) - 16.028$

46. $1.53(4y - 3) - 2.8(3y + 1) = .74(5y + 6) - 6.448$

Write each phrase as a mathematical expression. Use x as the variable. See Examples 6–9.

47. 8 plus a number

48. A number added to -6

49. -1 added to a number

50. The sum of a number and 12

51. A quantity increased by -18

52. The total of x and 12

53. 5 less than a number

54. A number decreased by 6

55. Subtract 9 from a number

56. 16 fewer than a number

57. The product of a number and 9

58. Double a number

59. Triple a number

60. Three fifths of a number

61. A number subtracted from its reciprocal

62. A number added to twice the number

63. The sum of a number and 6, divided by -4

64. The difference of a number and 7, divided by the number

65. The product of 8 and the sum of a number and 3

66. The quotient of -9 and the sum of a number plus 3

67. The quotient of a number and 6, subtracted from 7

68. Three times the quotient of a number and 2

69. Eight times the difference of a number and 8

70. The difference of a number and 2, multiplied by -7

Write the answer to each of the following problems as an algebraic expression.

71. Two numbers have a sum of 11. One of the numbers is q. Find the other number.

72. The product of two numbers is 9. One of the numbers is k. What is the other number?

73. Yesterday Walt bought x apples. Today he bought 7 apples. How many apples did he buy altogether?

74. Joann has 15 books. She donated p books to the library. How many books does she have left?

75. Mary is a years old. How old will she be in 12 years?

76. Tom has r quarters. Find the value of the quarters in cents.

77. A bank teller has t dollars, all in five-dollar bills. How many five-dollar bills does the teller have?

78. A plane ticket costs b dollars for an adult and d dollars for a child. Find the total cost for 3 adults and 2 children.

79. Suppose n represents an even integer. Find the next higher even integer.

80. Let x represent the first of three consecutive integers. Find the next two integers.

2.5   Word Problems

Objective

1 Translate sentences into equations and solve the equations.

This section discusses the solution of many common types of word problems. To get the solution to a word problem, first read the problem carefully and determine what facts are given and what must be found. Then go through the following six steps.

Solving a Word Problem

Step 1 Choose a variable to represent the numerical value that you are asked to find—the unknown number. *Write down* what the variable represents.
Step 2 *Write down* a mathematical expression using the variable for any other unknown quantities.
Step 3 Translate the problem into an equation.
Step 4 Solve the equation.
Step 5 Answer the question asked in the problem.
Step 6 Check your solution by using the original words of the problem.

1 The third step in solving a word problem is often the hardest. Begin to translate the problem into an equation by writing the given phrases as mathematical expressions. Since equal mathematical expressions are names for the same number, translate any words that mean *equal* or *same* as $=$. The $=$ sign leads to an equation to be solved.

Example 1 Translate "the product of 4, and a number decreased by 7 is 100" into an equation. Use x as the variable. Solve the equation.

Translate as follows.

The product of 4,	and a number	decreased by	7	is	100.
↓	↓	↓	↓	↓	↓
$4 \cdot$	$(x$	$-$	$7)$	$=$	100

Because of the comma in the given sentence, writing the equation as $4x - 7 = 100$ is incorrect. The equation $4x - 7$ corresponds to the statement "The product of 4 and a number, decreased by 7, is 100." Solve the equation, using the steps given in the previous section.

$$4(x - 7) = 100$$
$$4x - 28 = 100 \qquad \text{Distributive property}$$
$$4x = 128 \qquad \text{Add 28 to both sides}$$
$$x = 32 \qquad \text{Divide by 4}$$

Now check the answer by substituting 32 for x in the original problem. Decrease the number, 32, by 7; $32 - 7 = 25$. Multiply the result by 4; $25 \cdot 4 = 100$. The solution, 32, gives a correct result. ✛

Notice how the six steps given above are used in the solution of the next few word problems.

Example 2   At a concert, there were 25 more women than men. The total number of people at the concert was 139. Find the number of men.

Let	$x = $ the number of men;	(Step 1)
	$x + 25 = $ the number of women.	(Step 2)

Now write an equation.

The total	is	the number of men	plus	the number of women.
↓	↓	↓	↓	↓
139	=	x	+	$x + 25$ (Step 3)

Solve the equation.

$$139 = 2x + 25$$
$$139 - 25 = 2x + 25 - 25 \qquad \text{Subtract 25}$$
$$114 = 2x \qquad \text{Simplify}$$
$$57 = x \qquad \text{Divide by 2 (Step 4)}$$

There were 57 men at the concert.   (Step 5)
   Check the answer.
   If there were 57 men, there were $57 + 25 = 82$ women at the concert. The total number of people was $82 + 57 = 139$ as required.   (Step 6) ✛

Example 3   The owner of a small cafe found one day that the number of orders for tea was 1/3 the number of orders for coffee. If the total number of orders for the two drinks was 76, how many orders were placed for tea?
   Let x equal the number of orders for coffee. Then $(1/3)x$ equals the number of orders for tea. Use the fact that the total number of orders was 76 to write an equation.

The total	is	orders for coffee	plus	orders for tea.
↓	↓	↓	↓	↓
76	=	x	+	$\dfrac{1}{3}x$

Now solve the equation.

$$76 = \frac{4}{3}x$$

$$\frac{3}{4}(76) = \frac{3}{4}\left(\frac{4}{3}x\right)$$

$$57 = x$$

There were 57 orders for coffee and $(1/3)(57) = 19$ orders for tea, giving a total of $57 + 19 = 76$ orders, as required. ✚

The next example uses **consecutive odd integers,** which are odd integers in a row, such as 5 and 7, or 31 and 33. Consecutive odd integers always differ by two.

When you work with consecutive integers, the following information may be helpful.

For a pair of consecutive:	*Use these variables:*
integers,	$x, \quad x + 1$
odd integers,	$x, \quad x + 2$
even integers,	$x, \quad x + 2$

Example 4   If the smaller of two consecutive odd integers is doubled, the result is 7 more than the larger of the two integers. Find the smaller integer.

Let $x =$ the first odd integer;

$x + 2 =$ the next consecutive odd integer.

(Add 2 to get the next consecutive *odd* integer.) Now write an equation from the statement of the problem.

If the smaller is doubled,	the result is	7	more than	the larger.
↓	↓	↓	↓	↓
$2x$	$=$	7	$+$	$x + 2$

Solve the equation.

$$2x = 7 + x + 2$$
$$2x = 9 + x \qquad \text{Combine terms}$$
$$2x - x = 9 + x - x \qquad \text{Subtract } x$$
$$x = 9 \qquad \text{Combine terms}$$

The smaller integer is 9. The larger integer would be $9 + 2 = 11$. Check that twice the smaller is 7 more than the larger. ✚

Example 5   Ten less than three times the smallest of three consecutive integers is nine more than the sum of the other two.

Let x = the smallest integer;

$x + 1$ = the second integer,

$x + 2$ = the third integer.

Use the statement of the problem to write an equation.

Three times smallest	less 10	is	9	more than	the sum of the other two.
↓	↓	↓	↓	↓	↓
$3x$	$- 10$	$=$	9	$+$	$(x + 1) + (x + 2)$

Solve this equation.

$$3x - 10 = 9 + x + 1 + x + 2$$
$$3x - 10 = 2x + 12$$
$$x - 10 = 12$$
$$x = 22$$

The smallest integer is 22, the next is $22 + 1 = 23$, and the largest is $22 + 2 = 24$. Check these numbers in the words of the original problem.   ✚

2.5 Exercises

The steps for solving word problems are repeated here. Follow these steps in working all the problems in this exercise set.

Step 1   Choose a variable to represent the numerical value that you are asked to find—the unknown number. *Write down* what the variable represents.

Step 2   *Write down* a mathematical expression using the variable for any other unknown quantities.

Step 3   Translate the problem into an equation.

Step 4   Solve the equation.

Step 5   Answer the question asked in the problem.

Step 6   Check your solution by using the original words of the problem.

Solve the following problems. See Examples 1–3.

1. When 6 is added to four times a number, the result is 42. Find the number.

2. The sum of a number and 3 is multiplied by 4, giving 36 as a result. Find the number.

3. If the sum of a number and 8 is multiplied by -2, the result is -8. Find the number.

4. When 6 is subtracted from a number, the result is seven times the number. Find the number.

5. If 4 is subtracted from twice a number, the result is 4 less than the number. Find the number.

6. If 2 is added to the product of 7 and a number, the result is 8 more than the number. Find the number.

7. On an algebra test, the highest grade was 42 points more than the lowest grade. The sum of the two grades was 138. Find the lowest grade.

8. In a physical fitness test, Rolfe did 25 more push-ups than Chuck did. The total number of push-ups for both men was 173. Find the number of push-ups that Chuck did.

9. A pharmacist found that at the end of the day she had 4/3 as many prescriptions for antibiotics as she had for tranquilizers. She had 84 prescriptions altogether for these two types of drugs. How many did she have for tranquilizers?

10. Mark White gives glass-bottom boat rides in the Bahama Islands. One day he noticed that the boat contained 17 more men (counting himself) than women. The number of men was 57 less than twice the number of women. How many women were on the boat?

11. Joann McKillip runs a dairy farm. Last year her cow Bossie gave 375 gallons less than twice the amount produced by another cow, Bessie. The two cows gave 1464 gallons of milk. How many gallons of milk did Bessie give?

Solve the following problems. See Examples 4 and 5.

12. The difference between 6 times the larger and the smaller of two consecutive integers is 166. Find the integers.

13. When the smaller of two consecutive integers is added to three times the larger, the result is 43. Find the smaller integer.

14. If five times the smaller of two consecutive integers is added to three times the larger, the result is 59. Find the smaller integer.

15. Find two consecutive odd integers such that when twice the larger is added to the smaller, the result is 169.

16. When the smaller of two consecutive even integers is multiplied by 5, the result is 110 more than the larger. Find the integers.

17. If 9 is added to the largest of three consecutive odd integers, the answer equals the sum of the first and second integers. Find the integers.

18. If the first and third of three consecutive even integers are added, the result is 20 more than the second integer. Find the integers.

Solve the following problems.

19. In a given amount of time, Larry drove twice as far as Rick. Altogether they drove 90 miles. Find the number of miles driven by each.

20. If a number is multiplied by 4, with the result added to 9, the sum is 3 more than the number. Find the number.

21. The sum of two consecutive odd integers is 92. Find the integers.

22. Twice a number is added to the number, giving 90. Find the number.

23. The smallest of three consecutive integers is added to twice the largest, producing a result 15 less than four times the middle integer. Find the smallest integer.

24. Nevarez and Smith were opposing candidates in the school board election. Nevarez received 30 more votes than did Smith, with 516 total votes cast. How many votes did Smith receive?

25. The sum of a number and 8 is multiplied by 5, giving 60 as the answer. Find the number.

26. Rick is 8 years older than Steve. The sum of their ages is 46 years. How old is each now?

27. A carpenter has a board 44 inches long. He wishes to cut it into two pieces so that one piece will be 6 inches longer than the other. How long should the shorter piece be?

28. If the middle of three consecutive integers is added to 100, the result is 1 less than the sum of the third integer and twice the smallest. Find the smallest integer.

The following word problems are real "head-scratchers."

29. Kevin is three times as old as Bob. Three years ago the sum of their ages was 22 years. How old is each now? (*Hint:* First write an expression for the age of each now, then for the age of each three years ago.)

30. A store has 39 quarts of milk, some in pint cartons and some in quart cartons. There are six times as many quart cartons as pint cartons. How many quart cartons are there? (*Hint:* 1 quart = 2 pints.)

31. A table is three times as long as it is wide. If it were 3 feet shorter and 3 feet wider, it would be square (with all sides equal). How long and how wide is the table?

32. Elena works for $6 an hour. A total of 25% of her salary is deducted for taxes and insurance. How many hours must she work to take home $450?

33. Paula received a paycheck for $585 for her weekly wages less 10% deductions. How much was she paid before the deductions were made?

34. At the end of a day, the owner of a gift shop had $2394 in the cash register. This included sales tax of 5% on all sales. Find the amount of the sales.

Review Exercises   *Use the given values to evaluate each expression. See Section 1.4.*

35. $LW; L = 6, W = 7$

36. $rt; r = 25, t = 2.5$

37. $prt; p = 4000, r = .08, t = 2$

38. $\frac{1}{2}Bh; B = 27, h = 4$

39. $2L + 2W; L = 18, W = 12$

40. $\frac{1}{2}(B + b)h; B = 10, b = 7, h = 5$

2.6 Formulas

Objectives

1 Solve a formula for one variable given the values of the other variables.
2 Use a formula to solve a word problem.
3 Solve a formula for a specified variable.

Many word problems can be solved with formulas. Formulas exist for geometric figures such as squares and circles, for distance, for money earned on bank savings, and for converting English measurements to metric measurements, for example. A list of the formulas used in this book is given inside the front cover.

1   Given the values of all but one of the variables in a formula, the value of the remaining variable can be found by using the methods introduced in this chapter for solving equations.

Example 1   Find the value of the remaining variable in each of the following.

(a) $A = LW$; $A = 64$, $L = 10$

As shown in Figure 2.1, this formula gives the area of a rectangle with length L and width W. Substitute the given values into the formula and then solve for W.

$$A = LW$$
$$64 = 10W \qquad \text{Let } A = 64, L = 10$$
$$6.4 = W \qquad \text{Divide by 10}$$

Check that the width of the rectangle is 6.4.

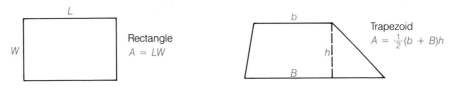

Rectangle $A = LW$	Trapezoid $A = \frac{1}{2}(b + B)h$
Figure 2.1	**Figure 2.2**

(b) $A = \dfrac{1}{2}(b + B)h$; $A = 210$, $B = 27$, $h = 10$

This formula gives the area of a trapezoid with parallel sides of length b and B and distance h between the parallel sides. See Figure 2.2. Again, begin by substituting the given values into the formula.

$$A = \frac{1}{2}(b + B)h$$
$$210 = \frac{1}{2}(b + 27)(10) \qquad A = 210, B = 27, h = 10$$

Now solve for b.

$$210 = \frac{1}{2}(10)(b + 27) \qquad \text{Commutative property}$$

$$210 = 5(b + 27)$$

$$210 = 5b + 135 \qquad \text{Distributive property}$$

$$75 = 5b \qquad \text{Subtract 135}$$

$$15 = b \qquad \text{Divide by 5}$$

Check that the length of the shorter parallel side, b, is 15. ✦

2 As the next examples show, formulas are often used to solve word problems.

Example 2 The perimeter of a square is 96 inches. Find the length of a side.

The list of formulas inside the front cover gives the formula $P = 4s$ for the **perimeter,** or the distance around a square, where s is the length of a side of a square. Here, the perimeter is given as 96 inches, so that $P = 96$. Substitute 96 for P in the formula.

$$P = 4s$$

$$96 = 4s \qquad \text{Let } P = 96$$

$$24 = s \qquad \text{Divide by 4}$$

Check this solution to see that each side of the square is 24 inches long. ✦

Example 3 The perimeter of a rectangle is 80 meters, and the length is 25 meters.* (See Figure 2.1.) Find the width of the rectangle.

The distance around a rectangle is called the perimeter of the rectangle. The formula for the perimeter of a rectangle is found by adding the lengths of the four sides.

$$P = L + L + W + W = 2L + 2W$$

Find the width by substituting 80 for P and 25 for L in the formula $P = 2L + 2W$.

$$P = 2L + 2W$$

$$80 = 2(25) + 2W \qquad P = 80, L = 25$$

$$80 = 50 + 2W \qquad \text{Multiply}$$

$$30 = 2W \qquad \text{Subtract 50}$$

$$15 = W \qquad \text{Divide by 2}$$

---

*A meter is a unit of length in the metric system, which is explained in Appendix C in the back of the book. No knowledge of the metric system is required for this example.

Check this result. If W is 15 and L is 25, the perimeter will be

$$2(15) + 2(25) = 30 + 50 = 80,$$

as required. The width of the rectangle is 15 meters.   ✦

Example 4   The area of a triangle is 126 square meters. The base of the triangle is 21 meters. Find the height.

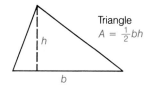

Triangle
$A = \frac{1}{2}bh$

Figure 2.3

The formula for the area of a triangle is $A = \dfrac{1}{2}bh$, where A is area, b is the base, and h is the height. See Figure 2.3. Substitute 126 for A and 21 for b in the formula.

$$A = \frac{1}{2}bh$$

$$126 = \frac{1}{2}(21)h \qquad A = 126, \, b = 21$$

$$126 = \frac{21}{2}h$$

To find h, multiply both sides by 2/21.

$$\frac{2}{21}(126) = \frac{2}{21} \cdot \frac{21}{2}h$$

$$12 = h \qquad \text{or} \qquad h = 12$$

The height of the triangle is 12 meters.   ✦

3   Sometimes it is necessary to solve a large number of problems that use the same formula. For example, a surveying class might need to solve several problems that involve the formula for the area of a rectangle, $A = LW$. Suppose that in each problem the area (A) and the length (L) of a rectangle are given and the width (W) must be found. Rather than solving for W each time the formula is used, it would be simpler to rewrite the *formula* so that it is solved for W. This process is called **solving for a specified variable.** As the following examples will show, solving a formula for a specified variable requires the same steps used earlier to solve equations with just one variable.

Example 5   Solve $A = LW$ for W.

Think of undoing what has been done to W. Since W is multiplied by L, undo the multiplication by dividing both sides of $A = LW$ by L.

$$A = LW$$

$$\frac{A}{L} = \frac{LW}{L} \qquad \text{Divide by } L$$

$$\frac{A}{L} = W \qquad \text{or} \qquad W = \frac{A}{L}$$

The formula is now solved for W.   ✚

The formula for converting temperatures given in degrees Celsius to degrees Fahrenheit is

$$F = \frac{9}{5}C + 32.$$

The next example shows how to solve this formula for C.

Example 6   Solve $F = \frac{9}{5}C + 32$ for C.

First undo the addition of 32 to $(9/5)C$ by subtracting 32 from both sides.

$$F = \frac{9}{5}C + 32$$

$$F - 32 = \frac{9}{5}C + 32 - 32 \qquad \text{Subtract 32}$$

$$F - 32 = \frac{9}{5}C$$

Now multiply both sides by 5/9. Use parentheses on the left.

$$\frac{5}{9}(F - 32) = \frac{5}{9} \cdot \frac{9}{5}C \qquad \text{Multiply by } \frac{5}{9}$$

$$\frac{5}{9}(F - 32) = C$$

This last result is the formula for converting temperatures from Fahrenheit to Celsius.   ✚

Example 7   Solve $A = \frac{1}{2}(b + B)h$ for B.

Begin by multiplying both sides by 2.

$$A = \frac{1}{2}(b + B)h$$

$$2A = 2 \cdot \frac{1}{2}(b + B)h$$

$$2A = (b + B)h$$

$$2A = bh + Bh \qquad \text{Distributive property}$$

Undo what was done to B by first subtracting bh on both sides. Then divide both sides by h.

$$2A - bh = Bh$$

$$\frac{2A - bh}{h} = B \qquad \text{or} \qquad B = \frac{2A - bh}{h}$$

The result can be written in a different form as follows.

$$B = \frac{2A - bh}{h} = \frac{2A}{h} - \frac{bh}{h} = \frac{2A}{h} - b$$

Either form is correct. ✛

2.6 Exercises

In the following exercises a formula is given, along with the values of all but one of the variables in the formula. Find the value of the variable that is not given. See Example 1.

1. $V = \frac{1}{3}Bh$; $B = 20, h = 9$

2. $V = \frac{1}{3}Bh$; $B = 82, h = 12$

3. $d = rt$; $d = 8, r = 2$

4. $d = rt$; $d = 100, t = 5$

5. $A = \frac{1}{2}bh$; $A = 20, b = 5$

6. $A = \frac{1}{2}bh$; $A = 30, b = 6$

7. $P = 2L + 2W$; $P = 40, W = 6$

8. $P = 2L + 2W$; $P = 180, L = 50$

9. $V = \frac{1}{3}Bh$; $V = 80, B = 24$

10. $V = \frac{1}{3}Bh$; $V = 52, h = 13$

11. $A = \frac{1}{2}(b + B)h$; $b = 6, B = 8, h = 3$

12. $A = \frac{1}{2}(b + B)h$; $b = 10, B = 12, h = 3$

13. $C = 2\pi r$; $C = 9.42, \pi = 3.14*$

14. $C = 2\pi r$; $C = 25.12, \pi = 3.14$

15. $A = \pi r^2$; $r = 9, \pi = 3.14$

16. $A = \pi r^2$; $r = 15, \pi = 3.14$

---

*Actually, π is approximately equal to 3.14, not exactly equal to 3.14.

17. $V = \frac{4}{3}\pi r^3$; $r = 3$, $\pi = 3.14$

18. $V = \frac{4}{3}\pi r^3$; $r = 6$, $\pi = 3.14$

19. $I = prt$; $I = 100$, $p = 500$, $r = .10$

20. $I = prt$; $I = 60$, $p = 150$, $r = .08$

21. $V = LWH$; $V = 150$, $L = 10$, $W = 5$

22. $V = LWH$; $V = 800$, $L = 40$, $W = 10$

23. $A = \frac{1}{2}(b + B)h$; $A = 42$, $b = 5$, $B = 7$

24. $A = \frac{1}{2}(b + B)h$; $A = 70$, $b = 15$, $B = 20$

25. $V = \frac{1}{3}\pi r^2 h$; $V = 9.42$, $\pi = 3.14$, $r = 3$

26. $V = \frac{1}{3}\pi r^2 h$; $V = 37.68$, $\pi = 3.14$, $r = 6$

Solve the given formulas for the indicated variables. See Examples 5–7.

27. $A = LW$;   for L

28. $d = rt$;   for t

29. $V = LWH$;   for H

30. $I = prt$;   for t

31. $C = 2\pi r$;   for r

32. $A = \frac{1}{2}bh$;   for b

33. $P = 2L + 2W$;   for W

34. $a + b + c = P$;   for b

35. $A = \frac{1}{2}(b + B)h$;   for b

36. $C = \frac{5}{9}(F - 32)$;   for F

37. $S = 2\pi rh + 2\pi r^2$;   for h

38. $A = p + prt$;   for r

39. $A = 4\pi r^2$;   for r^2

40. $c^2 = a^2 + b^2$;   for a^2

41. $V = \frac{1}{3}\pi r^2 h$;   for r^2

Write an equation for each word problem and then solve it. Check the solution in the words of the original problem. See Examples 2–4. The necessary formulas are given inside the front cover of the book. Use 3.14 as an approximation for π.

42. A rectangular classroom has an area of 61.2 square meters, and the width is 6 meters. Find the length.

43. The perimeter of a square picture is 80 centimeters. Find the length of a side.

44. The radius of a circular pond is 6 feet. Find the circumference (the distance around the pond).

45. The length of a rectangular map is 15 inches, and the perimeter is 50 inches. Find the width.

46. The perimeter of a triangular lot is 72 meters. One side is 16 meters, and another side is 32 meters. Find the third side.

47. The shorter base of a trapezoidal wall plaque is 16 centimeters, and the area is 108 square centimeters. The height is 6 centimeters. Find the length of the longer base.

48. The circumference of a circular clock face is 13.12 centimeters more than three times the radius. Find the radius of the face.

Solve each equation for x.

49. $y = 6 - 5x$

50. $y = 8 + 3x$

51. $2x + 3y = 9$

52. $5x - 4y = 12$

53. $3x - 5y = 15$

54. $-6x + 2y = 9$

55. $\dfrac{x}{2} + \dfrac{y}{3} = 8$

56. $\dfrac{x}{5} - \dfrac{y}{4} = 6$

57. $y = ax + b$

58. $y = mx + b$

59. $ax + by = c$

60. $mx + ny = k$

Review Exercises *Solve each equation. See Section 2.3.*

61. $6x = 12$

62. $9y = 36$

63. $3p = 8$

64. $2q = 11$

65. $-7z = 15$

66. $-5m = 3$

67. $-\dfrac{3}{5}x = 12$

68. $-\dfrac{2}{3}m = 4$

2.7 Ratios and Proportions

Objectives

1 Write ratios.

2 Decide whether proportions are true.

3 Solve proportions.

4 Solve word problems using proportions.

1 Ratios provide a way of comparing two numbers or quantities. A **ratio** is a quotient of two quantities with the same units. The ratio of the number a to the number b is written as

Ratio

$$a \text{ to } b, \qquad a : b, \qquad \text{or} \qquad \frac{a}{b}.$$

This last way of writing a ratio is the most common in algebra.

Example 1 Write a ratio for each word phrase.

(a) The ratio of 5 hours to 3 hours is

$$\frac{5}{3}.$$

(b) To find the ratio of 5 hours to 3 days, first convert 3 days to hours.

$$3 \text{ days} = 3 \cdot 24 = 72 \text{ hours}$$

The ratio of 5 hours to 3 days is thus

$$\frac{5}{72}.$$

2   A ratio is used to compare two numbers or amounts. A **proportion** says that two ratios are equal. For example,

$$\frac{3}{4} = \frac{15}{20}$$

is a proportion that say that the ratios 3/4 and 15/20 are equal. In the proportion

$$\frac{a}{b} = \frac{c}{d},$$

a, b, c, and d are the **terms** of the proportion. Beginning with the proportion

$$\frac{a}{b} = \frac{c}{d}$$

and multiplying both sides by the common denominator, bd, gives

$$bd \cdot \frac{a}{b} = bd \cdot \frac{c}{d}$$

$$ad = bc.$$

The products ad and bc are found by multiplying diagonally.

This is called **cross multiplication** and ad and bc are called **cross products.**

If $\dfrac{a}{b} = \dfrac{c}{d}$, then the cross products ad and bc are equal.

Also, if $ad = bc$, then $\dfrac{a}{b} = \dfrac{c}{d}$.

From the rule given above,

$$\text{if } \frac{a}{b} = \frac{c}{d} \text{ then } ad = bc.$$

However,

$$\text{if } \frac{a}{c} = \frac{b}{d}, \text{ then } ad = cb, \text{ or } ad = bc.$$

This means that the two proportions are equivalent, and

the proportion $\dfrac{a}{b} = \dfrac{c}{d}$ can always be written as $\dfrac{a}{c} = \dfrac{b}{d}$.

Sometimes one form is more convenient to work with than the other.

Example 2   Decide whether the following proportions are true or false.

(a) $\dfrac{3}{4} = \dfrac{15}{20}$

Check to see whether the cross products are equal.

$$4 \cdot 15 = 60$$

$$\frac{3}{4} = \frac{15}{20}$$

$$3 \cdot 20 = 60$$

The cross products are equal, so the proportion is true.

(b) $\dfrac{6}{7} = \dfrac{30}{32}$

The cross products are $6 \cdot 32 = 192$ and $7 \cdot 30 = 210$. The cross products are different, so the proportion is false.   ✛

3   Four numbers are used in a proportion. If any three of these numbers are known, the fourth can be found.

Example 3   (a) Find x in the proportion

$$\frac{63}{x} = \frac{9}{5}.$$

The cross products must be equal, so

$$63 \cdot 5 = 9x$$
$$315 = 9x.$$

Divide both sides by 9 to get

$$35 = x.$$

(b) Solve for r in the proportion $\dfrac{8}{5} = \dfrac{12}{r}$.

Set the cross products equal to each other.

$$8r = 5 \cdot 12$$
$$8r = 60$$
$$r = \frac{60}{8} = \frac{15}{2}$$ ✛

Example 4   Solve the equation

$$\frac{m - 2}{m + 1} = \frac{5}{3}.$$

Find the cross products, and set them equal to each other.

$$3(m - 2) = 5(m + 1)$$

$3m - 6 = 5m + 5$	Distributive property
$3m = 5m + 11$	Add 6
$-2m = 11$	Subtract $5m$
$m = -\dfrac{11}{2}$	Divide by -2 ❖

4   Proportions occur in many practical applications, as the next example shows.

Example 5   A hospital charges a patient $7.80 for 12 capsules. How much should it charge for 18 capsules?

Let x be the cost of 18 capsules. Set up a proportion; one ratio in the proportion can involve the number of capsules, and the other ratio can use the costs. Make sure that corresponding numbers appear in the numerator and the denominator.

$$\frac{\text{Cost of 18}}{\text{Cost of 12}} = \frac{18}{12}$$

$$\frac{x}{7.80} = \frac{18}{12}$$

Solve the proportion. Find the cross products, and set them equal to each other.

$12x = 18(7.80)$	
$12x = 140.40$	
$x = 11.70$	Divide by 12

The 18 capsules should cost $11.70. As shown above, this proportion could also be written as

$$\frac{18}{\text{Cost of 18}} = \frac{12}{\text{Cost of 12}}$$

$$\frac{18}{x} = \frac{12}{7.80}.$$

Find the cross products, and set them equal, to get

$$12x = 18(7.80)$$
$$12x = 140.40$$
$$x = 11.70,$$

the same answer as with the first method.   ❖

2.7   Exercises

Determine the following ratios. Write each ratio in lowest terms. See Example 1.

1. 30 miles to 20 miles
2. 50 feet to 90 feet
3. 72 dollars to 110 dollars
4. 120 people to 80 people
5. 6 feet to 5 yards
6. 10 yards to 8 feet
7. 30 inches to 4 feet
8. 100 inches to 5 yards
9. 12 minutes to 2 hours
10. 8 quarts to 5 pints
11. 4 dollars to 10 quarters
12. 35 dimes to 6 dollars
13. 20 hours to 5 days
14. 6 days to 9 hours
15. 80¢ to $3

Decide whether the following proportions are true. See Example 2.

16. $\dfrac{5}{8} = \dfrac{35}{56}$
17. $\dfrac{4}{7} = \dfrac{12}{21}$
18. $\dfrac{9}{10} = \dfrac{18}{20}$
19. $\dfrac{6}{8} = \dfrac{15}{20}$

20. $\dfrac{12}{18} = \dfrac{8}{12}$
21. $\dfrac{7}{10} = \dfrac{82}{120}$
22. $\dfrac{19}{30} = \dfrac{57}{90}$
23. $\dfrac{110}{18} = \dfrac{160}{27}$

24. $\dfrac{12.39}{8.91} = \dfrac{4.13}{2.97}$
25. $\dfrac{.612}{1.05} = \dfrac{1.0404}{1.785}$

26. $\dfrac{6.354}{7.823} = \dfrac{3.987}{4.096}$
27. $\dfrac{3.827}{2.4513} = \dfrac{8.0946}{5.7619}$

Solve the following equations. See Examples 3 and 4.

28. $\dfrac{7}{12} = \dfrac{a}{24}$
29. $\dfrac{35}{4} = \dfrac{k}{20}$
30. $\dfrac{z}{56} = \dfrac{7}{8}$
31. $\dfrac{m}{32} = \dfrac{3}{24}$

32. $\dfrac{6}{x} = \dfrac{4}{18}$
33. $\dfrac{z}{80} = \dfrac{20}{100}$
34. $\dfrac{25}{100} = \dfrac{8}{m}$
35. $\dfrac{2}{3} = \dfrac{y}{7}$

36. $\dfrac{5}{8} = \dfrac{m}{5}$
37. $\dfrac{5}{9} = \dfrac{z}{15}$
38. $\dfrac{3}{4} = \dfrac{r}{10}$
39. $\dfrac{7}{k} = \dfrac{8}{5}$

40. $\dfrac{5}{p} = \dfrac{3}{4}$
41. $\dfrac{m}{m-3} = \dfrac{5}{3}$
42. $\dfrac{r+1}{r} = \dfrac{1}{3}$
43. $\dfrac{3k-1}{k} = \dfrac{6}{7}$

Solve the following word problems involving proportions. See Example 5.

44. A certain lawn mower uses 3 tanks of gas to cut 10 acres of lawn. How many tanks of gas would be needed for 30 acres?

45. A Hershey bar contains 200 calories. How many bars would you need to eat to get 500 calories?

46. José can assemble 12 car parts in 40 minutes. How many minutes would he need to assemble 15 car parts?

47. If 2 pounds of fertilizer will cover 50 square feet of garden, how many pounds would be needed for 225 square feet?

48. The tax on a $20 item is $1. Find the tax on a $110 item.

49. On a road map, 3 inches represents 8 miles. How many inches would represent a distance of 24 miles?

50. A garden service charges $30 to install 50 square feet of sod. Find the charge to install 125 square feet.

51. If 9 pairs of jeans cost $121.50, find the cost of 5 pairs.

52. Suppose that 7 sacks of fertilizer cover 3325 square feet of lawn. Find the number of sacks needed for 7125 square feet.

53. The distance between two cities on a road map is 11 inches. Actually, the cities are 308 miles apart. The distance between two other cities is 15 inches. How far apart are these cities?

54. Twelve yards of material is needed for 5 dresses. How much material would be needed for 8 dresses?

55. The charge to move a load of freight 750 miles is $90. Find the charge to move the freight 1200 miles.

56. To ride a bus 80 miles cost $5. Find the charge to ride 180 miles.

Solve.

57. $\dfrac{3y - 2}{6y - 5} = \dfrac{5}{11}$

58. $\dfrac{2p + 7}{p - 1} = \dfrac{3}{4}$

59. $\dfrac{2r + 8}{4} = \dfrac{3r - 9}{3}$

60. $\dfrac{5k + 1}{6} = \dfrac{3k - 2}{3}$

61. The amount of material needed for 5 dresses is 7.2 yards less than the amount needed for 8 dresses. How much is needed for 8 dresses?

62. A bus ride of 180 miles costs $6.25 more than a ride of 80 miles. Find the cost to ride 80 miles.

63. The charge to move a load of freight 1200 miles is $36 less than twice the charge to move the freight 750 miles. Find both charges.

64. To prepare a medication, a nurse knows he must mix 4 ounces more than twice as much medicine with 50 ounces of water than he uses with 20 ounces. How much medicine should be used in each case?

Review Exercises   *Solve the following problems by choosing the proper formula. See Section 2.6.*

65. Tom travels 520 miles in 13 hours. How fast does he travel?

66. Joann goes 49 miles per hour, and covers 367.5 miles. How many hours did she travel?

67. Seamus earned $5700 in interest on a deposit of $19,000 at 12%. For how long was the money deposited?

68. Linda deposited $12,000 at 14% interest for 3 years. How much interest did she earn?

2.8 Further Word Problems

Objectives
Learn to solve word problems about

1 geometric figures;

2 money;

3 interest;

4 distance;

5 mixtures.

In this section, additional types of word problems are discussed. As you read through these examples, pay careful attention to the way each solution is begun by stating what the variable represents.

1 The first example shows how word problems result from discussions of geometric figures.

Example 1 A small shop is rectangular with the length 2 meters more than the width. The perimeter is 40 meters. (See Figure 2.4.) Find the width and length of the shop.

$$L = x + 2$$

$$W = x \qquad\qquad P = 40$$

Figure 2.4

Let $\qquad x =$ the width,

$\qquad x + 2 =$ the length.

The formula for the perimeter of a rectangle is $P = 2L + 2W$. Substitute x for W, $x + 2$ for L, and 40 for P.

$P = 2L = 2W$	
$40 = 2(x + 2) + 2x$	$P = 40, L = x + 2, W = x$
$40 = 2x + 4 + 2x$	Distributive property
$40 = 4x + 4$	Simplify
$36 = 4x$	Subtract 4 from both sides
$9 = x$	Divide by 4

The width of the shop is 9 meters, and the length is $9 + 2 = 11$ meters. Check these answers in the formula for the perimeter. Since $2(9) + 2(11) = 40$, the answers are correct.

2   The next example shows a word problem involving money.

Example 2   A bank teller has 25 more five-dollar bills than ten-dollar bills. The total value of the money is $200. How many of each type of bill does he have?

Let x = the number of ten-dollar bills,

$x + 25$ = the number of five-dollar bills.

The value of all the tens is given by the product of the number of tens (x) and the value per bill (10), so

$$\text{value of } \textbf{tens} = 10x.$$

In the same way,

$$\text{value of } \textbf{fives} = 5(x + 25).$$

The total value of all the money is $200.

value of fives	plus	value of tens	is	$200
↓	↓	↓	↓	↓
$5(x + 25)$	$+$	$10x$	$=$	200

Solve this equation.

$$5x + 125 + 10x = 200 \qquad \text{Distributive property}$$
$$15x + 125 = 200 \qquad \text{Simplify}$$
$$15x = 75 \qquad \text{Subtract 125}$$
$$x = 5 \qquad \text{Divide by 15}$$

Since x represents the number of tens, the teller has 5 tens and $5 + 25 = 30$ fives. Check that the value of this money is $5(\$10) + 30(\$5) = \$200$.   ✢

3   The next example uses the formula for interest, $I = prt$. The formula gives the amount of interest I earned by p dollars at a rate of interest r (in percent) for t years.

Example 3   Elizabeth Thornton receives an inheritance. She invests part of it at 9% and $2000 more than this amount at 10%. Altogether, she makes $1150 per year in interest. How much does she have invested at each rate?

Let x = the amount invested at 9%,

$x + 2000$ = the amount invested at 10%.

The formula for simple interest is $I = prt$. For the money at 9%, $p = x$, $r = 9\% = .09$, and $t = 1$, so that

$$\text{interest at } \textbf{9\%} = x(.09)(1) = .09x.$$

Also, $\text{interest at } \textbf{10\%} = .10(x + 2000).$

The total interest is $1150.

interest at 9%	plus	interest at 10%	is	total interest
↓	↓	↓	↓	↓
.09x	+	.10(x + 2000)	=	1150

Solve.

$$.09x + .10x + 200 = 1150 \qquad \text{Distributive property}$$
$$.19x + 200 = 1150 \qquad \text{Simplify}$$
$$.19x = 950 \qquad \text{Subtract 200}$$
$$x = 5000 \qquad \text{Divide by .19}$$

She has $5000 invested at 9% and $5000 + $2000 = $7000 invested at 10%. ✛

4   The next example uses the formula for distance, $d = rt$, where d represents distance, r is the rate or speed, and t is the time. The information given in this example is summarized in a chart. Making a chart is often a useful technique in problem solving.

Example 4   Two cars start from the same point at the same time and travel in the same direction at constant speeds of 34 and 45 miles per hour, respectively. In how many hours will they be 33 miles apart? (See Figure 2.5.)

Let t represent the unknown number of hours.

The distance traveled by the slower car is its rate multiplied by its time, or $34t$. The distance traveled by the faster car is $45t$. Summarize this information in a chart.

Car	r	t	d
Faster	45	t	$45t$
Slower	34	t	$34t$

$r \times t = d$

Difference is 33

Use the chart to draw the diagram in Figure 2.5.

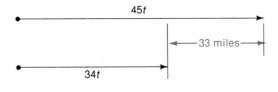

Figure 2.5

The quantities $45t$ and $34t$ represent different distances. Since the cars are 33 miles apart, the diagram suggests the equation

$$45t - 34t = 33.$$

Solve this equation.

$$11t = 33$$
$$t = 3$$

Check that the two cars will be 33 miles apart in 3 hours.   ✦

5   Algebra can be used to solve problems about mixing chemicals that occur in science and medicine. Again, in the solution of this example, a chart is used to help keep track of the information.

Example 5   A chemist needs to mix 20 liters of 40% salt solution with some 70% solution to get a mixture that is 50% salt. How many liters of the 70% solution should be used?

   Let $x =$ the number of liters of 70% solution that are needed. The amount of pure salt in this solution will be given by the product of the percent of strength and the number of liters of solution, or

$$\text{liters of pure salt in } x \text{ liters of 70\% solution} = .70x.$$

The amount of pure salt in the 20 liters of 40% solution is

$$\text{liters of pure salt in the 40\% solution} = .40(20) = 8.$$

The new solution will contain $20 + x$ liters of 50% solution. The amount of pure salt in this solution is

$$\text{liters of pure salt in the 50\% solution} = .50(20 + x).$$

The following chart shows all this information.

Strength	Liters of solution	Liters of pure salt	
40%	20	.40(20)	←
70%	x	.70x	← Sum must equal
50%	$20 + x$	.50(20 + x)	←

As the chart shows, the amount of pure salt before and after mixing the solutions must be equal, so the equation is

pure salt in 70%	plus	pure salt in 40%	is	pure salt in 50%
↓	↓	↓	↓	↓
.70x	+	.40(20)	=	.50(20 + x).

Solve for x.

$$.70x + 8 = 10 + .50x \qquad \text{Distributive property}$$
$$.20x + 8 = 10 \qquad \text{Subtract } .50x$$
$$.20x = 2 \qquad \text{Subtract 8}$$
$$x = 10 \qquad \text{Divide by } .20$$

Check this solution to see that the chemist needs to use 10 liters of 70% salt solution. ✚

2.8 Exercises

Work the following word problems. See Example 1.

1. The perimeter of a square is seven times the length of the side, decreased by 12. Find the length of a side.

2. The perimeter of a rectangle is 16 times the width. The length is 12 centimeters more than the width. Find the width of the rectangle.

3. The width of a rectangle is 1 less than the length. The perimeter is five times the length, decreased by 5. Find the length of the rectangle.

4. One side of a triangle is 10 centimeters longer than the shortest side. A third side is 20 centimeters longer than the shortest side. The perimeter of the triangle is 120 centimeters. Find the length of the shortest side of the triangle.

5. If the radius of a certain circle is tripled, with 8.2 centimeters then added, the result is the circumference of the circle. Find the radius of the circle. (Use 3.14 as an approximation for π.)

6. The circumference of a certain circle is five times the radius, increased by 2.56 meters. Find the radius of the circle. (Use 3.14 as an approximation for π.)

Work the following word problems. See Example 2.

7. A bank teller has some five-dollar bills and some twenty-dollar bills. The teller has 5 more of the twenties. The total value of the money is $725. Find the number of five-dollar bills that the teller has.

8. A woman has $1.70 in dimes and nickels; she has 2 more dimes than nickels. How many nickels does she have?

9. A stamp collector buys some 16¢ stamps and some 29¢ stamps, paying $8.68 for them. He buys 2 more 29¢ stamps than 16¢ stamps. How many 16¢ stamps does he buy?

10. For a retirement party, a person buys some 32¢ favors and some 50¢ favors, paying $46 in total. If she buys 10 more of the 50¢ favors, how many of the 32¢ favors were bought?

11. A cashier has a total of 126 bills, made up of fives and tens. The total value of the money is $840. How many of each kind does he have?

12. A convention manager finds that she has $1290, made up of twenties and fifties. She has a total of 42 bills. How many of each kind does the manager have?

Work the following word problems. See Example 3.

13. Dorothy Raymond inherited a sum of money from a relative. She deposits some of the money at 16%, and $4000 more than this amount at 12%. She earns $3840 in interest per year. Find the amount she has invested at 16%.

14. Adda McDowell invested some money at 18%, and $3000 less than that amount at 20%. The two investments produce a total of $3200 per year interest. How much is deposited at 18%?

15. Two investments produce an annual interest income of $4200. The amount invested at 14% is $6000 less than the amount invested at 10%. Find the amount invested at each rate.

16. Carol Foresman invested some money at 10%, and invested $5000 more than this at 14%. Her total annual income from these investments is $3100. How much does she have invested at each rate?

Work the following word problems. See Example 4.

17. Two trains leave a city at the same time. One travels north at 60 miles per hour, and the other travels south at 80 miles per hour. In how many hours will they be 280 miles apart?

Train	r	t	d
Northbound	60	t	
Southbound	80	t	

18. Two planes leave an airport at the same time. One flies east at 300 miles per hour, and the other flies west at 450 miles per hour. In how many hours will they be 2250 miles apart?

Plane	r	t	d
Eastbound	300	t	
Westbound	450	t	

19. From a point on a straight road, John and Fred ride bicycles in opposite directions. John rides 10 miles per hour and Fred rides 12 miles per hour. In how many hours will they be 55 miles apart?

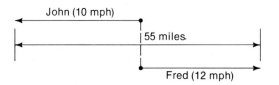

John (10 mph)

55 miles.

Fred (12 mph)

20. At a given hour, two steamboats leave a city in the same direction on a straight canal. One travels at 18 miles per hour, and the other travels at 25 miles per hour. In how many hours will the boats be 35 miles apart?

Work the following word problems. See Example 5.

21. How many gallons of 50% antifreeze must be mixed with 80 gallons of 20% antifreeze to get a mixture that is 40% antifreeze?

Strength	Gallons of solution	Gallons of pure antifreeze
50%	x	$.50x$
20%	80	
40%		

22. How many liters of 25% salt solution must be added to 80 liters of 40% solution to get a solution that is 30% salt?

Strength	Liters of solution	Liters of pure salt
25%	x	
40%	80	$.40(80)$
30%		

23. A certain metal is 40% tin. How many kilograms of this metal must be mixed with 80 kilograms of a metal that is 70% tin to get a metal that is 50% tin?

24. Ink worth $100 per barrel will be mixed with 30 barrels of ink worth $60 per barrel to get a mixture worth $75 per barrel. How many barrels of $100 ink should be used?

Work the following miscellaneous word problems.

25. Ann has saved $290 for a trip to Disneyland. Transportation will cost $56, tickets for park entrance and rides will cost $18 per day, and lodging and meals will cost $60 per day. How many days can she spend there?

26. At Irv's Burgerville, hamburgers cost 90 cents each, and a bag of french fries costs 40 cents. How many hamburgers and how many bags of french fries can Ted buy with $8.80 if he wants twice as many hamburgers as bags of french fries?

27. A boat travels upstream for 3 hours. The return trip requires 2 hours. If the speed of the current is 5 miles per hour, find the speed of the boat in still water.

28. In an automobile race, a driver was 120 miles from the finish line after 5 hours. Another driver, who was in a later race, traveled at the same speed as the first driver. After 3 hours, the second driver was 250 miles from the finish. Find the speed of each driver.

29. The longest side of a triangle is 5 meters longer than the shortest side, and the medium side is 2 meters longer than the shortest side. The perimeter of the triangle is 55 meters. Find the length of the shortest side of the triangle.

30. An actor invests his earnings in two ways: some goes into a 10% tax-free bond, and $5000 more than twice as much goes into an apartment house paying 20%. His total annual income from the investments is $9200. Find the amount he has invested at 10%.

31. A merchant wishes to mix candy worth $5 per pound with 40 pounds of candy worth $2 per pound to get a mixture that can be sold for $3 per pound. How many pounds of $5 candy should be used?

32. Two cars are 400 miles apart. Both start at the same time and travel toward one another. They meet 4 hours later. If the speed of one car is 20 miles per hour faster than the other, what is the speed of each car?

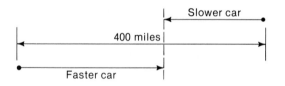

33. The length of a rectangle is 8 inches more than the width. The perimeter of the rectangle is 6 inches more than five times the width. Find the width of the rectangle.

34. With income earned by selling a patent, an engineer invests some money at 16%, and $3000 more than twice as much at 20%. The total annual income from the investments is $5080. Find the amount invested at 16%.

35. A pharmacist has 20 liters of a 10% drug solution. How many liters of 5% solution must be added to get a mixture that is 8%?

36. Two runners start from the same point and run in the same direction. One runner goes at 8 miles per hour, and the other runs at 11 miles per hour. In how many hours will the runners be 9 miles apart?

Review Exercises   *Place $<$ or $>$ in each blank to make a true statement. See Section 1.5.*

37. -7 _____ 3 **38.** -9 _____ 6 **39.** -11 _____ -4 **40.** -10 _____ -12

41. -8 _____ -10 **42.** -5 _____ -3 **43.** 5 _____ -9 **44.** 8 _____ -1

2.9   The Addition and Multiplication Properties of Inequality

Objectives

1 Graph intervals on a number line.

2 Learn the addition property of inequality.

3 Learn the multiplication property of inequality.

4 Solve inequalities.

5 Write inequalities to solve word problems.

6 Solve inequalities about numbers between other numbers.

Inequalities are statements with algebraic expressions related by

$<$ "is less than"

$\leq$ "is less than or equal to"

$>$ "is greater than"

$\geq$ "is greater than or equal to."

An inequality is solved by finding all real number solutions for it. For example, the solution of $x \leq 2$ includes all real numbers that are less than or equal to 2, and not just the integers less than or equal to 2. For example, -2.5, -1.7, -1, $7/4$, $1/2$, $\sqrt{2}$, and 2 are all real numbers less than or equal to 2, and are therefore solutions of $x \leq 2$.

1   A good way to show the solution of an inequality is by graphing. Graph all real numbers satisfying $x \leq 2$ by placing a dot at 2 on a number line and drawing an arrow extending from the dot to the left (to represent the fact that all numbers less than 2 are also part of the graph). The graph is shown in Figure 2.6.

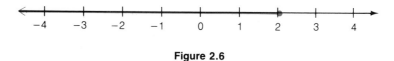

Figure 2.6

Example 1   Graph $x > -5$.

The statement $x > -5$ says that x can represent any number greater than -5, but x cannot equal -5 itself. Show this on a graph by placing an open circle at -5 and drawing an arrow to the right, as in Figure 2.7. The open circle at -5 shows that -5 is not part of the graph.   ✧

Figure 2.7

Example 2   Graph $-3 \le x < 2$.

The statement $-3 \le x < 2$ is read "-3 is less than or equal to x *and* x is less than 2." Graph this inequality by placing a solid dot at -3 (because -3 is part of the graph) and an open circle at 2 (because 2 is not part of the graph). Then draw a line segment between the two circles, as in Figure 2.8.   ✦

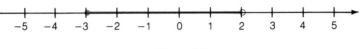

Figure 2.8

2   Inequalities such as $x + 4 \le 9$ can be solved in much the same way as equations. Consider the inequality $2 < 5$. If 4 is added to both sides of this inequality, the result is

$$2 + 4 < 5 + 4$$
$$6 < 9,$$

a true sentence. Now subtract 8 from both sides:

$$2 - 8 < 5 - 8$$
$$-6 < -3.$$

The result is again a true sentence. These examples suggest the following **addition property of inequality,** which states that the same number can be added to both sides of an inequality.

Addition Property of Inequality	For any real numbers A, B, and C, the inequalities $\qquad A < B \qquad$ and $\qquad A + C < B + C$ have exactly the same solutions. In words, the same number may be added on both sides of an inequality.

The addition property of inequality also works with $>$, $\le$, or $\ge$. Just as with the addition property of equality, the same number may also be *subtracted* on both sides of an inequality.

The following examples show how the addition property is used to solve inequalities.

Example 3   Solve the inequality $7 + 3k > 2k - 5$.

Use the addition property of inequality twice, once to get the terms containing k on one side of the inequality and a second time to get the integers together on the other side. (These steps can be done in either order.)

$$7 + 3k > 2k - 5$$

$$7 + 3k - 2k > 2k - 5 - 2k \qquad \text{Subtract } 2k$$

$$7 + k > -5 \qquad \text{Simplify}$$

$$7 + k - 7 > -5 - 7 \qquad \text{Subtract 7}$$

$$k > -12$$

The graph of the solution $k > -12$ is shown in Figure 2.9.   ✚

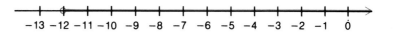

Figure 2.9

3   The addition property of inequality alone cannot be used to solve inequalities such as $4y \geq 28$. These inequalities require the multiplication property of inequality. To see how this property works, it will be helpful to look at some examples.

First, start with the inequality $3 < 7$ and multiply both sides by the positive number 2.

$$3 < 7$$

$$2(3) < 2(7) \qquad \text{Multiply both sides by 2}$$

$$6 < 14 \qquad \text{True}$$

Now multiply both sides of $3 < 7$ by the negative number -5.

$$3 < 7$$

$$-5(3) < -5(7) \qquad \text{Multiply both sides by } -5$$

$$-15 < -35 \qquad \text{False}$$

To get a true statement when multiplying both sides by -5 requires reversing the direction of the inequality symbol.

$$3 < 7$$

$$-5(3) > -5(7) \qquad \text{Multiply by } -5; \text{ reverse the symbol}$$

$$-15 > -35 \qquad \text{True}$$

Take the inequality $-6 < 2$ as another example. Multiply both sides by the positive number 4.

$$-6 < 2$$

$$4(-6) < 4(2) \qquad \text{Multiply by 4}$$

$$-24 < 8 \qquad \text{True}$$

Multiplying both sides of $-6 < 2$ by -5 and at the same time reversing the direction of the inequality symbol gives

$$-6 < 2$$
$$(-5)(-6) > (-5)(2) \qquad \text{Multiply by } -5, \text{ change } < \text{ to } >$$
$$30 > -10. \qquad \text{True}$$

In summary, the two parts of the **multiplication property of inequality** are stated below.

Multiplication Property of Inequality	For any real numbers A, B, and C ($C \neq 0$), (1) if C is *positive,* then the inequalities $\qquad A < B \qquad$ and $\qquad AC < BC$ have exactly the same solutions; (2) if C is *negative,* then the inequalities $\qquad A < B \qquad$ and $\qquad AC > BC$ have exactly the same solutions. In other words, both sides of an inequality may be multiplied by the same nonzero number. If the number is negative, reverse the inequality symbol.

The multiplication property of inequality works with $>$, $\leq$, or $\geq$, as well. The multiplication property of inequality also permits *division* of both sides of an inequality by the same nonzero number.

It is important to remember the differences in the multiplication property for positive and negative numbers.

(1) When both sides of an inequality are multiplied or divided by a positive number, the direction of the inequality symbol *does not change.* Adding or subtracting terms on both sides also does not change the symbol.
(2) When both sides of an inequality are multiplied or divided by a negative number, the direction of the symbol *does change. Reverse the symbol of inequality only when you multiply or divide both sides by a negative number.*

The next examples show how to solve inequalities with the multiplication property.

Example 4   Solve the inequality $3r < 18$.

Simplify this inequality by using the multiplication property of inequality and dividing both sides by 3. Since 3 is a positive number, the direction of the inequality symbol does not change.

$$3r < 18$$

$$\frac{3r}{3} < \frac{18}{3} \qquad \text{Divide by 3}$$

$$r < 6$$

The graph of this solution is shown in Figure 2.10.   ✛

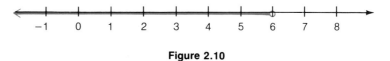

Figure 2.10

Example 5   Solve the inequality $-4t \geq 8$.

Here both sides of the inequality must be divided by -4, a negative number, which *does* change the direction of the inequality symbol.

$$-4t \geq 8$$

$$\frac{-4t}{-4} \leq \frac{8}{-4} \qquad \text{Divide by } -4; \text{ symbol reversed}$$

$$t \leq -2$$

The solution is graphed in Figure 2.11.   ✛

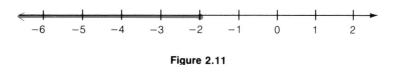

Figure 2.11

4   The steps in solving an inequality are summarized below. (Remember that $<$ can be replaced with $>$, $\leq$, or $\geq$ in this summary.)

Solving an Inequality	*Step 1* Use the associative, commutative, and distributive properties to combine like terms on each side of the inequality.
	Step 2 Use the addition property of inequality to simplify the inequality to one of the form $ax < b$, where a and b are real numbers.
	Step 3 Use the multiplication property of inequality to simplify further to an inequality of the form $x < c$ or $x > c$, where c is a real number.

Notice how these steps are used in the next example.

Example 6   Solve the inequality $3z + 2 - 5 > -z + 7 + 2z$.

Step 1   Combine like terms and simplify.

$$3z + 2 - 5 > -z + 7 + 2z$$
$$3z - 3 > z + 7$$

Step 2   Use the addition property of inequality.

$$3z - 3 + 3 > z + 7 + 3 \qquad \text{Add 3}$$
$$3z > z + 10$$
$$3z - z > z + 10 - z \qquad \text{Subtract } z$$
$$2z > 10$$

Step 3   Use the multiplication property of inequality.

$$\frac{2z}{2} > \frac{10}{2} \qquad \text{Divide by 2}$$
$$z > 5$$

Since 2 is positive, the direction of the inequality symbol was not changed in the third step. A graph of the solution is shown in Figure 2.12.   ✤

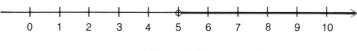

Figure 2.12

Example 7   Solve $5(k - 3) - 7k \geq 4(k - 3) + 9$.

Step 1   Simplify and combine like terms.

$$5(k - 3) - 7k \geq 4(k - 3) + 9$$
$$5k - 15 - 7k \geq 4k - 12 + 9 \qquad \text{Distributive property}$$
$$-2k - 15 \geq 4k - 3 \qquad \text{Combine like terms}$$

Step 2   Use the addition property.

$$-2k - 15 - 4k \geq 4k - 3 - 4k \qquad \text{Subtract } 4k$$
$$-6k - 15 \geq -3$$
$$-6k - 15 + 15 \geq -3 + 15 \qquad \text{Add 15}$$
$$-6k \geq 12$$

Step 3   Divide both sides by -6, a negative number. Change the direction of the inequality symbol.

$$\frac{-6k}{-6} \le \frac{12}{-6} \qquad \text{Divide by } -6$$

$$k \le -2$$

A graph of the solution is shown in Figure 2.13. ✛

Figure 2.13

5   The next example shows how inequalities can be used in the solution of a word problem.

Example 8   If 2 is added to five times a number, the result is greater than or equal to 5 more than four times the number. Find the number.

First translate the word problem into an inequality. Let x represent the number you want to find. The phrase, "2 is added to five times a number" is expressed as $5x + 2$. And "5 more than four times the number" is $4x + 5$. The two expressions are related by "is greater than or equal to."

$$5x + 2 \ge 4x + 5$$

Solve the inequality $5x + 2 \ge 4x + 5$, by subtracting $4x$ and 2 from both sides (in either order).

$$5x + 2 - 4x \ge 4x + 5 - 4x \qquad \text{Subtract } 4x$$

$$x + 2 \ge 5$$

$$x + 2 - 2 \ge 5 - 2 \qquad \text{Subtract } 2$$

$$x \ge 3$$

The solution is any number greater than or equal to 3. ✛

6   Inequalities that say that one number is *between* two other numbers also can be solved by using the addition and multiplication properties of inequality.

Example 9   **(a)** Solve $4 \le 3x - 5 < 6$.

First add 5 to each part.

$$4 \le 3x - 5 < 6$$

$$4 + 5 \le 3x - 5 + 5 < 6 + 5 \qquad \text{Add } 5$$

$$9 \le 3x < 11$$

Now divide each part by the positive number 3.

$$\frac{9}{3} \leq \frac{3x}{3} < \frac{11}{3} \qquad \text{Divide by 3}$$

$$3 \leq x < \frac{11}{3}$$

A graph of the solution is shown in Figure 2.14.

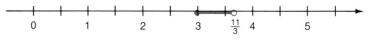

Figure 2.14

(b) Solve $-4 \leq \frac{2}{3}m - 1 < 8$.

First add 1 to each part.

$$-4 \leq \frac{2}{3}m - 1 < 8$$

$$-4 + 1 \leq \frac{2}{3}m - 1 + 1 < 8 + 1 \qquad \text{Add 1}$$

$$-3 \leq \frac{2}{3}m < 9$$

Multiply each part by 3/2, a positive number.

$$\frac{3}{2}(-3) \leq \frac{3}{2} \cdot \frac{2}{3}m < \frac{3}{2} \cdot 9 \qquad \text{Multiply by } \frac{3}{2}$$

$$-\frac{9}{2} \leq m < \frac{27}{2}$$

A graph of the solution is shown in Figure 2.15.

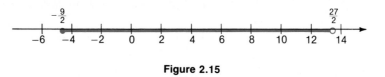

Figure 2.15

2.9 Exercises

Graph each inequality on a number line. See Examples 1 and 2.

1. $x \leq 4$ **2.** $k \geq -5$ **3.** $a < 3$ **4.** $p > 4$

5. $-2 \leq x \leq 5$ **6.** $8 \leq m \leq 10$ **7.** $3 \leq y < 5$ **8.** $0 < y \leq 10$

Solve each inequality. See Examples 3–5.

9. $a + 6 < 8$ **10.** $k - 4 < 2$ **11.** $z - 3 \geq -2$ **12.** $p + 2 \geq -6$

13. $-3 + k \geq 2$ **14.** $x + 5 > 5$ **15.** $3x < 27$ **16.** $5h \geq 20$

17. $-2k \leq 12$ **18.** $-3v > 6$ **19.** $-8y < -72$ **20.** $-9a \geq -63$

Solve each inequality and graph the solution. See Examples 6 and 7.

21. $3n + 5 \leq 2n - 6$ **22.** $5x - 2 < 4x - 5$

23. $2z - 8 > z - 3$ **24.** $4x + 6 \leq 3x - 5$

25. $4k + 1 \geq 2k - 9$ **26.** $5y + 3 < 2y + 12$

27. $3 + \dfrac{2}{3}r > \dfrac{5}{3}r - 27$ **28.** $8 + \dfrac{6}{5}t \leq \dfrac{8}{5}t + 12$

29. $4q + 1 - \dfrac{5}{3} < 8q + \dfrac{4}{3}$ **30.** $5x - \dfrac{2}{3} \leq \dfrac{2}{3}x + 6 - x$

31. $10p + 20 - p > p + 3 - 23$ **32.** $-3v + 6 + 3 - 2 > -5v - 19$

33. $-k + 4 + 5k \leq -1 + 3k + 5$ **34.** $6y - 2y - 4 + 7y > 3y - 4 + 7y$

35. $5 - (2 - r) \leq 3r + 5$ **36.** $-9 + (8 + y) > 7y - 4$

Solve each inequality and graph the solution. See Example 9.

37. $-5 \leq 2x - 3 \leq 9$ **38.** $-7 \leq 3x - 4 \leq 8$ **39.** $5 < 1 - 6m < 12$

40. $10 < 7p + 3 < 24$ **41.** $-1 \leq 1 - 5q \leq 16$ **42.** $-8 \leq 3r - 1 \leq -1$

43. $-12 \leq \dfrac{1}{2}z + 1 \leq 4$ **44.** $-6 \leq 3 - \dfrac{1}{3}a \leq 5$ **45.** $1 \leq 3 + \dfrac{2}{3}p \leq 7$

46. $2 < 6 + \dfrac{3}{4}y < 12$ **47.** $-7 \leq \dfrac{5}{4}r - 1 \leq -1$ **48.** $-12 \leq \dfrac{3}{7}a + 2 \leq -4$

Solve each word problem. See Example 8.

49. If half a number is added to 5, the result is greater than or equal to -3. Find all such numbers.

50. When four times a number is subtracted from 8, the result is less than 15. Find all numbers that satisfy this condition.

51. Twice a number added to three times the sum of the number and 2 is more than 17. Find the numbers that satisfy this condition.

52. If two-thirds of a number is added to -3, the result is no more than two times the number. Find all such numbers.

53. A student has test grades of 75 and 82. What must he score on a third test to have an average of 80 or higher?

54. In Exercise 53, if 100 is the highest score possible on the third test, how high an average (to the nearest tenth) can the student make? What is the lowest average possible for the three tests?

55. Mr. Odjakjian earned $200 at odd jobs during July, $300 during August, and $225 during September. If his average salary for the four months from July through October is to be at least $250, how much must he earn during October?

56. One side of a triangle is twice as long as a second side. The third side of the triangle is 17 feet long. The perimeter of the triangle cannot be more than 50 feet. Find the longest possible values for the other two sides of the triangle.

57. In order to qualify for a company pension plan, an employee must average at least $1000 per month in earnings. During the first four months of the year, an employee made $900, $1200, $1040, and $760. What amount of earnings during the fifth month will qualify the employee?

58. The perimeter of a rectangle must be no greater than 120 meters. The width of the rectangle must be 22 meters. Find the longest possible value for the length of the rectangle.

59. One side of a rectangle is 8 meters long. The area of the rectangle must be at least 240 square meters. Find the shortest possible length for the rectangle.

60. A triangle has a height of 20 meters. The area of the triangle must be less than or equal to 40 square meters. Find the longest possible length for the base of the triangle.

Solve each inequality.

61. $2(x - 5) + 3x < 4(x - 6) + 3$

62. $5(t + 3) - 6t \le 3(2t + 1) - 4t$

63. $5(2k + 3) - 2(k - 8) > 3(2k + 4) + k - 2$

64. $2(3z - 5) + 4(z + 6) \ge 2(3z + 2) + 3(z - 5)$

65. $3(p + 1) - 2(p - 4) \ge 5(2p - 3) + 2$

66. $-5(m - 3) + 4(m + 6) < 2(m - 3) + 4$

Review Exercises   *Evaluate each expression. See Section 1.3.*

67. $2 \cdot 2 \cdot 2 \cdot 2 \cdot 2$

68. $3 \cdot 3 \cdot 3$

69. $5 \cdot 5 \cdot 5 \cdot 5$

70. $4 \cdot 4 \cdot 4 \cdot 4$

71. $\dfrac{2}{3} \cdot \dfrac{2}{3} \cdot \dfrac{2}{3}$

72. $\dfrac{5}{8} \cdot \dfrac{5}{8}$

73. $(2 \cdot 2 \cdot 2)(2 \cdot 2 \cdot 2 \cdot 2)$

74. $(3 \cdot 3) \cdot (3 \cdot 3 \cdot 3)$

Simplify. See Sections 1.6 and 1.7.

75. $8 + (-5)$

76. $9 + (-12)$

77. $3 - (-2)$

78. $5 - (-7)$

79. $8 - 4 - (-9)$

80. $-6 - (-8)$

81. $-2 - [-4 - (3 - 2)]$

82. $1 - [-(8 - 9) - 2]$

Chapter 2 *Summary*

Addition Property of Equality	If A, B, and C are real numbers, then the equations $$A = B \quad \text{and} \quad A + C = B + C$$ have exactly the same solution. In words, the same number may be added to both sides of an equation.
Multiplication Property of Equality	If A, B, and C are real numbers, then the equations $$A = B \quad \text{and} \quad AC = BC$$ have exactly the same solution. (Assume that $C \neq 0$.) In words, both sides of an equation may be multiplied by the same nonzero number.
Addition Property of Inequality	For any real numbers A, B, and C, the inequalities $$A < B \quad \text{and} \quad A + C < B + C$$ have exactly the same solutions. In words, the same number may be added on both sides of an inequality.
Multiplication Property of Inequality	For any real numbers A, B, and C, (1) if C is *positive*, then the inequalities $$A < B \quad \text{and} \quad AC < BC$$ have exactly the same solutions; (2) if C is *negative*, then the inequalities $$A < B \quad \text{and} \quad AC > BC$$ have exactly the same solutions. (Assume that $C \neq 0$.) In words, both sides of an inequality may be multiplied by the same nonzero number. When you multiply by a negative number, reverse the inequality symbol.

Chapter 2 *Review Exercises*

❶ [2.1] *Combine terms whenever possible.*

1. $2m + 9m$

2. $15p^2 - 7p^2 + 8p^2$

3. $5p^2 - 4p + 6p + 11p^2$

4. $-2(3k - 5) + 2(k + 1)$

5. $7(2m + 3) - 2(8m - 4)$

6. $-(2k + 8) - (3k - 7)$

[2.2–2.4] *Solve each equation.*

7. $m - 5 = 1$

8. $y + 8 = -4$

9. $3k + 1 = 2k + 8$

10. $5k = 4k + \dfrac{2}{3}$

11. $(4r - 2) - (3r + 1) = 8$

12. $3(2y - 5) = 2 + 5y$

13. $7k = 35$ **14.** $12r = -48$ **15.** $2p - 7p + 8p = 15$

16. $\dfrac{m}{12} = -1$ **17.** $\dfrac{5}{8}k = 8$ **18.** $12m + 11 = 59$

19. $-(m + 2) = -3m - 11$

[2.5] *Solve the word problems.*

20. If 7 is added to six times a number, the result is 22. Find the number.

21. If 4 is subtracted from twice a number, the result is 8. Find the number.

22. The sum of a number and 5 is multiplied by 6, giving 72 as a result. Find the number.

23. In a marathon, Susan ran 2/3 as far as Linda. In all, the two people ran 30 miles. How many miles did Susan run?

24. Charles has 15 more college units than Tom. Altogether, the two men have 95 units. How many units does Tom have?

[2.6] *A formula is given in the following exercises, along with the values of some of the variables. Find the value of the variable that is not given.*

25. $A = \dfrac{1}{2}bh$; $A = 22, b = 4$ **26.** $A = \dfrac{1}{2}(b + B)h$; $b = 9, B = 12, h = 8$

27. $C = 2\pi r$; $C = 12.56, \pi = 3.14$ **28.** $V = \dfrac{4}{3}\pi r^3$; $\pi = 3.14, r = 1$

Solve each formula for the specified variable.

29. $A = LW$;   for W **30.** $A = \dfrac{1}{2}(b + B)h$;   for h

[2.7] *Write the following ratios and reduce them to lowest terms.*

31. 50 centimeters to 30 centimeters   **32.** 6 days to 1 week

33. 45 inches to 5 feet   **34.** 2 months to 3 years

Decide whether or not the following proportions are true.

35. $\dfrac{15}{18} = \dfrac{45}{54}$ **36.** $\dfrac{11}{19} = \dfrac{55}{105}$ **37.** $\dfrac{38}{51} = \dfrac{722}{1020}$ **38.** $\dfrac{29}{72} = \dfrac{899}{2232}$

Solve each proportion.

39. $\dfrac{p}{5} = \dfrac{21}{30}$ **40.** $\dfrac{2 + m}{2 - m} = \dfrac{3}{4}$

Solve each problem.

41. The distance between two towns is 520 miles. On a map the towns are 2.6 centimeters apart. How far apart are towns that are 1.5 centimeters apart on the map?

42. A recipe that serves 6 people calls for 2 1/2 cups of ground beef. How much ground beef should be used to prepare the recipe for 16 people?

[2.8] *Solve the word problems.*

43. The perimeter of a rectangle is ten times the short side. The length is 9 meters more than the width. Find the width of the rectangle.

44. The longest side of a triangle is 11 meters longer than the shortest side. The medium side is 15 meters long. The perimeter of the triangle is 46 meters. Find the length of the shortest side of the triangle.

45. A person has $250 in fives and tens. He has twice as many tens as fives. How many fives does he have?

46. For a party, Joann bought some 15¢ candy and some 30¢ candy, paying $15 in all. If there are 25 more pieces of 15¢ candy, how many pieces of 30¢ did she buy?

47. How many liters of 40% salt solution should be added to 72 liters of 80% salt solution to get a 70% mixture?

48. A nurse must mix 15 liters of a 10% solution of a drug with some 60% solution to get a 20% mixture. How many liters of the 60% solution would be needed?

[2.9] *Graph each inequality on a number line.*

49. $m \geq -2$ **50.** $a < -3$ **51.** $-5 \leq p < 6$ **52.** $1 \leq m < 4$

Solve each inequality.

53. $y + 5 \geq 2$ **54.** $5y > 4y + 8$

55. $9(k - 5) - (3 + 8k) \geq 5$ **56.** $3(2z + 5) + 4(8 + 3z) \leq 5(3z + 2) + 2z$

57. $-6 \leq x + 2 \leq 0$ **58.** $3 < y - 4 < 5$

59. $6k \geq -18$ **60.** $-11y < 22$

61. $2 - 4p + 7p + 8 < 6p - 5p$ **62.** $-(y + 2) + 3(2y - 7) \leq 4 - 5y$

63. $-3 \leq 2m + 1 \leq 4$ **64.** $9 < 3m + 5 \leq 20$

❷ *Solve each of the following.*

65. $\dfrac{y}{y - 5} = \dfrac{7}{2}$ **66.** $I = prt;$ for r

67. $-z > -4$ **68.** $\dfrac{3}{5} = \dfrac{k}{12}$

69. $2k - 5 = 4k + 7$ **70.** $7m + 3 \leq 5m - 9$

71. $a + b + c = P;$ for a **72.** $2 - 3(y - 5) = 4 + y$

73. If 1 quart of oil must be mixed with 24 quarts of gasoline, how much oil would be needed for 192 quarts of gasoline?

74. Two trains are 390 miles apart. They start at the same time and travel toward one another, meeting 3 hours later. If the speed of one train is 30 miles per hour faster than the speed of the other train, find the speed of each train.

75. The area of a triangle is 25 square meters. The base is 10 meters in length. Find the height.

76. On a test in geometry, the highest grade was 35 points more than the lowest. The sum of the highest and lowest grades was 157. Find the lowest score.

77. The perimeter of a square cannot be greater than 200 meters. Find the longest possible value for the length of a side.

78. The distance between two cities on a road map is 16 centimeters. The two cities are actually 150 kilometers apart. The distance on the map between two other cities is 40 centimeters. How far apart are these cities?

79. If three times the smaller of two consecutive odd integers is added to twice the larger, the result is 99. Find the smaller integer.

80. One side of a triangle is 3 centimeters longer than the shortest side. The third side is twice as long as the shortest side. If the perimeter of the triangle cannot exceed 39 centimeters, find the maximum possible length for the shortest side.

81. An inheritance is invested two ways—some at 5% and $6000 more at 10%. The total annual income from interest is $2100. Find the amount invested at each rate.

82. The shorter base of a trapezoid is 42 centimeters long, and the longer base is 48 centimeters long. The area of the trapezoid is 360 square centimeters. Find the height of the trapezoid.

Chapter 2 *Test*

Simplify by combining like terms.

1. $9r + 3r - 4r - r - 8r$

2. $3z - 7z + 8 - 9 - (-5) + 4z$

3. $4(2m - 1) - (m + 5) - 3(m - 5)$

Solve each equation.

4. $3(a + 12) = 1 - 2(a - 5)$

5. $4k - 6k + 8(k - 3) = -2(k + 12)$

6. $\dfrac{5}{4}m = -3$

7. $\dfrac{1}{2}p + \dfrac{1}{3} = \dfrac{5}{2}p - \dfrac{4}{3}$

8. $-(y + 3) + 2y - 5 = 4 - 3y$

9. Three times a number is subtracted from 4, giving a result 20 more than the number. Find the number.

10. Vern paid $57 more to tune up his Bronco than his Oldsmobile. He paid $257 in all. How much did it cost for the tune-up on the Oldsmobile?

11. Solve the formula $V = LWH$;   for W.

12. Solve the formula $A = \frac{1}{2}(b + B)h$;   for b.

13. Is the proportion $\frac{15}{79} = \frac{465}{2449}$ true or not?

Solve each proportion.

14. $\frac{z}{16} = \frac{3}{48}$

15. $\frac{y + 5}{y - 2} = \frac{1}{4}$

16. If 11 hamburgers cost $6.05, find the cost of 32 hamburgers.

Write an equation for each problem, and then solve it.

17. Sam has some ten-dollar bills and some twenty-dollar bills. He has 10 more twenties than tens. The money has a total value of $800. Find the number of ten-dollar bills that he has.

18. Trains A and B leave the same station at the same time, traveling in opposite directions. Train A travels at 50 miles per hour, and Train B travels at 70 miles per hour. After how many hours will the trains be 360 miles apart?

19. A woman invests some money at 12%, with $6000 more than this amount invested at 15%. Her total annual income from interest is $3870. How much is invested at 12%?

20. How many liters of a 20% chemical solution must be mixed with 30 liters of a 60% solution to get a 50% mixture?

Solve each inequality. Graph each solution.

21. $-2m < -14$

22. $5(k - 2) + 3 \le 2(k - 3) + 2k$

23. $-4r + 2(r - 3) \ge 5r - (3 + 6r) - 7$

24. $-8 < 3k - 2 \le 12$

Exponents and Polynomials

Objectives

1 Use exponents.

2 Use the product rule for exponents.

Use the following power rules for exponents.

3 $(a^m)^n = a^{mn}$

4 $(ab)^m = a^m b^m$

5 $\left(\dfrac{a}{b}\right)^m = \dfrac{a^m}{b^m}$

1 In Chapter 1, exponents were used to write products of repeated factors:

$$5^2 = 5 \cdot 5 = 25,$$
$$4^3 = 4 \cdot 4 \cdot 4 = 64,$$
$$9^1 = 9,$$

and so on. In the expression 5^2, the number 5 is the **base** and 2 is the **exponent.** The expression 5^2 is called an **exponential expression.**

Example 1 Evaluate each exponential expression. Name the base and the exponent.

	Base	*Exponent*
(a) $5^4 = 5 \cdot 5 \cdot 5 \cdot 5 = 625$	5	4
(b) $-5^4 = -1 \cdot 5^4 = -1 \cdot (5 \cdot 5 \cdot 5 \cdot 5) = -625$	5	4
(c) $(-5)^4 = (-5)(-5)(-5)(-5) = 625$	-5	4

It is important to understand the differences between parts (b) and (c) of Example 1. In -5^4 the lack of parentheses shows that the exponent 4 refers only to the base 5, and not -5; in $(-5)^4$ the parentheses show that the exponent 4 refers to the base -5. In summary,

$-a^n$ and $(-a)^n$ are not necessarily the same.			
Expression	*Base*	*Exponent*	*Example*
$-a^n$	a	n	$-3^2 = -(3 \cdot 3) = -9$
$(-a)^n$	$-a$	n	$(-3)^2 = (-3)(-3) = 9$

2 By the definition of exponents,

$$2^4 \cdot 2^3 = (2 \cdot 2 \cdot 2 \cdot 2)(2 \cdot 2 \cdot 2)$$
$$= 2 \cdot 2 \cdot 2 \cdot 2 \cdot 2 \cdot 2 \cdot 2$$
$$= 2^7.$$

Also, $$6^2 \cdot 6^3 = (6 \cdot 6)(6 \cdot 6 \cdot 6)$$
$$= 6 \cdot 6 \cdot 6 \cdot 6 \cdot 6$$
$$= 6^5.$$

Generalizing from these examples, $2^4 \cdot 2^3 = 2^{4+3} = 2^7$ and $6^2 \cdot 6^3 = 6^{2+3} = 6^5$, gives the **product rule for exponents.**

For any positive integers m and n, $a^m \cdot a^n = a^{m+n}$.

The bases must be the same before the product rule for exponents can be applied. (A summary of the rules for exponents is given at the end of the section.)

Example 2 Use the product rule for exponents to find each product.

(a) $6^3 \cdot 6^5 = 6^{3+5} = 6^8$ by the product rule.

(b) $(-4)^7(-4)^2 = (-4)^{7+2} = (-4)^9$

(c) There is no shortcut for finding the product $2^3 \cdot 3^2$, since the bases are different.

(d) $x^2 \cdot x = x^2 \cdot x^1 = x^{2+1} = x^3$

(e) $m^4 m^3 m^5 = m^{4+3+5} = m^{12}$

Example 3 Multiply $2x^3$ and $3x^7$.

Since $2x^3$ means $2 \cdot x^3$ and $3x^7$ means $3 \cdot x^7$, use the associative and commutative properties to get

$$2x^3 \cdot 3x^7 = 2 \cdot 3 \cdot x^3 \cdot x^7 = 6x^{10}. \quad \clubsuit$$

Be careful to understand the difference between *adding* and *multiplying* terms. For example,

$$8x^3 + 5x^3 = 13x^3, \quad \text{but} \quad (8x^3)(5x^3) = 8 \cdot 5x^{3+3} = 40x^6.$$

3 Simplify an expression such as $(8^3)^2$ with the product rule for exponents.

$$(8^3)^2 = (8^3)(8^3)$$
$$= 8^{3+3}$$
$$= 8^6$$

The exponents in $(8^3)^2$ are multiplied to give the exponent in 8^6: $3 \cdot 2 = 6$. As another example,

$$(5^2)^3 = 5^2 \cdot 5^2 \cdot 5^2$$
$$= 5^{2+2+2}$$
$$= 5^6,$$

and $2 \cdot 3 = 6$. These examples suggest **power rule (a) for exponents.**

For any positive integers m and n, $(a^m)^n = a^{mn}$.

Example 4 Use power rule (a) for exponents to simplify each expression.

(a) $(2^5)^3 = 2^{5 \cdot 3} = 2^{15}$

(b) $(5^7)^2 = 5^{7(2)} = 5^{14}$

(c) $(x^2)^5 = x^{2(5)} = x^{10}$

(d) $(n^3)^2 = n^{3(2)} = n^6 \quad \clubsuit$

4 The properties studied in Chapter 1 can be used to develop two more rules for exponents. Using the definition of an exponential expression and the commutative and associative properties,

$$(4 \cdot 8)^3 = (4 \cdot 8)(4 \cdot 8)(4 \cdot 8)$$
$$= 4 \cdot 4 \cdot 4 \cdot 8 \cdot 8 \cdot 8$$
$$= 4^3 \cdot 8^3.$$

This example suggests **power rule (b) for exponents.**

For any positive integer m, $(ab)^m = a^m b^m$.

Example 5   Use power rule (b) for exponents to simplify each expression.

(a) $5(pq)^2 = 5(p^2q^2) = 5p^2q^2$

(b) $(3xy)^2 = 3^2x^2y^2 = 9x^2y^2$

(c) $(2m^2p^3)^4 = 2^4(m^2)^4(p^3)^4 = 2^4m^8p^{12}$ ✛

---

5   Since the quotient a/b can be written as $a\left(\dfrac{1}{b}\right)$, the rule above, together with some of the properties of real numbers, gives **power rule (c) for exponents.**

$$\text{For any positive integer } m, \qquad \left(\frac{a}{b}\right)^m = \frac{a^m}{b^m} \qquad (b \neq 0).$$

Example 6   Use power rule (c) for exponents to simplify each expression.

(a) $\left(\dfrac{2}{3}\right)^5 = \dfrac{2^5}{3^5}$

(b) $\left(\dfrac{m}{n}\right)^3 = \dfrac{m^3}{n^3}$ ✛

    As shown in the next example, more than one rule may be needed to simplify an expression with exponents.

Example 7   Simplify each expression.

(a) $\left(\dfrac{2}{3}\right)^2 \cdot 2^3$

$$\left(\frac{2}{3}\right)^2 \cdot 2^3 = \frac{2^2}{3^2} \cdot \frac{2^3}{1} \qquad \text{Power rule (c)}$$

$$= \frac{2^2 \cdot 2^3}{3^2 \cdot 1}$$

$$= \frac{2^5}{3^2} \qquad \text{Product rule}$$

(b) $(5x)^3(5x)^4$

$$(5x)^3(5x)^4 = (5x)^7 \qquad \text{Product rule}$$
$$= 5^7x^7 \qquad \text{Power rule (b)}$$

(c) $(2x^2y^3)^4(3xy^2)^3$

$$(2x^2y^3)^4(3xy^2)^3 = 2^4(x^2)^4(y^3)^4 \cdot 3^3x^3(y^2)^3 \qquad \text{Power rule (b)}$$
$$= 2^4 \cdot 3^3x^8y^{12}x^3y^6 \qquad \text{Power rule (a)}$$
$$= 16 \cdot 27x^{11}y^{18} \qquad \text{Product rule}$$
$$= 432x^{11}y^{18} \quad ✛$$

The rules for exponents discussed in this section are summarized below.

Rules for Exponents

For positive integers m and n:

Product rule $a^m \cdot a^n = a^{m+n}$

Power rules (a) $(a^m)^n = a^{mn}$

(b) $(ab)^m = a^m b^m$

(c) $\left(\dfrac{a}{b}\right)^m = \dfrac{a^m}{b^m} \ (b \neq 0)$

3.1 Exercises

Identify the base and exponent for each exponential expression. See Example 1.

1. 5^{12}　　　　　**2.** $(3m)^4$　　　　　**3.** -2^4　　　　　**4.** -125^3

5. $(-24)^2$　　　　**6.** $-(-3)^5$　　　　**7.** $-r^5$　　　　**8.** $5y^3$

Write each expression using exponents.

9. $3 \cdot 3 \cdot 3 \cdot 3 \cdot 3$　　　　**10.** $4 \cdot 4 \cdot 4$　　　　**11.** $(-2)(-2)(-2)(-2)(-2)$

12. $(-1)(-1)(-1)(-1)$　　**13.** $\dfrac{1}{(-2)(-2)(-2)}$　　**14.** $\dfrac{1}{2 \cdot 2 \cdot 2 \cdot 2 \cdot 2}$

15. $p \cdot p \cdot p \cdot p \cdot p$　　　**16.** $k \cdot k \cdot k$　　　**17.** $\dfrac{1}{y \cdot y \cdot y \cdot y}$

18. $\dfrac{1}{a \cdot a \cdot a \cdot a \cdot a \cdot a}$　　**19.** $(-2z)(-2z)(-2z)$　　**20.** $(-3m)(-3m)(-3m)(-3m)$

Evaluate each expression. For example, $5^2 + 5^3 = 25 + 125 = 150$.

21. $3^2 + 3^4$　　　　**22.** $2^8 - 2^6$　　　　**23.** $4^2 + 4^3$　　　　**24.** $3^3 + 3^4$

25. $2^2 + 2^5$　　　　**26.** $4^2 + 4^1$　　　　**27.** $(-4)^2 - (-2)^2$　　**28.** $(-2)^3 - (-3)^2$

Use the product rule to simplify each expression. Write each answer in exponential form. See Examples 2 and 3.

29. $4^2 \cdot 4^3$　　　**30.** $3^5 \cdot 3^4$　　　**31.** $3^4 \cdot 3^7$　　　**32.** $2^5 \cdot 2^{15}$

33. $4^3 \cdot 4^5 \cdot 4^{10}$　**34.** $2^3 \cdot 2^4 \cdot 2^6$　**35.** $(-3)^3(-3)^2$　**36.** $(-4)^5(-4)^3$

37. $y^3 \cdot y^4 \cdot y^7$　**38.** $a^8 \cdot a^5 \cdot a^2$　**39.** $r \cdot r^5 \cdot r^4 \cdot r^7$　**40.** $m^9 \cdot m \cdot m^5 \cdot m^8$

41. $(-9r^3)(7r^6)$　**42.** $(8a^9)(-3a^{14})$　**43.** $(2p^4)(5p^9)$　**44.** $(3q^8)(7q^5)$

In each of the following exercises, first add the given terms; then start over and multiply them. See Example 3.

45. $4m^3, \ 9m^3$　　**46.** $8y^2, \ 7y^2$　　**47.** $-12p, \ 11p$　　**48.** $3q^4, \ 5q^4$

49. $7r, \ 3r, \ 5r$　　**50.** $9a^3, \ 2a^3, \ 3a^3$　**51.** $-5a^2, \ 3a^2$　　**52.** $6r^4, \ -8r^4$

Use the power rules for exponents to simplify each expression. Write each answer in exponential form. See Examples 4–6.

53. $(6^3)^2$　　　　**54.** $(8^4)^6$　　　　**55.** $(9^3)^2$　　　　**56.** $(2^3)^4$

57. $(-5^2)^4$ **58.** $(-2^3)^2$ **59.** $(-4^2)^3$ **60.** $(-3^5)^3$

61. $(5m)^3$ **62.** $(2xy)^4$ **63.** $(-2pq)^4$ **64.** $(-3ab)^5$

65. $\left(\dfrac{-3x^5}{4}\right)^2$ **66.** $\left(\dfrac{4m^3n^2}{5}\right)^4$ **67.** $\left(\dfrac{5a^2b}{c^4}\right)^3$ **68.** $\left(\dfrac{2y^2z^4}{w^3}\right)^5$

Simplify each expression. See Example 7.

69. $\left(\dfrac{4}{3}\right)^5 \cdot (4)^3$ **70.** $\left(\dfrac{2}{5}\right)^2 \cdot \left(\dfrac{2}{5}\right)^3$ **71.** $\left(\dfrac{3}{4}x\right)^5$ **72.** $\left(\dfrac{5}{3}y\right)^2$

73. $(3m)^2(3m)^5$ **74.** $(-5p)^4(-5p)^2$ **75.** $(8z)^6(8z)^3$ **76.** $(2p)^4(2p)$

77. $(2m^2n)^3(mn^2)$ **78.** $(3p^2q^2)^3(q^4)$ **79.** $(5ab^2)^5(5a^3b)^2$ **80.** $(-r^3s)^4(r^2s)^3$

Simplify each expression. Assume that all variables represent positive integers.

81. $5^r \cdot 5^{7r}$ **82.** $6^{5p} \cdot 6^p$ **83.** $2^{3q} \cdot 2^q \cdot 2^{4q}$ **84.** $3^{5x} \cdot 3^{7x} \cdot 3^x$

85. $(2m)^p$ **86.** $(3k)^t$ **87.** $(4r^2 5^x)^y$ **88.** $(2p^3q^y)^w$

89. $\left(\dfrac{4^2}{3^3}\right)^r$ **90.** $\left(\dfrac{5^3}{7^4}\right)^k$ **91.** $\left(\dfrac{2p^m}{q^r}\right)^n$ **92.** $\left(\dfrac{4r^a}{r^b}\right)^c$

Review Exercises *Evaluate each expression when (a) $x = 2$ and (b) $x = -3$. See Sections 1.6–1.8.*

93. $-x^2 + 4x - 7$ **94.** $-x^2 - 6x + 2$ **95.** $-x^3 + x^2 - 4$ **96.** $8 - 6x - x^3$

3.2 The Quotient Rule and Integer Exponents

Objectives

1 Use zero as an exponent.

2 Use negative numbers as exponents.

3 Use the quotient rule for exponents.

4 Use combinations of rules.

5 Use variables as exponents.

1 Section 3.1 gave the product rule for exponents. The rule for division with exponents is similar to the product rule for exponents. For example,

$$\frac{6^5}{6^2} = \frac{6 \cdot 6 \cdot 6 \cdot 6 \cdot 6}{6 \cdot 6} = 6 \cdot 6 \cdot 6 = 6^3,$$

and $5 - 2 = 3$. Also,

$$\frac{m^4}{m^2} = \frac{m \cdot m \cdot m \cdot m}{m \cdot m} = m \cdot m = m^2,$$

and $4 - 2 = 2$. Generalizing from these examples, the difference of the exponents gives the exponent of the answer.

If the exponents in the numerator and denominator are equal, then, for example,

$$\frac{6^5}{6^5} = \frac{6 \cdot 6 \cdot 6 \cdot 6 \cdot 6}{6 \cdot 6 \cdot 6 \cdot 6 \cdot 6} = 1.$$

If, however, the exponents are subtracted as above,

$$\frac{6^5}{6^5} = 6^{5-5} = 6^0.$$

This means that $6^0 = 1$. Based on this,

for any nonzero real number a, $a^0 = 1$.

Example 1   Evaluate each exponential expression.

(a) $60^0 = 1$

(b) $(-60)^0 = 1$

(c) $-(60^0) = -(1) = -1$ ✚

Notice the difference between parts (b) and (c) of Example 1. In Example 1(b) the base is -60 and the exponent is 0. Any nonzero base raised to a zero exponent is 1. But in Example 1(c), the base is 60. Then $60^0 = 1$, so that $-60^0 = -1$.

2   The discussion at the beginning of the section showed that

$$\frac{6^5}{6^2} = 6^3,$$

where the bottom exponent was subtracted from the top exponent. If the bottom exponent were larger than the top exponent, subtracting would result in a negative exponent.

For example,

$$\frac{6^2}{6^5} = 6^{2-5} = 6^{-3}.$$

On the other hand,

$$\frac{6^2}{6^5} = \frac{6 \cdot 6}{6 \cdot 6 \cdot 6 \cdot 6 \cdot 6} = \frac{1}{6 \cdot 6 \cdot 6} = \frac{1}{6^3},$$

so that

$$6^{-3} = \frac{1}{6^3}.$$

This example suggests that **negative exponents** be defined as follows.

For any nonzero real number a and any integer n, $a^{-n} = \dfrac{1}{a^n}$.

By definition, a^{-n} and a^n are reciprocals, since

$$a^n \cdot a^{-n} = a^n \cdot \frac{1}{a^n} = 1.$$

The definition of a^{-n} also can be written as

$$a^{-n} = \frac{1}{a^n} = \left(\frac{1}{a}\right)^n.$$

For example, using the last result above,

$$6^{-3} = \left(\frac{1}{6}\right)^3 \quad \text{and} \quad \left(\frac{1}{3}\right)^{-2} = 3^2.$$

Example 2   Simplify by using the definition of negative exponents.

(a) $3^{-2} = \dfrac{1}{3^2} = \dfrac{1}{9}$

(b) $5^{-3} = \dfrac{1}{5^3} = \dfrac{1}{125}$

(c) $\left(\dfrac{1}{2}\right)^{-3} = 2^3 = 8$

(d) $\left(\dfrac{2}{5}\right)^{-4} = \left(\dfrac{5}{2}\right)^4$

(e) $4^{-1} - 2^{-1} = \dfrac{1}{4} - \dfrac{1}{2} = \dfrac{1}{4} - \dfrac{2}{4} = -\dfrac{1}{4}$

(f) $p^{-2} = \dfrac{1}{p^2}, \quad p \neq 0$

(g) $\dfrac{1}{x^{-4}}, \quad x \neq 0$

Since any real number power of 1 is 1,

$$\frac{1}{x^{-4}} = \frac{1^{-4}}{x^{-4}} = \left(\frac{1}{x}\right)^{-4} = x^4.$$

(h) $\dfrac{2}{x^{-4}}, \quad x \neq 0$

Write $\dfrac{2}{x^{-4}}$ as $2 \cdot \dfrac{1}{x^{-4}}$. Then use the result from part (g).

$$\frac{2}{x^{-4}} = \frac{2}{1} \cdot \frac{1}{x^{-4}} = 2 \cdot x^4 \text{ or } 2x^4 \quad \blacklozenge$$

A negative exponent does not indicate a negative number; negative exponents lead to reciprocals.

Expression	Example	
a^{-n}	$3^{-2} = \dfrac{1}{3^2} = \dfrac{1}{9}$	Not negative
$-a^{-n}$	$-3^{-2} = -\dfrac{1}{3^2} = -\dfrac{1}{9}$	Negative

Since exponential expressions with negative exponents can be written with positive exponents, the rules for exponents given in Section 3.1 are also true for negative exponents.

3 Now that zero and negative exponents have been defined, the **quotient rule for exponents** can be given.

For any nonzero real number a and any integers m and n,

$$\frac{a^m}{a^n} = a^{m-n}.$$

Example 3 Simplify by using the quotient rule for exponents. Write answers with positive exponents.

(a) $\dfrac{5^8}{5^6} = 5^{8-6} = 5^2$

(b) $\dfrac{4^2}{4^9} = 4^{2-9} = 4^{-7} = \dfrac{1}{4^7}$

(c) $\dfrac{5^{-3}}{5^{-7}} = 5^{-3-(-7)} = 5^4$

(d) $\dfrac{q^5}{q^{-3}} = q^{5-(-3)} = q^8,\ q \neq 0$

(e) $\dfrac{3^2 x^5}{3^4 x^3} = \dfrac{3^2}{3^4} \cdot \dfrac{x^5}{x^3} = 3^{2-4} \cdot x^{5-3}$

$\qquad = 3^{-2} x^2 = \dfrac{x^2}{3^2},\ x \neq 0$

Sometimes numerical expressions with small exponents, such as 3^2, are evaluated. Doing that would give the result as $x^2/9$. ✦

4 The next example shows the use of more than one rule to simplify an expression.

Example 4   Use a combination of the rules for exponents to simplify each expression.

(a) $\dfrac{(4^2)^3}{4^5}$

Use power rule (a) and then the quotient rule.

$$\dfrac{(4^2)^3}{4^5} = \dfrac{4^6}{4^5} = 4^{6-5} = 4^1 = 4$$

(b) $(2x)^3(2x)^2$

Use the product rule first. Then use power rule (b).

$$(2x)^3(2x)^2 = (2x)^5 = 2^5 x^5 \quad \text{or} \quad 32x^5$$

(c) $\left(\dfrac{2x^3}{5}\right)^{-4}$

By the definition of a negative exponent and power rules (b) and (c),

$$\left(\dfrac{2x^3}{5}\right)^{-4} = \left(\dfrac{5}{2x^3}\right)^4 = \dfrac{5^4}{2^4 x^{12}}.$$

(d) $\left(\dfrac{3x^{-2}}{4^{-1}y^3}\right)^{-3} = \dfrac{3^{-3}x^6}{4^3 y^{-9}} \qquad$ Power rules

$$= \dfrac{\dfrac{1}{3^3} \cdot x^6}{4^3 \cdot \dfrac{1}{y^9}} = \dfrac{\dfrac{x^6}{3^3}}{\dfrac{4^3}{y^9}} = \dfrac{x^6}{3^3} \cdot \dfrac{y^9}{4^3} = \dfrac{x^6 y^9}{3^3 \cdot 4^3} \quad \clubsuit$$

Since the steps can be done in several different orders, there are many equally good ways to simplify a problem like Example 4(d).

The definitions and rules for exponents given in this section and the last are summarized below.

Definitions and Rules for Exponents

For any integers m and n:

Product rule	$a^m \cdot a^n = a^{m+n}$	
Zero exponent	$a^0 = 1$	$(a \neq 0)$
Negative exponent	$a^{-n} = \dfrac{1}{a^n}$	$(a \neq 0)$
Quotient rule	$\dfrac{a^m}{a^n} = a^{m-n}$	$(a \neq 0)$
Power rules (a)	$(a^m)^n = a^{mn}$	
(b)	$(ab)^m = a^m b^m$	
(c)	$\left(\dfrac{a}{b}\right)^m = \dfrac{a^m}{b^m}$	$(b \neq 0)$

5 All the rules given in the box also apply when variables are used as exponents, as long as the variables represent only integers, an assumption we shall make throughout this chapter.

Example 5 Simplify the following.

(a) $3x^k \cdot 2x^5 = 3 \cdot 2 \cdot x^k \cdot x^5 = 6x^{k+5}$

(b) $\dfrac{y^{7z}}{y^{3z}} = y^{7z-3z} = y^{4z}$

(c) $(2a^4)^r = 2^r \cdot a^{4r} = 2^r a^{4r}$

(d) $z^q \cdot z^{5q} \cdot z^{-4q} = z^{q+5q+(-4q)} = z^{2q}$ ✦

3.2 Exercises

Evaluate each expression. See Examples 1 and 2.

1. $4^0 + 5^0$

2. $3^0 + 8^0$

3. $(-9)^0 + 9^0$

4. $(-8)^0 + (-8)^0$

5. 3^{-3}

6. 2^{-5}

7. $(-12)^{-1}$

8. $(-6)^{-2}$

9. $\left(\dfrac{1}{2}\right)^{-5}$

10. $\left(\dfrac{1}{5}\right)^{-2}$

11. $\left(\dfrac{1}{2}\right)^{-1}$

12. $\left(\dfrac{5}{4}\right)^{-2}$

13. $2^{-1} + 3^{-1}$

14. $3^{-1} - 4^{-1}$

15. $5^{-1} + 4^{-1}$

16. $3^{-1} + 6^{-1}$

17. $(.98)^{-2}$

18. $(1.76)^{-2}$

19. $(3.918)^{-3}$

20. $(.162)^{-3}$

Use the quotient rule to simplify each expression. Write each answer with positive exponents. Assume that all variables represent nonzero real numbers. See Example 3.

21. $\dfrac{4^7}{4^2}$

22. $\dfrac{11^5}{11^3}$

23. $\dfrac{8^3}{8^9}$

24. $\dfrac{5^4}{5^{10}}$

25. $\dfrac{6^{-4}}{6^2}$

26. $\dfrac{7^{-5}}{7^3}$

27. $\dfrac{14^{-2}}{14^{-5}}$

28. $\dfrac{6^{-3}}{6^{-8}}$

29. $\dfrac{x^6}{x^{-9}}$

30. $\dfrac{y^2}{y^{-5}}$

31. $\dfrac{z}{z^{-1}}$

32. $\dfrac{r^{-1}}{r}$

33. $\dfrac{1}{2^{-5}}$

34. $\dfrac{1}{4^{-3}}$

35. $\dfrac{2}{k^{-2}}$

36. $\dfrac{5}{p^{-5}}$

Simplify each expression. Give answers with only positive exponents. Assume that all variables represent nonzero numbers. See Examples 3 and 4.

37. $\dfrac{4^3 \cdot 4^{-5}}{4^7}$

38. $\dfrac{2^5 \cdot 2^{-4}}{2^{-1}}$

39. $\dfrac{5^4}{5^{-3} \cdot 5^{-2}}$

40. $\dfrac{8^6}{8^{-2} \cdot 8^5}$

41. $\dfrac{64^6}{32^6}$

42. $\dfrac{81^5}{9^5}$

43. $\dfrac{(3x)^5}{x^5}$

44. $\dfrac{m^8}{(2m)^8}$

45. $\left(\dfrac{5m^{-2}}{m^{-1}}\right)^2$

46. $\left(\dfrac{4x^3}{x^{-1}}\right)^{-1}$

47. $\dfrac{x^7(x^8)^{-2}}{(x^{-2})^3}$

48. $\dfrac{(m^3)^2(m^{-2})^4}{(m^{-1})^6}$

49. $\dfrac{3^{-1}a^{-2}}{3^2 a^{-4}}$ **50.** $\dfrac{2^{-5}p^{-3}}{2^{-7}p^5}$ **51.** $\dfrac{4k^{-3}m^5}{4^{-1}k^{-7}m^{-3}}$ **52.** $\dfrac{6^{-1}y^{-2}z^5}{6^2 y^{-1}z^{-2}}$

Simplify each expression. Assume that all variables represent nonzero integers.
Give answers with positive exponents only. See Example 5.

53. $5^r \cdot 5^{7r} \cdot 5^{-2r}$ **54.** $6^{5p} \cdot 6^p \cdot 6^{-2p}$ **55.** $x^{-a} \cdot x^{-3a} \cdot x^{-7a}$ **56.** $a^{4m} \cdot q^{-6m} \cdot q^{-3m}$

57. $\dfrac{a^{6y}}{a^{2y}}$ **58.** $\dfrac{p^{9m}}{p^{5m}}$ **59.** $\dfrac{q^{-3k}}{q^{-8k}}$ **60.** $\dfrac{z^{-7m}}{z^{-12m}}$

61. $(6 \cdot p^{-3})^{-y}$ **62.** $(2 \cdot a^{-p})^4$ **63.** $\left(\dfrac{3m^{-2}}{p^{-1}}\right)^{-q}$ **64.** $\left(\dfrac{7z^{-1}}{x^{-5}}\right)^{-m}$

Evaluate each expression.

65. $2 \cdot 3^{-1} + 4 \cdot 2^{-1}$ **66.** $5 \cdot 4^{-2} + 3 \cdot 2^{-3}$ **67.** $-4 \cdot 2^{-2} + 3 \cdot 2^{-3}$ **68.** $4 \cdot 3^{-1} - 2 \cdot 3^{-2}$

Simplify each expression. Give answers with only positive integers. Assume all
variables are nonzero real numbers.

69. $\dfrac{(4a^2b^3)^{-2}(2ab^{-1})^3}{(a^3b)^{-4}}$ **70.** $\dfrac{(m^6n)^{-2}(m^2n^{-2})^3}{m^{-1}n^{-2}}$

71. $\dfrac{(2y^{-1}z^2)^2(3y^{-2}z^{-3})^3}{(y^3z^2)^{-1}}$ **72.** $\dfrac{(3p^{-2}q^3)^2(5p^{-1}q^{-4})^{-1}}{(p^2q^{-2})^{-3}}$

73. $\dfrac{(9^{-1}z^{-2}x)^{-1}(4z^2x^4)^{-2}}{(5z^{-2}x^{-3})^2}$ **74.** $\dfrac{(4^{-1}a^{-1}b^{-2})^{-2}(5a^{-3}b^4)^{-2}}{(3a^{-3}b^{-5})^2}$

Review Exercises *Evaluate each of the following.*

75. $10(6427)$ **76.** $100(72.69)$ **77.** $1000(1.23)$ **78.** $10,000(26.94)$

79. $34 \div 10$ **80.** $6501 \div 100$ **81.** $237 \div 1000$ **82.** $42 \div 10,000$

3.3 An Application of Exponents: Scientific Notation

Objectives

1 Express numbers in scientific notation.

2 Convert numbers in scientific notation to numbers without exponents.

3 Use scientific notation in calculations.

1 One example of the use of exponents comes from science. The numbers in science are often extremely large (such as the distance from the earth to the sun, which is 93,000,000 miles) or extremely small (the wavelength of yellow green light is approximately .0000006 meters). Because of the difficulty of working with many zeros, scientists often express such numbers with exponents. Each number is written as the product of a number a, where $1 \le a < 10$, and some power of 10, that is, as $a \times 10^n$. This form is called **scientific notation.** When a number

is multiplied by a power of 10, such as 10^1, $10^2 = 100$, $10^3 = 1000$, $10^{-1} = .1$, $10^{-2} = .01$, $10^{-3} = .001$, and so on, the net effect is to move the decimal point to the right if it is a positive power and to the left if it is a negative power. This is shown in the examples below. (In work with scientific notation, the times symbol, $\times$, is used.)

$23.19 \times 10^1 = 23.19 \times 10 = 231.9$ decimal moves 1 place to the right

$23.19 \times 10^2 = 23.19 \times 100 = 2319.$ decimal moves 2 places to the right

$23.19 \times 10^3 = 23.19 \times 1000 = 23190.$ decimal moves 3 places to the right

$23.19 \times 10^{-1} = 23.19 \times .1 = 2.319$ decimal moves 1 place to the left

$23.19 \times 10^{-2} = 23.19 \times .01 = .2319$ decimal moves 2 places to the left

$23.19 \times 10^{-3} = 23.19 \times .001 = .02319$ decimal moves 3 places to the left

A number in scientific notation is always written with the decimal point after the first nonzero integer and then multiplied by the appropriate power of 10. For example, 35 is written 3.5×10^1, or 3.5×10; 56,200 is written 5.62×10^4, since

$$56,200 = 5.62 \times 10,000 = 5.62 \times 10^4.$$

The following box shows the steps involved in writing a number in scientific notation.

Writing a Number in Scientific Notation	*Step 1* Place a caret, $\wedge$, to the right of the first nonzero digit. *Step 2* Count the number of places from the caret to the actual decimal point. *Step 3* The number of places in Step 2 is the absolute value of the exponent on 10. *Step 4* The exponent on 10 is positive if you made the number smaller by moving the decimal point. The exponent will be negative if you made the number larger by moving the decimal point.

Example 1   Write each number in scientific notation.

(a) 93,000,000

Place a caret after the first nonzero digit.

9$_\wedge$3,000,000

└─Caret to right of first nonzero digit

Count from the caret to the decimal point.

$$9{\wedge}3{,}000{,}000 \qquad \text{7 places}$$

Since moving the decimal point to the right made the number smaller, the exponent on 10 is positive, and $93{,}000{,}000 = 9.3 \times 10^{7}$.

(b) $63{,}200{,}000{,}000 = 6.32 \times 10^{10}$

(c) .00462

Place a caret to the right of the first nonzero digit.

$$.004{\wedge}62$$

Count from the caret to the decimal point.

$$.004{\wedge}62 \qquad \text{3 places}$$

Since moving the decimal point made the number larger, the exponent will be negative.

$$.00462 = 4.62 \times 10^{-3}$$

(d) $.0000762 = 7.62 \times 10^{-5}$ ✛

2 To convert a number from scientific notation to a number without exponents, use the steps given below.

Converting from Scientific Notation	*Step 1* Count from the decimal point the same number of places as the exponent on 10, attaching additional zeros as necessary.
	Step 2 Move to the right if the exponent on 10 is positive; move to the left if the exponent is negative.

Example 2   Write each number without exponents.

(a) 6.2×10^{3}

Moving the decimal point 3 places to the right, because the exponent on 10 is positive, gives

$$6.2 \times 10^{3} = 6.2 \times 1000 = 6200.$$

(b) $4.283 \times 10^{5} = 4.28300 = 428{,}300 \qquad \text{5 places to the right}$

(c) $7.04 \times 10^{-3} = .00704$

Here the decimal point was moved 3 places left; because the exponent on 10 is negative. ✛

3 The next example shows how scientific notation can be used with products and quotients.

Example 3   Write each number without exponents.

(a) $(6 \times 10^3)(5 \times 10^{-4})$

First use the commutative and associative properties.

$$(6 \times 10^3)(5 \times 10^{-4}) = (6 \times 5)(10^3 \times 10^{-4})$$

Now use the rules for exponents.

$$(6 \times 5)(10^3 \times 10^{-4}) = 30 \times 10^{-1}$$

Then express the result without exponents, as 3.0.

(b) $\dfrac{4 \times 10^{-5}}{2 \times 10^3} = \dfrac{4}{2} \times \dfrac{10^{-5}}{10^3} = 2 \times 10^{-8} = .00000002$ ✚

3.3   Exercises

Express each number in scientific notation. See Example 1.

1. 6,835,000,000
2. 321,000,000,000,000
3. 8,360,000,000,000
4. 6850
5. 215
6. 683
7. 25,000
8. 110,000,000
9. .035
10. .005
11. .0101
12. .0000006
13. .000012
14. .000000982

Write each number without exponents. See Examples 2 and 3.

15. 8.1×10^9
16. 3.5×10^2
17. 9.132×10^6
18. 2.14×10^0
19. 3.24×10^8
20. 4.35×10^4
21. 3.2×10^{-4}
22. 5.76×10^{-5}
23. 4.1×10^{-2}
24. 1.79×10^{-3}
25. $(2 \times 10^8) \times (4 \times 10^{-3})$
26. $(5 \times 10^4) \times (3 \times 10^{-2})$
27. $(4 \times 10^{-1}) \times (1 \times 10^{-5})$
28. $(6 \times 10^{-5}) \times (2 \times 10^4)$
29. $(7 \times 10^3) \times (2 \times 10^2) \times (3 \times 10^{-4})$
30. $(3 \times 10^{-5}) \times (3 \times 10^2) \times (5 \times 10^{-2})$
31. $(1.2 \times 10^2) \times (5 \times 10^{-3}) \times (2.4 \times 10^3)$
32. $(4.6 \times 10^{-3}) \times (2 \times 10^{-1}) \times (4 \times 10^5)$
33. $\dfrac{9 \times 10^5}{3 \times 10^{-1}}$
34. $\dfrac{12 \times 10^{-4}}{4 \times 10^4}$
35. $\dfrac{4 \times 10^{-3}}{2 \times 10^{-2}}$
36. $\dfrac{5 \times 10^{-1}}{1 \times 10^{-5}}$
37. $\dfrac{2.6 \times 10^5}{2 \times 10^2}$
38. $\dfrac{9.5 \times 10^{-1}}{5 \times 10^3}$
39. $\dfrac{7.2 \times 10^{-3} \times 1.6 \times 10^5}{4 \times 10^{-2} \times 3.6 \times 10^9}$
40. $\dfrac{8.7 \times 10^{-2} \times 1.2 \times 10^{-6}}{3 \times 10^{-4} \times 2.9 \times 10^{11}}$

Write the numbers in each statement in scientific notation.

41. Light visible to the human eye has a wavelength between .0004 millimeters and .0008 millimeters.

42. In the ocean, the amount of oxygen per cubic mile of water is 4,037,000,000 tons, and the amount of radium is .0003 tons.

43. Each tide in the Bay of Fundy carries more than 3,680,000,000,000,000 cubic feet of water into the bay.

44. The mean (average) diameter of the sun is about 865,000 miles.

Write the numbers in each statement without exponents.

45. Many ocean trenches have a depth of 3.5×10^4 feet.

46. The average life span of a human is 1×10^9 seconds.

47. There are 1×10^3 cubic millimeters in 6.102×10^{-2} cubic inches.

48. In the food chain that links the largest sea creature, the whale, to the smallest, the diatom, 4×10^{14} diatoms sustain a medium-sized whale for only a few hours.

Review Exercises   *Simplify. See Section 2.1.*

49. $7m + 8m$ **50.** $9y + 11y$ **51.** $12p - 9p + p$

52. $8x - 9x + 10x$ **53.** $5(2m - 1) - 3m$ **54.** $-7 + 2(2x + 5)$

Evaluate each expression when $x = -4$. See Sections 1.6–1.8.

55. $2x - 5$ **56.** $3x + 7$ **57.** $x^2 + x^3$ **58.** $2x^2 + 4x$

3.4 Polynomials

Objectives

1. Identify terms and coefficients.
2. Combine like terms.
3. Know the vocabulary for polynomials.
4. Evaluate polynomials.
5. Add polynomials.
6. Subtract polynomials.

1   Recall that in an expression such as

$$4x^3 + 6x^2 + 5x,$$

the quantities $4x^3$, $6x^2$, and $5x$ are called **terms.** In the term $4x^3$, the number 4 is called the **numerical coefficient,** or simply the coefficient, of x^3. In the same way, 6 is the coefficient of x^2 in the term $6x^2$, and 5 is the coefficient of x in the term $5x$. In a term with more than two factors, the **coefficient** of one factor is the product of the other factors.

Example 1   Name the (numerical) coefficient of each term in these expressions.

(a) $4x^3$

The coefficient is 4.

(b) $x - 6x^4$

The coefficient of x is 1 because $x = 1 \cdot x$. The coefficient of x^4 is -6 since $x - 6x^4$ can be written as the sum $x + (-6x^4)$.

(c) $5 - v^3$

The coefficient of the term 5 is 5 since $5 = 5v^0$. By writing $5 - v^3$ as a sum, $5 + (-v^3)$, or $5 + (-1v^3)$, the coefficient of v^3 can be identified as -1.   ✛

2   Recall that **like terms** have exactly the same combination of variables with the same exponents. Only the coefficients may be different. Examples of like terms are

$$19m^5 \quad \text{and} \quad 14m^5,$$
$$6y^9, \quad -37y^9, \quad \text{and} \quad y^9,$$
$$3pq \quad \text{and} \quad -2pq,$$
$$2xy^2 \quad \text{and} \quad -xy^2.$$

As shown in Chapter 2, to combine like terms, use the distributive property.

Example 2   Simplify each expression by using the distributive property.

(a) $-4x^3 + 6x^3 = (-4 + 6)x^3 = 2x^3$

(b) $9x^6 - 14x^6 + x^6 = (9 - 14 + 1)x^6 = -4x^6$

(c) $12m^2 + 5m + 4m^2 = (12 + 4)m^2 + 5m = 16m^2 + 5m$

(d) $3x^2y + 4x^2y - x^2y = (3 + 4 - 1)x^2y = 6x^2y$ ✛

Example 3(c) shows that is it not possible to combine $16m^2$ and $5m$. These two terms are unlike because the exponents on the variables are different. **Unlike terms** have different variables or different exponents on the same variables.

3   *Polynomials* are basic to algebra. A **polynomial in x** is an expression containing the sum of only a finite number of terms of the form ax^n, for any real number a and any whole number n. For example,

$$16x^8 - 7x^6 + 5x^4 - 3x^2 + 4$$

is a polynomial in x (the 4 can be written as $4x^0$). This polynomial is written in **descending powers** of the variable, since the exponents on x decrease from left to right. On the other hand,

$$2x^3 - x^2 + \frac{4}{x}$$

is not a polynomial in x, since $4/x$ is not a *product, ax^n,* for a *whole number n.* (Of course, we could define *polynomial* using any variable or variables, and not just x.)

The **degree** of a term is the sum of the exponents on the variables. For example, $3x^4$ has degree 4, while $6x^{17}$ has degree 17. The term $5x$ has degree 1, -7 has degree 0 (since -7 can be written as $-7x^0$), and $2x^2y$ has degree $2 + 1 = 3$ (y has an exponent of 1). The **degree of a polynomial** in one variable is the highest exponent found in any nonzero term of the polynomial. For example, $3x^4 - 5x^2 + 6$ is of degree 4, the polynomial $5x + 7$ is of degree 1, and 3 (or $3x^0$) is of degree 0.

Three types of polynomials are very common and are given special names. A polynomial with exactly three terms is called a **trinomial.** (*Tri-* means "three," as in *tri*angle.) Examples are

$$9m^3 - 4m^2 + 6, \qquad 19y^2 + 8y + 5, \qquad \text{and} \qquad -3m^5 - 9m^2 + 2.$$

A polynomial with exactly two terms is called a **binomial.** (*Bi-* means "two," as in *bi*cycle.) Examples are

$$-9x^4 + 9x^3, \qquad 8m^2 + 6m, \qquad \text{and} \qquad 3m^5 - 9m^2.$$

A polynomial with only one term is called a **monomial.** (*Mon(o)-* means "one," as in *mono*rail.) Examples are

$$9m, \qquad -6y^5, \qquad a^2, \qquad \text{and} \qquad 6.$$

Example 3   For each polynomial, first simplify if possible by combining like terms. Then give the degree and tell whether it is a monomial, a binomial, a trinomial, or none of these.

(a) $2x^3 + 5$

The polynomial cannot be simplified. The degree is 3. The polynomial is a binomial.

(b) $4x - 5x + 2x$

Add like terms to simplify: $4x - 5x + 2x = x$, a monomial of degree 1.   ✚

4   A polynomial represents different numbers for different values of the variable, as shown in the next examples.

Example 4   Find the value of $3x^4 + 5x^3 - 4x - 4$ when $x = -2$ and when $x = 3$.

First, substitute -2 for x.

$$3x^4 + 5x^3 - 4x - 4 = 3(-2)^4 + 5(-2)^3 - 4(-2) - 4$$

$$= 3 \cdot 16 + 5 \cdot (-8) + 8 - 4$$

$$= 48 - 40 + 8 - 4$$

$$= 12$$

Next, replace x with 3.

$$3x^4 + 5x^3 - 4x - 4 = 3 \cdot 3^4 + 5 \cdot 3^3 - 4 \cdot 3 - 4$$
$$= 3 \cdot 81 + 5 \cdot 27 - 12 - 4$$
$$= 362 \quad \clubsuit$$

5   Polynomials may be added, subtracted, multiplied, and divided. Polynomial addition and subtraction are explained in the rest of this section.

Adding Polynomials

> To add two polynomials, add like terms.

Example 5   Add $6x^3 - 4x^2 + 3$ and $-2x^3 + 7x^2 - 5$.
Write like terms in columns.

$$\begin{array}{r} 6x^3 - 4x^2 + 3 \\ -2x^3 + 7x^2 - 5 \\ \hline \end{array}$$

Now add, column by column.

$$\begin{array}{ccc} 6x^3 & -4x^2 & 3 \\ -2x^3 & 7x^2 & -5 \\ \hline 4x^3 & 3x^2 & -2 \end{array}$$

Add the three sums together.

$$4x^3 + 3x^2 + (-2) = 4x^3 + 3x^2 - 2 \quad \clubsuit$$

The polynomials in Example 5 also could be added horizontally as shown below.

Example 6   Add $6x^3 - 4x^2 + 3$ and $-2x^3 + 7x^2 - 5$.
Write the sum.

$$(6x^3 - 4x^2 + 3) + (-2x^3 + 7x^2 - 5)$$

Use the associative and commutative properties to rewrite this sum with the parentheses removed and with the subtractions changed to additions of inverses.

$$6x^3 + (-4x^2) + 3 + (-2x^3) + 7x^2 + (-5)$$

Place like terms together.

$$6x^3 + (-2x^3) + (-4x^2) + 7x^2 + 3 + (-5)$$

Combine like terms to get

$$4x^3 + 3x^2 + (-2), \quad \text{or simply} \quad 4x^3 + 3x^2 - 2,$$

the same answer found in Example 5. $\clubsuit$

6   Earlier, the difference $x - y$ was defined as $x + (-y)$. (Find the difference $x - y$ by adding x and the opposite of y.) For example,

$$7 - 2 = 7 + (-2) = 5 \quad \text{and} \quad -8 - (-2) = -8 + 2 = -6.$$

A similar method is used to subtract polynomials.

Example 7   Subtract: $(5x - 2) - (3x - 8)$.

By the definition of subtraction,

$$(5x - 2) - (3x - 8) = (5x - 2) + [-(3x - 8)].$$

As in Chapter 1, the distributive property gives

$$-(3x - 8) = -1(3x - 8) = -3x + 8.$$

Now

$$(5x - 2) - (3x - 8) = (5x - 2) + (-3x + 8) = 2x + 6. \quad \clubsuit$$

In summary, polynomials are subtracted as follows.

Subtracting Polynomials	Subtract two polynomials by changing all the signs on the second polynomial and adding the result to the first polynomial.

Example 8   Subtract $6x^3 - 4x^2 + 2$ from $11x^3 + 2x^2 - 8$.

Write the problem.

$$(11x^3 + 2x^2 - 8) - (6x^3 - 4x^2 + 2)$$

Change all the signs on the second polynomial and add.

$$(11x^3 + 2x^2 - 8) + (-6x^3 + 4x^2 - 2) = 5x^3 + 6x^2 - 10$$

Check a subtraction problem by using the fact that if $a - b = c$, then $a = b + c$. For example, $6 - 2 = 4$. Check by writing $6 = 2 + 4$, which is correct. Check the polynomial subtraction above by adding $6x^3 - 4x^2 + 2$ and $5x^3 + 6x^2 - 10$. Since the sum is $11x^3 + 2x^2 - 8$, the subtraction was performed correctly. $\clubsuit$

Subtraction also can be done in columns.

Example 9   Use the method of subtracting by columns to find

$$(14y^3 - 6y^2 + 2y - 5) - (2y^3 - 7y^2 - 4y + 6).$$

Arrange like terms in columns.

$$\begin{array}{r} 14y^3 - 6y^2 + 2y - 5 \\ \underline{2y^3 - 7y^2 - 4y + 6} \end{array}$$

Change all signs in the second row, and then add.

$$\begin{array}{r} 14y^3 - 6y^2 + 2y - 5 \\ -2y^3 + 7y^2 + 4y - 6 \quad \text{All signs changed} \\ \hline 12y^3 + y^2 + 6y - 11 \quad \text{Add} \end{array}$$

Either the horizontal or the vertical method may be used for adding and subtracting polynomials.

3.4 Exercises

In each polynomial, combine terms where possible. Write the results in descending powers of the variable. See Example 2.

1. $2r^5 + (-3r^5)$

2. $-19y^2 + 9y^2$

3. $3x^5 + 2x^5 - 4x^5$

4. $6x^3 - 8x^7 - 9x^3$

5. $-4p^7 + 8p^7 - 5p^7$

6. $-3a^8 + 4a^8 - 3a^8 + 2a^8$

7. $4y^2 + 3y^2 - 2y^2 + y^2$

8. $3r^5 - 8r^5 + r^5 - 2r^5$

9. $-5p^5 + 8p^5 - 2p^5 - p^5$

10. $6k^3 - 9k^3 + 8k^3 - 2k^3$

11. $y^4 + 8y^4 - 9y^2 + 6y^2 + 10y^2$

12. $11a^2 - 10a^2 + 2a^2 - a^6 + 2a^6$

13. $4z^5 - 9z^3 + 8z^2 + 10z^5$

14. $-9m^3 + 2m^3 - 11m^3 + 15m^2 - 9m$

15. $-.823q^2 + 1.725q - .374 + 1.994q^2 - .324q + .122$

16. $5.893r^3 - 2.776r^2 + 5.409r - 6.783r^3 + 1.437r - r^2$

For each polynomial, first simplify, if possible; then give the degree of the polynomial and tell whether it is (a) a monomial, (b) a binomial, (c) a trinomial, (d) none of these. See Example 3.

17. $5x^4 - 8x$

18. $4y - 8y$

19. $23x^9 - \frac{1}{2}x^2 + x$

20. $2m^7 - 3m^6 + 2m^5 + m$

21. $x^8 + 3x^7 - 5x^4$

22. $2x - 2x^2$

23. $\frac{3}{5}x^5 + \frac{2}{5}x^5$

24. $\frac{9}{11}x^2$

25. $2m^8 - 5m^9$

Tell whether each statement is true always, sometimes, *or* never.

26. A binomial is a polynomial.

27. A polynomial is a trinomial.

28. A trinomial is a binomial.

29. A monomial has no coefficient.

30. A binomial is a trinomial.

31. A polynomial of degree 4 has 4 terms.

Find the value of each polynomial when (a) $x = 2$ and when (b) $x = -1$. See Example 4.

32. $2x^2 - 4x$

33. $8x + 5x^2 + 2$

34. $2x^5 - 4x^4 + 5x^3 - x^2$

35. $2x^2 + 5x + 1$

36. $-3x^2 + 14x - 2$

37. $-2x^2 + 3$

38. $-5x^2 + 4x + 5$

39. $-x^2 - 8x$

40. $-x^2 + 7x + 2$

Add or subtract as indicated. See Examples 5 and 9.

41. Add.

$$3m^2 + 5m$$
$$2m^2 - 2m$$

42. Add.

$$4a^3 - 4a^2$$
$$6a^3 + 5a^2$$

43. Subtract.

$$12x^4 - x^2$$
$$\underline{8x^4 + 3x^2}$$

44. Subtract.

$$2a + 5d$$
$$3a - 6d$$

45. Subtract.

$$2n^5 - 5n^3 + 6$$
$$3n^5 + 7n^3 + 8$$

46. Subtract.

$$3r^2 - 4r + 2$$
$$7r^2 + 2r - 3$$

47. Add.

$$9m^3 - 5m^2 + 4m - 8$$
$$3m^3 + 6m^2 + 8m - 6$$

48. Add.

$$12r^5 + 11r^4 - 7r^3 - 2r^2 - 5r - 3$$
$$-8r^5 - 10r^4 + 3r^3 + 2r^2 - 5r + 7$$

49. Subtract.

$$5a^4 - 3a^3 + 2a^2$$
$$\underline{a^3 - a^2 + a - 1}$$

50. Add.

$$3w^2 - 5w + 2$$
$$4w^2 + 6w - 5$$
$$8w^2 + 7w - 2$$

Perform the indicated operations. See Examples 6–8.

51. $(3r^2 + 5r - 6) + (2r - 5r^2)$

52. $(8m^2 - 7m) - (3m^2 + 7m)$

53. $(x^2 + x) - (3x^2 + 2x - 1)$

54. $(3x^2 + 2x + 5) + (8x^2 - 5x - 4)$

55. $(16x^3 - x^2 + 3x) + (-12x^3 + 3x^2 + 2x)$

56. $(-2b^6 + 3b^4 - b^2) - (-b^6 + 2b^4 + 2b^2)$

57. $(7y^4 + 3y^2 + 2y) - (-18y^4 - 5y^2 - y)$

58. $(3x^2 + 2x + 5) - (-7x^2 - 8x + 2) + (3x^2 - 4x + 7)$

Write each statement as an equation or an inequality. Do not solve.

59. When $4 + x^2$ is added to $-9x + 2$, the result is larger than 8.

60. When $6 + 3x$ is subtracted from $5 + 2x$, the difference is larger than $8x + x^2$.

61. The sum of $5 + x^2$ and $3 - 2x$ is not equal to 5.

62. The sum of $3 - 2x + x^2$ and $8 - 9x + 3x^2$ is negative.

Perform the indicated operations in Exercises 63–66.

63. $[(8m^2 + 4m - 7) - (2m^2 - 5m + 2)] - (m^2 + m + 1)$

64. $[(9b^3 - 4b^2 + 3b + 2) - (-2b^3 - 3b^2 + b)] - (8b^3 + 6b + 4)$

65. $(3.127m^2 - 5.148m - 3.947) - (-.259m^2 + 7.125m - 8.9)$

66. $(-4.009k^2 + 3.176k + 4.1) - (1.795k^2 - .165k - .9935)$

67. Subtract $9x^2 - 6x + 5$ from $3x^2 - 2$.

68. Add $x^3 + 4x^2 - 2x + 3$ to $2x^3 - 8x^2 - 5$.

69. Find the sum of $10x^4 - 3x^3 + 2x + 1$ and $3x^3 - 6x^2 + 8x - 2$.

70. Find the difference when $9x^4 + 3x^2 + 5$ is subtracted from $8x^4 - 2x^3 + x - 1$.

71. $p(2p)$

72. $3k(5k)$

73. $5x^2(2x)$

74. $9r^3(2r)$

75. $7m^3(8m^2)$

76. $4y^5(7y^2)$

77. $6p^5(5p^4)$

78. $2z^8(5z^3)$

3.5 Multiplication of Polynomials

Objectives

1 Multiply a monomial and a polynomial.

2 Multiply two polynomials.

3 Multiply vertically.

1   As shown earlier, the product of two monomials is found by using the rules for exponents and the commutative and associative properties. For example,

$$(-8m^6)(-9m^4) = (-8)(-9)(m^6)(m^4) = 72m^{6+4} = 72m^{10}.$$

As mentioned earlier, it is important not to confuse the *addition* of terms with the *multiplication* of terms. For example,

$$7q^5 + 2q^5 = 9q^5, \quad \text{but} \quad (7q^5)(2q^5) = 7 \cdot 2q^{5+5} = 14q^{10}.$$

Find the product of a monomial and a polynomial with more than one term by first using the distributive property and then the method shown above.

Example 1   Use the distributive property to find each product.

(a) $4x^2(3x + 5)$

$$4x^2(3x + 5) = (4x^2)(3x) + (4x^2)(5) = 12x^3 + 20x^2$$

(b) $-8m^3(4m^3 + 3m^2 + 2m - 1)$

$$-8m^3(4m^3 + 3m^2 + 2m - 1)$$
$$= (-8m^3)(4m^3) + (-8m^3)(3m^2) + (-8m^3)(2m) + (-8m^3)(-1)$$
$$= -32m^6 - 24m^5 - 16m^4 + 8m^3 \quad \clubsuit$$

2   The distributive property is used repeatedly to find the product of any two polynomials. To find the product of the polynomials $x + 1$ and $x - 4$, think of $x - 4$ as a single quantity and use the distributive property as follows.

$$(x + 1)(x - 4) = x(x - 4) + 1(x - 4)$$

Now use the distributive property twice to find $x(x - 4)$ and $1(x - 4)$.

$$x(x - 4) + 1(x - 4) = x(x) + x(-4) + 1(x) + 1(-4)$$
$$= x^2 - 4x + x - 4$$
$$= x^2 - 3x - 4$$

A rule for multiplying any two polynomials is given below.

| **Multiplying Polynomials** | Multiply two polynomials by multiplying each term of the second polynomial by each term of the first polynomial and adding the products. |

Example 2   Find the product $(2x + 1)(3x + 5)$.
Use the distributive property.

$$
\begin{aligned}
(2x + 1)(3x + 5) &= 2x(3x + 5) + 1(3x + 5) \\
&= 2x(3x) + 2x(5) + 1(3x) + 1(5) \\
&= 6x^2 + 10x + 3x + 5 \\
&= 6x^2 + 13x + 5 \quad \clubsuit
\end{aligned}
$$

In the second step in Example 2, notice that each term of the second polynomial is multiplied by each term of the first polynomial. This shortcut is used in the next example.

Example 3   Find the product of $4m^3 - 2m^2 + 4m$ and $m^2 + 5$.
Multiply each term of the second polynomial by each term of the first. (Either polynomial can be written first in the product.)

$$
\begin{aligned}
&(m^2 + 5)(4m^3 - 2m^2 + 4m) \\
&= m^2(4m^3) - m^2(2m^2) + m^2(4m) + 5(4m^3) - 5(2m^2) + 5(4m) \\
&= 4m^5 - 2m^4 + 4m^3 + 20m^3 - 10m^2 + 20m
\end{aligned}
$$

Now combine like terms.

$$
= 4m^5 - 2m^4 + 24m^3 - 10m^2 + 20m \quad \clubsuit
$$

Example 4   Find $(x + 5)^3$.

$$
\begin{aligned}
(x + 5)^3 &= (x + 5)(x + 5)(x + 5) \\
&= (x^2 + 5x + 5x + 25)(x + 5) \\
&= (x^2 + 10x + 25)(x + 5) \\
&= x^3 + 5x^2 + 10x^2 + 50x + 25x + 125 \\
&= x^3 + 15x^2 + 75x + 125 \qquad \text{Combine terms} \quad \clubsuit
\end{aligned}
$$

3   Polynomial multiplication can also be performed vertically in a way that is similar to multiplication with whole numbers.

Example 5   Multiply $2x^2 + 4x + 1$ by $3x + 5$.
Start with the following.

$$
\begin{array}{r}
2x^2 + 4x + 1 \\
3x + 5 \\
\hline
\end{array}
$$

It is not necessary to line up terms in columns, because any terms may be multiplied (not just like terms). Begin by multiplying each of the terms in the top row by 5.

Step 1
$$
\begin{array}{r}
2x^2 + 4x + 1 \\
3x + 5 \\
\hline
10x^2 + 20x + 5 \quad\quad 5(2x^2 + 4x + 1)
\end{array}
$$

Notice how this process is similar to multiplication of whole numbers. Now multiply each term in the top row by $3x$. Be careful to place the like terms in columns, since the final step will involve addition (as in multiplying two whole numbers).

Step 2
$$
\begin{array}{r}
2x^2 + 4x + 1 \\
3x + 5 \\
\hline
10x^2 + 20x + 5 \\
6x^3 + 12x^2 + \quad 3x \quad\quad 3x(2x^2 + 4x + 1)
\end{array}
$$

Step 3 Add like terms.

$$
\begin{array}{r}
2x^2 + 4x + 1 \\
3x + 5 \\
\hline
10x^2 + 20x + 5 \\
6x^3 + 12x^2 + \quad 3x \\
\hline
6x^3 + 22x^2 + 23x + 5
\end{array}
$$

The product is $6x^3 + 22x^2 + 23x + 5$. ✤

The next example shows how to find the product of polynomials having more than one variable.

Example 6 Find the product of $3p - 5q$ and $2p + 7q$.

$$
\begin{array}{r}
3p - 5q \\
2p + 7q \\
\hline
21pq - 35q^2 \quad \leftarrow 7q(3p - 5q) \\
6p^2 - 10pq \quad\quad\quad \leftarrow 2p(3p - 5q) \\
\hline
6p^2 + 11pq - 35q^2 \quad ✤
\end{array}
$$

3.5 Exercises

Find each product. See Example 1.

1. $(-4x^5)(8x^2)$

2. $(-3x^7)(2x^5)$

3. $(5y^4)(3y^7)$

4. $(10p^2)(5p^3)$

5. $(15a^4)(2a^5)$

6. $(-3m^6)(-5m^4)$

7. $(5p)(3q^2)$

8. $(4a^3)(3b^2)$

9. $(-6m^3)(3n^2)$

10. $(9r^3)(-2s^2)$

11. $2m(3m + 2)$

12. $-5p(6 - 3p)$

13. $3p(-2p^3 + 4p^2)$

14. $4x(3 + 2x + 5x^3)$

15. $-8z(2z + 3z^2 + 3z^3)$

16. $7y(3 + 5y^2 - 2y^3)$

17. $2y^3(3 + 2y + 5y^4)$

18. $-2m^4(3m^2 + 5m + 6)$

19. $-4r^3(-7r^2 + 8r - 9)$

20. $-9a^5(-3a^6 - 2a^4 + 8a^2)$

21. $3a^2(2a^2 - 4ab + 5b^2)$

22. $4z^3(8z^2 + 5zy - 3y^2)$

23. $7m^3n^2(3m^2 + 2mn - n^3)$

24. $2p^2q(3p^2q^2 - 5p + 2q^2)$

Find each product. See Examples 2 and 6.

25. $(m + 7)(m + 5)$

26. $(n - 1)(n + 4)$

27. $(x + 5)(x - 5)$

28. $(y + 8)(y - 8)$

29. $(2x + 3)(6x - 4)$

30. $(4m + 3)(4m + 3)$

31. $(3x - 2)(3x - 2)$

32. $(b + 8)(6b - 2)$

33. $(5a + 1)(2a + 7)$

34. $(8 - 3a)(2 + a)$

35. $(6 - 5m)(2 + 3m)$

36. $(-4 + k)(2 - k)$

37. $(5 - 3x)(4 + x)$

38. $(2m - 3n)(m + 5n)$

39. $(4a + 3b)(2a - b)$

40. $(5r + 3s)(2r - 7s)$

41. $(3x + 2y)(5x - 3y)$

42. $(2x - 5)(x + 3)$

Find each product. See Examples 3 and 5.

43. $(6x + 1)(2x^2 + 4x + 1)$

44. $(9y - 2)(8y^2 - 6y + 1)$

45. $(9a + 2)(9a^2 + a + 1)$

46. $(2r - 1)(3r^2 + 4r - 4)$

47. $(4m + 3)(5m^3 - 4m^2 + m - 5)$

48. $(y + 4)(3y^4 - 2y^2 + 1)$

49. $(2x - 1)(3x^5 - 2x^3 + x^2 - 2x + 3)$

50. $(2a + 3)(a^4 - a^3 + a^2 - a + 1)$

51. $(5x^2 + 2x + 1)(x^2 - 3x + 5)$

52. $(2m^2 + m - 3)(m^2 - 4m + 5)$

Find each product. For Exercises 53–64, see Example 4.

53. $(x + 7)^2$

54. $(m + 6)^2$

55. $(a - 4)^2$

56. $(b - 10)^2$

57. $(2p - 5)^2$

58. $(3m + 1)^2$

59. $(5k + 3q)^2$

60. $(8m - 3n)^2$

61. $(m - 5)^3$

62. $(p + 3)^3$

63. $(2a + 1)^3$

64. $(3m - 1)^3$

65. $7(4m - 3)(2m + 1)$

66. $-4r(3r + 2)(2r - 5)$

67. $-3a(3a + 1)(a - 4)$

68. $(k + 1)^4$

69. $(3r - 2s)^4$

70. $(2z + 5y)^4$

71. $3p^3(2p^2 + 5p)(p^3 + 2p + 1)$

72. $5k^2(k^2 - k + 4)(k^3 - 3)$

73. $-2x^5(3x^2 + 2x - 5)(4x + 2)$

74. $-4x^3(3x^4 + 2x^2 - x)(-2x + 1)$

Review Exercises Find two numbers having the given sum and product. See Sections 1.6 and 1.8.

	Product	Sum
75.	6	5
77.	-21	-4
79.	-54	3

	Product	Sum
76.	8	6
78.	-35	-2
80.	-96	4

3.6   Products of Binomials

Objectives

1 Multiply binomials by the FOIL method.

2 Square binomials.

3 Find the product of the sum and difference of two terms.

1 The methods introduced in the last section can be used to find the product of any two polynomials. They are the only practical methods for multiplying polynomials with three or more terms. However, many of the polynomials to be multiplied are both binomials (with only two terms) so this section discusses a shortcut that eliminates the need to write all the steps. To develop this shortcut, multiply $x + 3$ and $x + 5$ using the distributive property.

$$
\begin{aligned}
(x + 3)(x + 5) &= x(x + 5) + 3(x + 5) \\
&= x(x) + x(5) + 3(x) + 3(5) \\
&= x^2 + 5x + 3x + 15 \\
&= x^2 + 8x + 15
\end{aligned}
$$

The first term in the second line, $(x)(x)$, is the product of the first terms of the two binomials.

$(x + 3)(x + 5)$ Multiply the first terms: $(x)(x)$

The term $(x)(5)$ is the product of the first term of the first binomial and the last term of the second binomial. This is the **outer product.**

$(x + 3)(x + 5)$ Multiply the outer terms: $(x)(5)$

The term $(3)(x)$ is the product of the last term of the first binomial and the first term of the second binomial. The product of these middle terms is the **inner product.**

$(x + 3)(x + 5)$ Multiply the inner terms: $(3)(x)$

Finally, $(3)(5)$ is the product of the last terms of the two binomials.

$(x + 3)(x + 5)$ Multiply the last terms: $(3)(5)$

In the third step of the multiplication above, the inner product and the outer product are added. This step should be performed mentally, so that the three terms of the answer can be written without extra steps as

$$(x + 3)(x + 5) = x^2 + 8x + 15.$$

A summary of these steps is given below. This procedure is sometimes called the **FOIL method,** which comes from the abbreviation for *first, outer, inner, last.*

Multiplying Binomials by the FOIL Method	*Step 1* Multiply the two first terms of the binomials to get the first term of the answer. *Step 2* Find the outer product and the inner product and mentally add them, when possible, to get the middle term of the answer. *Step 3* Multiply the two last terms of the binomials to get the last term of the answer.

Example 1 Find the product $(x + 8)(x - 6)$ by the FOIL method.

Step 1 **F** Multiply the *first* terms.

$$x(x) = x^2$$

Step 2 **O** Find the product of the *outer* terms.

$$x(-6) = -6x$$

I Find the product of the *inner* terms.

$$8(x) = 8x$$

Add the outer and inner products mentally.

$$-6x + 8x = 2x$$

Step 3 **L** Multiply the *last* terms.

$$8(-6) = -48$$

The product of $x + 8$ and $x - 6$ is found by adding the terms found in the three steps above, so

$$(x + 8)(x - 6) = x^2 - 6x + 8x - 48 = x^2 + 2x - 48.$$

As a shortcut, this product can be found in the following manner.

Example 2 Multiply $9x - 2$ and $3x + 1$.

First $(9x - 2)\ 3x + 1)$ $\quad 27x^2$
Outer $(9x - 2)(3x + 1)$ $\quad 9x$
Inner $(9x - 2)(3x + 1)$ $\quad -6x$
Last $(9x - 2)(3x + 1)$ $\quad -2$

$$\begin{array}{cccc} & \text{F} & \text{O} & \text{I} & \text{L} \end{array}$$
$$(9x - 2)(3x + 1) = 27x^2 + 9x - 6x - 2$$
$$= 27x^2 + 3x - 2$$

Example 3   Find the following products.

$$\overset{\text{F}\quad\text{O}\quad\text{I}\quad\text{L}}{\text{(a) } (2k + 5y)(k + 3y) = (2k)(k) + (2k)(3y) + (5y)(k) + (5y)(3y)}$$
$$= 2k^2 + 6ky + 5ky + 15y^2$$
$$= 2k^2 + 11ky + 15y^2$$

(b) $(7p + 2q)(3p - q) = 21p^2 - pq - 2q^2$ ✚

2   Certain special types of binomial products occur so often that the form of the answers should be memorized. For example, to find the **square of a binomial** quickly, use the method shown in Example 4.

Example 4   Find $(2m + 3)^2$.

Squaring $2m + 3$ by the FOIL method gives

$$(2m + 3)(2m + 3) = 4m^2 + 12m + 9.$$ ✚

The result has the square of both the first and the last terms of the binomial:

$$(2m)^2 = 4m^2 \quad \text{and} \quad 3^2 = 9.$$

The middle term is twice the product of the two terms of the binomial, since both the outer and inner products are $(2m)(3)$:

$$(2m)(3) + (2m)(3) = 2(2m)(3) = 12m.$$

This example suggests the following rule.

Square of a Binomial	The square of a binomial is equal to the square of the first term, plus twice the product of the two terms, plus the square of the last term:
	$$(x + y)^2 = x^2 + 2xy + y^2,$$
or	$$(x - y)^2 = x^2 - 2xy + y^2.$$

Example 5   Use the formula to square each binomial.

(a) $(5z - 1)^2 = (5z)^2 - 2(5z)(1) + (1)^2 = 25z^2 - 10z + 1$
Recall that $(5z)^2 = 5^2z^2 = 25z^2$.

(b) $(3b + 5r)^2 = (3b)^2 + 2(3b)(5r) + (5r)^2 = 9b^2 + 30br + 25r^2$

(c) $(2a - 9x)^2 = 4a^2 - 36ax + 81x^2$

(d) $\left(4m + \dfrac{1}{2}\right)^2 = (4m)^2 + 2(4m)\left(\dfrac{1}{2}\right) + \left(\dfrac{1}{2}\right)^2 = 16m^2 + 4m + \dfrac{1}{4}$ ✚

3   Binomial products of the form $(x + y)(x - y)$ also occur frequently. In these products, one binomial is the sum of two terms, and the other is the difference of the same two terms. As an example, the product of $a + 2$ and $a - 2$ is

$$(a + 2)(a - 2) = a^2 - 2a + 2a - 4 = a^2 - 4.$$

Using the FOIL method, the product of $x + y$ and $x - y$ is the difference of two squares.

Product of the Sum and Difference of Two Terms	$$(x + y)(x - y) = x^2 - y^2$$

Example 6   Find each product.

(a) $(5m + 3)(5m - 3)$

Use the formula for the sum and difference of two terms.

$$(5m + 3)(5m - 3) = (5m)^2 - 3^2 = 25m^2 - 9$$

(b) $(4x + y)(4x - y) = (4x)^2 - y^2 = 16x^2 - y^2$

(c) $\left(z - \dfrac{1}{4} \right)\left(z + \dfrac{1}{4} \right) = z^2 - \dfrac{1}{16}$ ✚

The product formulas of this section will be very useful in later work, particularly in Chapter 4. Therefore, it is important to memorize the formulas and practice using them.

3.6   Exercises

Find each product. See Examples 1–3.

1. $(r - 1)(r + 3)$
2. $(x + 2)(x - 5)$
3. $(x - 7)(x - 3)$
4. $(r + 3)(r + 6)$
5. $(2x - 1)(3x + 2)$
6. $(4y - 5)(2y + 1)$
7. $(6z + 5)(z - 3)$
8. $(3x - 1)(2x + 3)$
9. $(2r - 1)(4r + 3)$
10. $(11m - 10)(10m + 11)$
11. $(4 + 5x)(5 - 4x)$
12. $(8 + 3x)(2 - x)$
13. $(-3 + 2r)(4 + r)$
14. $(-5 + 6z)(2 - z)$
15. $(-3 + a)(-5 - 2a)$
16. $(-6 - 3y)(1 - 4y)$
17. $(p + 3q)(p + q)$
18. $(2r - 3s)(3r + s)$
19. $(5y + z)(2y - z)$
20. $(9m + 4k)(2m - 3k)$
21. $(8y - 9z)(y + 5z)$
22. $(4r + 9s)(-2r + 5s)$
23. $(7m + 11n)(3m - 8n)$
24. $(2.13y + 4.06)(1.92y - 3.9)$
25. $(8.17m - 2.4)(3.5m + 1.8)$

Find each square. See Examples 4 and 5.

26. $(m + 2)^2$ **27.** $(x + 8)^2$ **28.** $(r - 3)^2$

29. $(z - 5)^2$ **30.** $(x + 2y)^2$ **31.** $(3m - p)^2$

32. $(5p + 2q)^2$ **33.** $(8a - 3b)^2$ **34.** $(4a + 5b)^2$

35. $(9y + z)^2$ **36.** $(.85r + .23s)^2$ **37.** $(.67m - .17k)^2$

Find the following products. See Example 6.

38. $(a + 8)(a - 8)$ **39.** $(k + 5)(k - 5)$ **40.** $(2 + p)(2 - p)$

41. $(4 - 3t)(4 + 3t)$ **42.** $(2m + 5)(2m - 5)$ **43.** $(2b + 5)(2b - 5)$

44. $(3x + 4y)(3x - 4y)$ **45.** $(6a - p)(6a + p)$ **46.** $(5y + 3x)(5y - 3x)$

47. $(.48q + .37r)(.48q - .37r)$ **48.** $(.26a + .15b)(.26a - .15b)$

Write each statement as an equation or an inequality, using x to represent the unknown number. Do not solve.

49. The square of 3 more than a number is 5.

50. The square of the sum of a number and 6 is less than 3.

51. When 3 plus a number is multiplied by the number less 4, the result is greater than 7.

52. Twice a number plus 4, multiplied by 6 times the number less 5, gives 8.

Find the products in Exercises 53–62.

53. $\left(3p + \frac{4}{5}q\right)\left(2p - \frac{5}{3}q\right)$ **54.** $\left(-x + \frac{2}{3}y\right)\left(3x - \frac{3}{4}y\right)$

55. $\left(2z - \frac{5}{2}x\right)^2$ **56.** $\left(6a - \frac{3}{2}b\right)^2$

57. $(.2x - 3y)^2$ **58.** $(m - .5q)^2$

59. $\left(2m - \frac{5}{3}\right)\left(2m + \frac{5}{3}\right)$ **60.** $\left(3a - \frac{4}{5}\right)\left(3a + \frac{4}{5}\right)$

61. $(7y^2 + 10z)(7y^2 - 10z)$ **62.** $(6x + 5y^2)(6x - 5y^2)$

63. Let $a = 2$ and $b = 5$. Evaluate $(a + b)^2$ and $a^2 + b^2$. Does $(a + b)^2 = a^2 + b^2$?

64. Let $p = 7$ and $q = 3$. Evaluate $(p - q)^2$ and $p^2 - q^2$. Does $(p - q)^2 = p^2 - q^2$?

Review Exercises Simplify. See Section 3.2.

65. $\dfrac{9y^4}{3y}$ **66.** $\dfrac{8p^2}{2p}$ **67.** $\dfrac{5m^7}{m^3}$ **68.** $\dfrac{-10p^4}{2p^3}$

69. $\dfrac{-8z^5}{10z^7}$ **70.** $\dfrac{20k^3}{50k^5}$ **71.** $\dfrac{36r^5s^7}{24r^9s}$ **72.** $\dfrac{-18p^7q^{10}}{32p^2q^{12}}$

3.7   Dividing a Polynomial by a Monomial

Objective

1 Divide a polynomial by a monomial.

1   The quotient rule for exponents is used to divide a monomial by another monomial. For example,

$$\frac{12x^2}{6x} = 2x, \qquad \frac{25m^5}{5m^2} = 5m^3, \qquad \text{and} \qquad \frac{30a^2b^8}{15a^3b^3} = \frac{2b^5}{a}.$$

Dividing a Polynomial by a Monomial	Divide a polynomial by a monomial by dividing each term of the polynomial by the monomial: $$\frac{a + b}{c} = \frac{a}{c} + \frac{b}{c} \qquad (c \neq 0).$$

Example 1 Divide $5m^5 - 10m^3$ by $5m^2$.

Use the rule above, with $+$ replace by $-$.

$$\frac{5m^5 - 10m^3}{5m^2} = \frac{5m^5}{5m^2} - \frac{10m^3}{5m^2} = m^3 - 2m$$

Check by multiplication.

$$5m^2(m^3 - 2m) = 5m^5 - 10m^3$$

Since division by 0 is meaningless, the quotient

$$\frac{5m^5 - 10m^3}{5m^2}$$

has no value if $m = 0$. In the rest of the chapter, assume that no denominators are 0.   ✦

Example 2 Divide $\dfrac{16a^5 - 12a^4 + 8a^2}{4a^3}$.

Divide each term of $16a^5 - 12a^4 + 8a^2$ by $4a^3$.

$$\frac{16a^5 - 12a^4 + 8a^2}{4a^3} = \frac{16a^5}{4a^3} - \frac{12a^4}{4a^3} + \frac{8a^2}{4a^3}$$

$$= 4a^2 - 3a + \frac{2}{a}$$

The result is not a polynomial because of the expression $\frac{2}{a}$, which has a variable in the denominator. While the sum, difference, and product of two polynomials are always polynomials, the quotient of two polynomials may not be.

Again, check by multiplying.

$$4a^3\left(4a^2 - 3a + \frac{2}{a}\right) = 4a^3(4a^2) - 4a^3(3a) + 4a^3\left(\frac{2}{a}\right)$$

$$= 16a^5 - 12a^4 + 8a^2. \quad \clubsuit$$

Example 3   Divide $-7x^3 + 12x^4 - 4 + x$ by $4x$.

The polynomial should be written in descending order before dividing. Write it as $12x^4 - 7x^3 + x - 4$; then divide by $4x$.

$$\frac{12x^4 - 7x^3 + x - 4}{4x} = \frac{12x^4}{4x} - \frac{7x^3}{4x} + \frac{x}{4x} - \frac{4}{4x}$$

$$= 3x^3 - \frac{7x^2}{4} + \frac{1}{4} - \frac{1}{x}$$

$$= 3x^3 - \frac{7}{4}x^2 + \frac{1}{4} - \frac{1}{x}$$

Check by multiplication. $\clubsuit$

Example 4   Divide $180y^{10} - 150y^8 + 120y^6 - 90y^4 + 100y$ by $-30y^2$.

Using the methods of this section,

$$\frac{180y^{10} - 150y^8 + 120y^6 - 90y^4 + 100y}{-30y^2}$$

$$= \frac{180y^{10}}{-30y^2} - \frac{150y^8}{-30y^2} + \frac{120y^6}{-30y^2} - \frac{90y^4}{-30y^2} + \frac{100y}{-30y^2}$$

$$= -6y^8 + 5y^6 - 4y^4 + 3y^2 - \frac{10}{3y}.$$

Check by multiplying this result by $-30y^2$. $\clubsuit$

3.7 Exercises

Divide.

1. $\dfrac{4x^2}{2x}$

2. $\dfrac{8m^7}{2m}$

3. $\dfrac{10a^3}{5a}$

4. $\dfrac{36p^8}{4p^3}$

5. $\dfrac{27k^4m^5}{3km^6}$

6. $\dfrac{18x^5y^6}{3x^2y^2}$

7. $\dfrac{-15m^3p^2}{5mp^4}$

8. $\dfrac{-32a^4b^5}{4a^5b}$

Divide each polynomial by 2m. See Examples 1–4.

9. $60m^4 - 20m^2$

10. $120m^6 - 60m^3 + 80m^2$

11. $10m^5 - 16m^2 + 8m^3$

12. $6m^5 - 4m^3 + 2m^2$

13. $8m^5 - 4m^3 + 4m^2$

14. $8m^3 - 4m^2 + 6m$

15. $2m^5 - 4m^2 + 8m$

16. $m^2 + m + 1$

17. $2m^2 - 2m + 5$

Divide each polynomial by $-5q$. See Examples 1–4.

18. $-30q^4 + 15q^3 - 10q$

19. $25q^3 - 10q^2 + 5q$

20. $15q^6 - 30q^4 + 5q^2 - q$

21. $-40q^5 + 35q^3 - 20$

22. $q^5 - 10q^3 + 8q^2$

23. $3q^4 + 5q^2 - 6q$

24. $q^3 + q^2 - 9q + 1$

25. $5q^3 - 3q^2 + 2q - 7$

26. $3q^5 - 2q^4 + 10q^2$

Divide each polynomial by $3x^2$. See Examples 1–4.

27. $3x^4 + 9x^3$

28. $15x^2 - 9x^3$

29. $12x^4 + 3x^2 - 3x^3$

30. $45x^3 + 15x^2 - 9x^5$

31. $36x + 24x^2 + 3x^3$

32. $4x^4 - 3x^3 + 2x$

33. $x^3 + 6x^2 - x$

34. $-3x^4 + 6x^5 + 2 + 9x^2$

Perform each division in Exercises 35–42. See Examples 1–4.

35. $\dfrac{8k^4 - 12k^3 - 2k^2 + 7k - 3}{2k}$

36. $\dfrac{27r^4 - 36r^3 - 6r^2 + 26r - 2}{3r}$

37. $\dfrac{100p^5 - 50p^4 + 30p^3 - 30p}{-10p^2}$

38. $\dfrac{2m^5 - 6m^4 + 8m^2}{-2m^3}$

39. $(16y^5 - 8y^2 + 12y) \div (4y^2)$

40. $(20a^4 - 15a^5 + 25a^3) \div (15a^4)$

41. $(120x^{11} - 60x^{10} + 140x^9 - 100x^8) \div (10x^{12})$

42. $(5 + x + 6x^2 + 8x^3) \div (3x^4)$

43. Evaluate $\dfrac{5y + 6}{2}$ when $y = 2$. Evaluate $5y + 3$ when $y = 2$. Does

$\dfrac{5y + 6}{2} = 5y + 3$?

44. Evaluate $\dfrac{10r + 7}{5}$ when $r = 1$. Evaluate $2r + 7$ when $r = 1$. Does

$\dfrac{10r + 7}{5} = 2r + 7$?

Solve each problem.

45. What polynomial, when divided by $3x^2$, yields $4x^3 + 3x^2 - 4x + 2$ as a quotient?

46. What polynomial, when divided by $4m^3$, yields $-6m^2 + 4m$ as a quotient?

47. The quotient of a certain polynomial and $-7y^2$ is $9y^2 + 3y + 5 - 2/y$. Find the polynomial.

48. The quotient of a certain polynomial and a is $2a^2 + 3a + 5$. Find the polynomial.

In the following problems, choose the appropriate formula from the list of formulas inside the front cover and use it to write an expression that solves the problem.

49. The area of a rectangular building is $10x^5 + 15x^4 + 40x^3$. One dimension is $5x^3$. What is the other dimension?

50. A vehicle travels $3d^3 + 27d^2 + 15d$ miles in $3d$ hours. Find its rate (or speed).

51. The volume of a box is $60x^4 + 70x^3 - 30x^2$. The length is $2x^2$ and the width is $5x$. Find the height.

52. The interest on a loan of p dollars is $2p^4 - 2p^3 + 3p^2 - 3$ and the interest rate is $3p^2$. Find the time.

Review Exercises   *Find each product. See Section 3.5.*

53. $x(2x^2 - 5x + 1)$ **54.** $m(3m^3 - 2m^2 + 5)$ **55.** $-4k(5k^3 - 3k^2 + 2k)$

56. $-7z(z^2 - 5z + 6)$ **57.** $3m^5(2m^5 - 4m^3 + m^2)$ **58.** $2p^3(8p^2 - 4p + 9)$

Subtract. See Section 3.4.

59. $8x - 5$
$3x - 7$

60. $5a - 4$
$7a - 2$

61. $6x^2 - 4$
$8x^2 + 7$

62. $3k^2 - 4k + 1$
$4k^2 + 7k - 5$

63. $9x^2 - 4x - 7$
$2x^2 + 6x - 8$

64. $4m^2 - 5m - 1$
$3m^2 + 2m - 5$

3.8   The Quotient of Two Polynomials

Objective

1 Divide a polynomial by a polynomial.

1   A method of "long division" is used to divide a polynomial by a polynomial (other than a monomial). This method is similar to the method of long division used for two whole numbers. For comparison, the division of whole numbers is shown alongside the division of polynomials. Both polynomials should be written with descending powers before beginning the division process.

Step 1

Divide 27 into 6696.

Divide $2x + 3$ into
$8x^3 - 4x^2 - 14x + 15$.

$$27\overline{)6696} \qquad 2x + 3\overline{)8x^3 - 4x^2 - 14x + 15}$$

Step 2

27 divides into 66 **2** times;
$2 \cdot 27 = 54$.

$2x$ divides into $8x^3$ **$4x^2$** times;
$4x^2(2x + 3) = 8x^3 + 12x^2$.

$$\begin{array}{r} 2 \\ 27\overline{)6696} \\ \underline{54} \end{array} \qquad \begin{array}{r} 4x^2 \\ 2x + 3\overline{)8x^3 - 4x^2 - 14x + 15} \\ \underline{8x^3 + 12x^2} \end{array}$$

Step 3

Subtract; then bring down the next term.

$$\begin{array}{r} 2 \\ 27\overline{)6696} \\ 54\downarrow \\ \hline 129 \end{array}$$

Subtract; then bring down the next term.

$$\begin{array}{r} 4x^2 \\ 2x + 3\overline{)8x^3 - 4x^2 - 14x + 15} \\ 8x^3 + 12x^2 \downarrow \\ \hline -16x^2 - 14x \end{array}$$

(To subtract two polynomials, change the sign of the second and then add.)

Step 4

27 divides into 129 **4** times; $4 \cdot 27 = 108$.

$$\begin{array}{r} 24 \\ 27\overline{)6696} \\ 54 \\ \hline 129 \\ 108 \end{array}$$

$2x$ divides into $-16x^2$ **$-8x$** times; $-8x(2x + 3) = -16x^2 - 24x$.

$$\begin{array}{r} 4x^2 - 8x \\ 2x + 3\overline{)8x^3 - 4x^2 - 14x + 15} \\ 8x^3 + 12x^2 \\ \hline -16x^2 - 14x \\ -16x^2 - 24x \end{array}$$

Step 5

Subtract; then bring down the next term.

$$\begin{array}{r} 24 \\ 27\overline{)6696} \\ 54 \\ \hline 129 \\ 108\downarrow \\ \hline 216 \end{array}$$

Subtract; then bring down the next term.

$$\begin{array}{r} 4x^2 - 8x \\ 2x + 3\overline{)8x^3 - 4x^2 - 14x + 15} \\ 8x^3 + 12x^2 \\ \hline -16x^2 - 14x \\ -16x^2 - 24x \downarrow \\ \hline 10x + 15 \end{array}$$

Step 6

27 divides into 216 **8** times; $8 \cdot 27 = 216$.

$$\begin{array}{r} 248 \\ 27\overline{)6696} \\ 54 \\ \hline 129 \\ 108 \\ \hline 216 \\ 216 \end{array}$$

$2x$ divides into $10x$ **5** times; $5(2x + 3) = 10x + 15$.

$$\begin{array}{r} 4x^2 - 8x + 5 \\ 2x + 3\overline{)8x^3 - 4x^2 - 14x + 15} \\ 8x^3 + 12x^2 \\ \hline -16x^2 - 14x \\ -16x^2 - 24x \\ \hline 10x + 15 \\ 10x + 15 \end{array}$$

6696 divided by 27 is 248. There is no remainder.

$8x^3 - 4x^2 - 14x + 15$ divided by $2x + 3$ is $4x^2 - 8x + 5$. There is no remainder.

Step 7

Check by multiplication.   Check by multiplication.
$$27 \cdot 248 = 6696$$ $$(2x + 3)(4x^2 - 8x + 5)$$
$$= 8x^3 - 4x^2 - 14x + 15$$

Notice that at each step in the polynomial division process, the *first* term was divided into the *first* term.

Example 1   Divide $4x^3 - 4x^2 + 5x - 8$ by $2x - 1$.

$$
\begin{array}{r}
2x^2 - x + 2 \\
2x - 1 \overline{)4x^3 - 4x^2 + 5x - 8} \\
\underline{4x^3 - 2x^2} \\
-2x^2 + 5x \\
\underline{-2x^2 + x} \\
4x - 8 \\
\underline{4x - 2} \\
-6
\end{array}
$$

Step 1 $2x$ divides into $4x^3$ $(2x^2)$ times; $2x^2(2x - 1) = 4x^3 - 2x^2$.

Step 2   Subtract; bring down the next term.

Step 3 $2x$ divides into $-2x^2$ $(-x)$ times; $-x(2x - 1) = -2x^2 + x$.

Step 4   Subtract; bring down the next term.

Step 5 $2x$ divides into $4x$ **2** times; $2(2x - 1) = 4x - 2$.

Step 6   Subtract. The remainder is -6.

Thus, $2x - 1$ divides into $4x^3 - 4x^2 + 5x - 8$ with a quotient of $2x^2 - x + 2$ and a remainder of -6. The result is not a polynomial because of the remainder.

$$\frac{4x^3 - 4x^2 + 5x - 8}{2x - 1} = 2x^2 - x + 2 + \frac{-6}{2x - 1}$$

Step 7   Check by multiplication.

$$(2x - 1)\left(2x^2 - x + 2 + \frac{-6}{2x - 1}\right) = 4x^3 - 4x^2 + 5x - 8 \quad \clubsuit$$

Example 2   Divide $x^3 + 2x - 3$ by $x - 1$.

Here the polynomial $x^3 + 2x - 3$ is missing the x^2 term. When terms are missing, use 0 as the coefficient for the missing terms.

$$x^3 + 2x - 3 = x^3 + 0x^2 + 2x - 3$$

Now divide.

$$
\begin{array}{r}
x^2 + x + 3 \\
x - 1 \overline{)x^3 + 0x^2 + 2x - 3} \\
\underline{x^3 - x^2} \\
x^2 + 2x \\
\underline{x^2 - x} \\
3x - 3 \\
\underline{3x - 3}
\end{array}
$$

The remainder is 0. The quotient is $x^2 + x + 3$. Check by multiplication.

$$(x^2 + x + 3)(x - 1) = x^3 + 2x - 3 \quad \clubsuit$$

Example 3   Divide $x^4 + 2x^3 + 2x^2 - x - 1$ by $x^2 + 1$.

Since $x^2 + 1$ has a missing x term, write it as $x^2 + 0x + 1$. Then proceed through the division process.

$$
\begin{array}{r}
x^2 + 2x + 1 \\
x^2 + 0x + 1 \overline{)x^4 + 2x^3 + 2x^2 - x - 1} \\
\underline{x^4 + 0x^3 + x^2 } \\
2x^3 + x^2 - x \\
\underline{2x^3 + 0x^2 + 2x } \\
x^2 - 3x - 1 \\
\underline{x^2 + 0x + 1} \\
-3x - 2
\end{array}
$$

When the result of subtracting ($-3x - 2$, in this case) is a polynomial of smaller degree than the divisor ($x^2 + 0x + 1$), that polynomial is the remainder. Write the result as

$$x^2 + 2x + 1 + \frac{-3x - 2}{x^2 + 1}. \quad \clubsuit$$

3.8   Exercises

Perform each division. (Be sure each polynomial is written in descending powers of the variable.) See Examples 1 and 2.

1. $\dfrac{x^2 - x - 6}{x - 3}$

2. $\dfrac{m^2 - 2m - 24}{m + 4}$

3. $\dfrac{2y^2 + 9y - 35}{y + 7}$

4. $\dfrac{y^2 + 2y + 1}{y + 1}$

5. $\dfrac{p^2 + 2p - 20}{p + 6}$

6. $\dfrac{x^2 + 11x + 16}{x + 8}$

7. $\dfrac{r^2 - 8r + 15}{r - 3}$

8. $\dfrac{t^2 - 3t - 10}{t - 5}$

9. $\dfrac{12m^2 - 20m + 3}{2m - 3}$

10. $\dfrac{9w^2 + 6w + 10}{3w - 2}$

11. $\dfrac{2x^2 + 3}{2x + 4}$

12. $\dfrac{4m^2 + 5}{2m - 6}$

13. $\dfrac{12r^2 - 17r + 5}{3r - 5}$

14. $\dfrac{15y^2 + 11y - 60}{3y + 7}$

15. $\dfrac{14k^2 + 19k - 30}{7k - 8}$

16. $\dfrac{15m^2 + 34m + 28}{5m + 3}$

17. $\dfrac{2x^3 + 3x - x^2 + 2}{2x + 1}$

18. $\dfrac{-11t^2 + 12t^3 + 18 + 9t}{4t + 3}$

19. $\dfrac{8k^4 - 12k^3 - 2k^2 + 7k - 6}{2k - 3}$

20. $\dfrac{27r^4 - 36r^3 - 6r^2 + 26r - 24}{3r - 4}$

21. $\dfrac{3y^3 + y^2 + 4y + 1}{y + 1}$

22. $\dfrac{2r^3 - 5r^2 - 6r + 15}{r - 3}$

23. $\dfrac{3k^3 - 4k^2 - 6k + 10}{k - 2}$

24. $\dfrac{5z^3 - z^2 + 10z + 2}{z + 2}$

25. $\dfrac{6p^4 - 16p^3 + 15p^2 - 5p + 10}{3p + 1}$

26. $\dfrac{6r^4 - 11r^3 - r^2 + 16r - 8}{2r - 3}$

Perform each division. See Examples 2 and 3.

27. $\dfrac{-6x - x^2 + x^4}{x^2 - 2}$

28. $\dfrac{5 - 2r^2 + r^4}{r^2 - 1}$

29. $\dfrac{-m^3 - 9m^4 - 22m^2 + 4m^5 - 15}{4m^2 - m - 3}$

30. $\dfrac{6x^4 - x^3 + x^2 + 2x^5 + 5 - x}{2x^2 + 2x + 1}$

31. $\dfrac{y^3 + 1}{y + 1}$

32. $\dfrac{x^4 - 1}{x^2 - 1}$

33. $\dfrac{a^4 - 1}{a^2 + 1}$

34. $\dfrac{p^5 - 2}{p^2 - 1}$

35. $(2y^2 - 5y - 3) \div (2y + 4)$

36. $(3a^2 - 11a + 16) \div (2a + 6)$

37. $(9w^2 + 5w + 10) \div (3w - 2)$

38. $(4m^4 - 3m^3 + m^2 - m) \div (4m^2 + 3)$

39. $(6p^5 + 4p^4 - p^3 + 3p + 2) \div (3p^3 + 2p)$

40. $(2x^5 + 6x^4 - x^3 - x^2 + 5) \div (2x^2 + 2x + 1)$

Review Exercises   *List all positive integer factors of each number. See Section 1.9.*

41. 12   **42.** 18   **43.** 20   **44.** 36

45. 50   **46.** 70   **47.** 29   **48.** 37

Multiply. See Section 3.5.

49. $2m(m^2 - m + 1)$

50. $4k(2k^2 - 5k + 6)$

51. $-y^2(3y^4 + 2y^3 - 1)$

52. $-p^3(3p^5 - 8p^4 - 6p^3)$

Chapter 3 *Summary*

Definitions and Rules for Exponents

For any integers m and n:

Product rule $a^m \cdot a^n = a^{m+n}$

Zero exponent $a^0 = 1$ $(a \neq 0)$

Negative exponent $a^{-n} = \dfrac{1}{a^n}$ $(a \neq 0)$

Quotient rule $\dfrac{a^m}{a^n} = a^{m-n}$ $(a \neq 0)$

Power rules   (a) $(a^m)^n = a^{mn}$

              (b) $(ab)^m = a^m b^m$

              (c) $\left(\dfrac{a}{b}\right)^m = \dfrac{a^m}{b^m}$ $(b \neq 0)$

Multiplying Binomials by the FOIL Method	*Step 1* Multiply the two first terms of the binomials to get the first term of the answer. *Step 2* Find the outer product and the inner product and mentally add them, when possible, to get the middle term of the answer. *Step 3* Multiply the two last terms of the binomials to get the last term of the answer.
Square of a Binomial	The square of a binomial is equal to the square of the first term, plus twice the product of the two terms, plus the square of the last term: $$(x + y)^2 = x^2 + 2xy + y^2$$ or $$(x - y)^2 = x^2 - 2xy + y^2.$$
Product of the Sum and Difference of Two Terms	$$(x + y)(x - y) = x^2 - y^2$$

Chapter 3   *Review Exercises*

[3.1] *Simplify. Assume all variables are nonzero real numbers.*

1. $4^2 + 4^3$

2. $(4)^5(4)^3$

3. $\left(\dfrac{3}{4}\right)^4$

4. $(3x^2)^3$

5. $(ab^2c^3)^4$

6. $\left(\dfrac{2m^3n}{p^2}\right)^2$

[3.2] *Simplify. Write each answer with only positive exponents. Assume that all variables are nonzero real numbers.*

7. -7^{-2}

8. $2^{-1} + 4^{-1}$

9. $(5^{-2})^{-4}$

10. $(-3)^7 \cdot (-3)^3$

11. $\dfrac{x^{-7}}{x^{-9}}$

12. $\dfrac{y^4 \cdot y^{-2}}{y^{-5}}$

13. $(6r^{-2})^{-1}$

14. $(3p)^4(3p^{-7})$

15. $\dfrac{(6r^{-1})^2 \cdot (2r^{-4})}{r^{-5}(r^2)^{-3}}$

[3.3] *Write each number in scientific notation.*

16. 15,800,000

17. .0004251

18. .0000976

19. .784

Write each number without exponents.

20. 1.2×10^4

21. 4.253×10^{-4}

22. $(6 \times 10^4) \times (1.5 \times 10^3)$

23. $(2 \times 10^{-3}) \times (4 \times 10^5)$

24. $\dfrac{9 \times 10^{-2}}{3 \times 10^2}$

25. $\dfrac{8 \times 10^4}{2 \times 10^{-2}}$

[3.4] *Combine terms where possible in the following polynomials. Write the answers in descending powers of the variable. Give the degree of the answer.*

26. $9m^2 + 11m^2 + 2m^2$

27. $-4p + p^3 - p^2 + 8p + 2$

28. $12a^5 - 9a^4 + 8a^3 + 2a^2 - a + 3$

29. $-7y^5 - 8y^4 - y^5 + y^4 + 9y$

Add or subtract as indicated.

30. Add.

$$-2a^3 + 5a^2$$
$$\underline{3a^3 - a^2}$$

31. Subtract.

$$6y^2 - 8y + 2$$
$$\underline{5y^2 + 2y - 7}$$

32. Subtract.

$$-12k^4 - 8k^2 + 7k$$
$$\underline{k^4 + 7k^2 - 11k}$$

33. $(2m^3 - 8m^2 + 4) + (3m^3 + 2m^2 - 7)$

34. $(12r^4 - 7r^3 + 2r^2) - (5r^4 - 3r^3 + 2r^2)$

[3.5] *Find each product.*

35. $2y^2(-11y^2 + 2y + 9)$

36. $-m^5(8m^2 - 10m + 6)$

37. $(a + 2)(a^2 - 4a + 1)$

38. $(3r - 2)(2r^2 + 4r - 3)$

39. $(5p^2 + 3p)(p^3 - p^2 + 5)$

40. $(r + 2)^3$

[3.6] *Find each product.*

41. $(a + 3b)(2a - b)$

42. $(6k + 5q)(2k - 7q)$

43. $(a + 4)^2$

44. $(8z - 3y)^2$

45. $(6m - 5)(6m + 5)$

46. $(5a + 6b)(5a - 6b)$

[3.7] *Perform each division.*

47. $\dfrac{-15y^4}{9y^2}$

48. $\dfrac{6y^4 - 12y^2 + 18y}{6y}$

49. $(-10m^4 + 5m^3 + 6m^2) \div (5m^2)$

50. $(25y^3 - 8y^2 + 15y) \div (10y^3)$

[3.8] *Perform each division.*

51. $\dfrac{2r^2 + 3r - 14}{r - 2}$

52. $\dfrac{10a^3 + 9a^2 - 14a + 9}{5a - 3}$

53. $\dfrac{2k^4 + 3k^3 + 9k^2 - 8}{2k^2 + k + 1}$

54. $\dfrac{m^4 + 4m^3 - 5m^2 - 12m + 6}{m^2 - 3}$

❷ *Simplify. Perform the indicated operations. Write with positive exponents only. Assume all variables represent nonzero real numbers.*

55. $5^0 + 7^0$

56. $(12a + 1)(12a - 1)$

57. 2^{-4}

58. $(8^{-3})^4$

59. $\dfrac{12m^2 - 11m - 10}{3m - 5}$

60. $\dfrac{2p^3 - 6p^2 + 5p}{2p^2}$

61. $(3k - 6)(2k^2 + 4k + 1)$

62. $\dfrac{r^9 \cdot r^{-5}}{r^{-2} \cdot r^{-7}}$

63. $(2r + 5s)^2$

64. $(-5y^2 + 3y - 11) + (4y^2 - 7y + 15)$

65. $(2r + 5)(5r - 2)$

66. $\dfrac{2y^3 + 17y^2 + 37y + 7}{2y + 7}$

67. $(6p^2 - p - 8) - (-4p^2 + 2p - 3)$

68. $(-7 + 2k)^2$

69. $(5^2)^4$

70. $(3k + 1)(2k - 3)$

Chapter 3   *Test*

Evaluate each expression.

1. $\left(\dfrac{4}{3}\right)^3$

2. $(5)^{-2}$

Simplify. Write each answer using only positive exponents.

3. $5^2 \cdot 5^6 \cdot 5$

4. $\dfrac{(8^3)^2 \cdot 8^4}{8^3 \cdot 8^9}$

5. $\left(\dfrac{6q^{-2}}{q^{-3}}\right)^{-1}$ $(q \neq 0)$

6. $\dfrac{(2p^2)^3(4p^4)^2}{(2p^5)^2}$ $(p \neq 0)$

7. Write .000379 in scientific notation.

Write each number without exponents.

8. $(6 \times 10^{-4}) \times (1.5 \times 10^6)$

9. $\dfrac{5.6 \times 10^{-7}}{1.4 \times 10^{-3}}$

For each polynomial, combine terms; then give the degree of the polynomial. Finally, select the most specific description from this list: (a) trinomial, (b) binomial, (c) monomial, (d) none of these.

10. $3x^2 + 6x - 4x^2$

11. $11m^3 - m^2 + m^4 + m^4 - 7m^2$

12. Subtract.

$$-5m^3 + 2m^2 - 7m + 3$$
$$\underline{8m^3 - 5m^2 + 9m - 5}$$

Perform the indicated operation.

13. $(2x^5 - 4x + 7) - (x^5 + x^2 - 2x - 5)$

14. $(y^2 - 5y - 3) + (3y^2 + 2y) - (y^2 - y - 1)$

15. $6m^2(m^3 + 2m^2 - 3m + 7)$

16. $(5m + 6)(3m - 11)$

17. $(2t + 5s)(7t - 4s)$

18. $(3k + 2)(2k^2 - 7k + 8)$

19. $(2k + 5m)^2$

20. $\left(2r + \dfrac{1}{2}t\right)^2$

21. $(6p^2 - 5r)(6p^2 + 5r)$

22. $\dfrac{-15y^5 - 8y^4 + 6y^3}{3y^2}$

23. $(10r^3 + 25r^2 - 15r + 8) \div (5r^3)$

24. $\dfrac{3x^3 - 2x^2 - 6x - 4}{x - 2}$

25. $\dfrac{10r^4 + 4r^3 - 25r^2 - 6r + 20}{2r^2 - 3}$

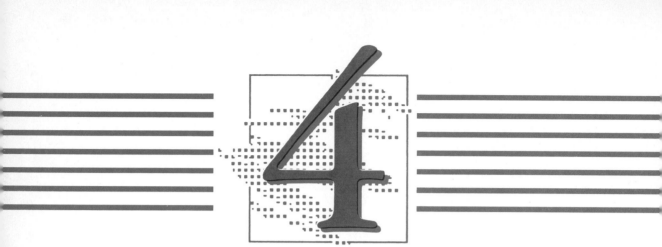

Factoring and Applications

Objectives

1 Find the greatest common factor of a list of integers.

2 Factor out the greatest common factor.

3 Factor by grouping.

Prime factors were discussed in Chapter 1. Recall, to **factor** means to write a quantity as a product. That is, factoring is the opposite of multiplication. For example,

Multiplication	*Factoring*
$6 \cdot 2 = 12,$	$12 = 6 \cdot 2.$
↑ ↑ ↑	↑ ↑ ↑
Factors Product	Product Factors

Other factored forms of 12 are $(-6)(-2)$, $3 \cdot 4$, $(-3)(-4)$, $12 \cdot 1$, and $(-12)(-1)$. More than two factors may be used, so another factored form of 12 is $2 \cdot 2 \cdot 3$. The positive integer factors of 12 are

$$1, 2, 3, 4, 6, 12.$$

1   An integer that is a factor of two or more integers is called a **common factor** of those integers. For example, 6 is a common factor of 18 and 24 since 6 is a factor of both 18 and 24. Other common factors of 18 and 24 are 1, 2, and 3. The **greatest common factor** of a list of integers is the largest common factor of those integers. Thus, 6 is the greatest common factor of 18 and 24, since it is the largest of the common factors of these numbers.

Find the greatest common factor of a list of numbers as follows.

Finding the Greatest Common Factor	
	Step 1 Write each number in prime factored form.
	Step 2 List each different prime factor that is in every prime factorization.
	Step 3 Use as exponents on the prime factors the *smallest* exponent from the prime factored forms. (If a prime does not appear in one of the prime factored forms, it cannot appear in the greatest common factor.)
	Step 4 Multiply together the primes of Step 3. If there are no primes left after Step 3, the greatest common factor is 1.

Example 1   Find the greatest common factor for each list of numbers.

(a) 30, 45

First write each number in prime factored form.

$$30 = 2 \cdot 3 \cdot 5 \qquad 45 = 3^2 \cdot 5$$

Now, take each prime the *least* number of times it appears in all the factored forms. There is no 2 in the prime factored form of 45, so there will be no 2 in the greatest common factor. The least number of times 3 appears in all the factored forms is 1, and the least number of times 5 appears is also 1. From this, the greatest common factor is

$$3^1 \cdot 5^1 = 15$$

(b) 72, 120, 432

Find the prime factored form of each number.

$$72 = 2^3 \cdot 3^2 \qquad 120 = 2^3 \cdot 3 \cdot 5 \qquad 432 = 2^4 \cdot 3^3$$

The least number of times 2 appears in all the factored forms is 3, and the least number of times 3 appears is 1. There is no 5 in the prime factored form of either 72 or 432, so the greatest common factor is

$$2^3 \cdot 3 = 24.$$

(c) 10, 11, 14

Write the prime factored form of each number.

$$10 = 2 \cdot 5 \qquad 11 = 11 \qquad 14 = 2 \cdot 7$$

There are no primes common to all three numbers, so the greatest common factor is 1.   ✥

The greatest common factor also can be found for a list of variables. For example, the terms x^4, x^5, x^6, and x^7 have x^4 as the greatest common factor, because 4 is the smallest exponent on x.

The exponent on a variable in the greatest common factor is the *smallest* exponent that appears on the factors.

Example 2   Find the greatest common factor for each list of terms.

(a) $21m^7$, $-18m^6$, $45m^8$, $-24m^5$

First, 3 is the greatest common factor of the coefficients 21, -18, 45, and -24. The smallest exponent on m is 5, so the greatest common factor of the terms is $3m^5$.

(b) x^4y^2, x^7y^5, x^3y^7, y^{15}

There is no x in the last term, y^{15}, so x will not appear in the greatest common factor. There is a y in each term, however, and 2 is the smallest exponent on y. The greatest common factor is y^2.   ✤

2   The idea of a greatest common factor can be used to write a polynomial in factored form. For example, the polynomial

$$3m + 12$$

consists of the two terms $3m$ and 12. The greatest common factor for these two terms is 3. Write $3m + 12$ so that each term is a product with 3 as one factor.

$$3m + 12 = 3 \cdot m + 3 \cdot 4$$

Now use the distributive property.

$$3m + 12 = 3 \cdot m + 3 \cdot 4 = 3(m + 4)$$

The factored form of $3m + 12$ is $3(m + 4)$. This process is called **factoring out the greatest common factor.**

Example 3   Factor out the greatest common factor.

(a) $20m^5 + 10m^4 + 15m^3$

The greatest common factor for the terms of this polynomial is $5m^3$.

$$20m^5 + 10m^4 + 15m^3 = (5m^3)(4m^2) + (5m^3)(2m) + (5m^3)3$$
$$= 5m^3(4m^2 + 2m + 3)$$

Check this work by multiplying $5m^3$ and $4m^2 + 2m + 3$. You should get the original polynomial as your answer.

(b) $x^5 + x^3 = (x^3)x^2 + (x^3)1 = x^3(x^2 + 1)$

Do not forget the 1 here; always be sure that the factored form can be multiplied out to yield the original polynomial.

(c) $20m^7p^2 - 36m^3p^4 = 4m^3p^2(5m^4 - 9p^2)$

(d) $a(a + 3) + 4(a + 3)$

The binomial $a + 3$ is the greatest common factor here.

$$a(a + 3) + 4(a + 3) = (a + 3)(a + 4)$$ ✛

> Always look for a greatest common factor as the first step in factoring a polynomial.

3   Common factors are used in **factoring by grouping,** as explained in the next example.

Example 4   Factor by grouping.

(a) $2x + 6 + ax + 3a$

The first two terms have a common factor of 2, and the last two terms have a common factor of a.

$$2x + 6 + ax + 3a = 2(x + 3) + a(x + 3)$$

Now $x + 3$ is a common factor.

$$2x + 6 + ax + 3a = 2(x + 3) + a(x + 3)$$
$$= (x + 3)(2 + a)$$

Note that the goal in factoring by grouping is to get a common factor, $x + 3$ here, so that the last step is possible.

(b) $m^2 + 6m + 2m + 12 = m(m + 6) + 2(m + 6)$
$$= (m + 6)(m + 2)$$

(c) $6y^2 - 21y - 8y + 28 = 3y(2y - 7) - 4(2y - 7)$
$$= (2y - 7)(3y - 4)$$

Since the quantities in parentheses in the second step must be the same, it was necessary here to factor out -4 rather than 4.   ✛

Use these steps when factoring four terms by grouping.

Factoring by Grouping

Step 1   Write the four terms so that the first two have a common factor and the last two have a common factor.

Step 2   Use the distributive property to factor each group of two terms.

Step 3   If possible, factor a common binomial factor from the results of Step 2.

Step 4   If Step 2 does not result in a common binomial factor, try grouping the terms of the original polynomial in a different way.

4.1 Exercises

Find the greatest common factor for each list of terms. See Examples 1 and 2.

1. $12y$, 24

2. $72m$, 12

3. $30p^2$, $20p^3$, $40p^5$

4. $14r^5$, $28r^2$, $56r^8$

5. $18r$, $32y$, $11z$

6. $45m^2$, $12n$, $7p^2$

7. $18m^2n^2$, $36m^4n^5$, $12m^3n$

8. $50p^5r^2$, $25p^4r^7$, $30p^7r^8$

9. $32y^4x^5$, $24y^7x$, $36y^3x^5$

Complete the factoring.

10. $18 = 9(\quad)$

11. $3x^2 = 3x(\quad)$

12. $8x^3 = 8x(\quad)$

13. $9m^4 = 3m^2(\quad)$

14. $12p^5 = 6p^3(\quad)$

15. $-8z^9 = 4z^5(\quad)$

16. $-15k^{11} = -5k^8(\quad)$

17. $x^2y^3 = xy(\quad)$

18. $a^3b^2 = a^2b(\quad)$

19. $27a^3b^2 = 9a^2b(\quad)$

20. $14x^4y^3 = 2xy(\quad)$

21. $-16m^3n^3 = 4mn^2(\quad)$

Factor out the greatest common factor. See Example 3.

22. $5x + 25$

23. $8k - 64$

24. $10p + 20p^2$

25. $49a^2 + 14a$

26. $21m^5 - 14m^4$

27. $65y^9 - 35y^5$

28. $100a^4 - 16a^2$

29. $121p^5 - 33p^4$

30. $8p^2 - 4p^4$

31. $11z^2 - 100$

32. $12z^2 - 11y^4$

33. $19y^3p^2 + 38y^2p^3$

34. $100m^5 - 50m^3 + 100m^2$

35. $13y^6 + 26y^5 - 39y^3$

36. $5x^4 + 25x^3 - 20x^2$

37. $45q^4p^5 - 36qp^6 + 81q^2p^3$

Factor by grouping. See Example 4.

38. $m^2 + 2m + 5m + 10$

39. $a^2 - 2a + 5a - 10$

40. $y^2 - 6y + 4y - 24$

41. $7z^2 + 3zm + 14zm + 6m^2$

42. $8k^2 + 6kq + 12kq + 9q^2$

43. $5m^2 + 15mp - 2mp - 6p^2$

44. $18r^2 + 12ry - 3ry - 2y^2$

45. $3a^3 + 3ab^2 + 2a^2b + 2b^3$

Factor out the greatest common factor.

46. $a^5 + 2a^5b + 3a^5b^2 - 4a^5b^3$

47. $a^3b^5 - a^2b^7 + ab^3$

48. $m^6n^5 - 2m^5 + 5m^3n^5$

49. $125z^5a^3 - 60z^4a^5 + 85z^3a^4$

50. $30a^2m^2 + 60a^3m + 180a^3m^2$

51. $33y^8 - 44y^{12} + 77y^3 + 11y^4$

52. $26g^6h^4 + 13g^3h^4 - 39g^4h^3$

53. $36a^4b^3 + 32a^5b^2 - 48a^6b^3$

Review Exercises   *Find each product. See Section 3.6.*

54. $(x + 2)(x + 5)$

55. $(m + 3)(m + 4)$

56. $(y + 7)(y - 3)$

57. $(r + 5)(r - 9)$

58. $(q - 5)(q - 7)$

59. $(y - 4)(y - 8)$

60. $(p - 10)(p + 10)$

61. $(a + 7)(a - 7)$

62. $(m - 12)(m + 12)$

4.2   *Factoring Trinomials*

Objectives

1 Factor trinomials with a coefficient of 1 for the squared term.

2 Factor such polynomials after factoring out the greatest common factor.

1 The product of the polynomials $k - 3$ and $k + 1$ is

$$(k - 3)(k + 1) = k^2 - 2k - 3.$$

The polynomial $k^2 - 2k - 3$ can be rewritten as the product $(k - 3)(k + 1)$. This product is called the **factored form** of $k^2 - 2k - 3$, and the process of finding the factored form is called **factoring.** The discussion of factoring in this section is limited to trinomials like $x^2 - 2x - 24$ or $y^2 + 2y - 15$, where the coefficient of the squared term is 1.

When factoring polynomials with only integer coefficients, use only integers for the numerical factors. For example, $x^2 + 5x + 6$ can be factored by finding integers a and b such that

$$x^2 + 5x + 6 = (x + {}^a)(x + {}^b).$$

To find these integers a and $b,$ first find the product of the two terms on the right-hand side of the equation:

$$(x + a)(x + b) = x^2 + ax + bx + ab$$

By the distributive property,

$$x^2 + ax + bx + ab = x^2 + (a + b)x + ab.$$

By this result, $x^2 + 5x + 6$ can be factored by finding integers a and b having a sum of 5 and a product of 6

$$x^2 + {}^5x + 6 = x^2 + ({}^a + {}^b)x + {}^{ab.}$$

Sum of a and b is 5

└Product of a and b is 6

Since infinitely many pairs of integers have a sum of 5, it is best to begin by listing those pairs of integers whose product is 6. Both 5 and 6 are positive, so only pairs in which both integers are positive need be considered.

Product	Sum
$1 \cdot 6 = 6$	$1 + 6 = 7$
$2 \cdot 3 = 6$	$2 + 3 = 5$

Both pairs have a product of 6, but only the pair 2 and 3 has a sum of 5. So 2 and 3 are the required integers, and

$$x^2 + 5x + 6 = (x + {}^2)(x + {}^3).$$

Check by multiplying the binomials. Make sure that the sum of the outer and inner products produces the correct middle term.

$$(x + 2)(x + 3) = x^2 + 5x + 6$$

$$\begin{array}{c} 2x \\ \underline{3x} \\ 5x \end{array}$$

This method of factoring can be used only for trinomials having the coefficient of the squared term equal to 1. Methods for factoring other trinomials will be given in the next section.

Example 1   Factor $m^2 + 9m + 14$.

Look for two integers whose product is 14 and whose sum is 9. List the pairs of integers whose product is 14. Then examine the sums. Again, only positive integers are needed because all signs are positive.

$$14, 1 \qquad 14 + 1 = 15$$
$$7, 2 \qquad 7 + 2 = 9 \qquad \text{Sum is 9}$$

From the list, 7 and 2 are the required integers, since $7 \cdot 2 = 14$ and $7 + 2 = 9$. Thus $m^2 + 9m + 14 = (m + 2)(m + 7)$. This answer also could have been written $(m + 7)(m + 2)$. Because of the commutative property of multiplication, the order of the factors does not matter.   ✚

Example 2   Factor $p^2 - 2p - 15$.

Find two integers whose product is -15 and whose sum is -2. If these numbers do not come to mind right away, find them (if they exist) by listing all the pairs of integers whose product is -15. Because of the minus signs in $p^2 - 2p - 15$, negative integers as well as positive ones must be considered.

$$15, -1 \qquad 15 + (-1) = 14$$
$$5, -3 \qquad 5 + (-3) = 2$$
$$-15, 1 \qquad -15 + 1 = -14$$
$$-5, 3 \qquad -5 + 3 = -2 \qquad \text{Sum is } -2$$

The necessary integers are -5 and 3, and

$$p^2 - 2p - 15 = (p - 5)(p + 3). \quad ✚$$

Example 3   **(a)** Factor $x^2 - 5x + 12$.

First, list all pairs of integers whose product is 12. Both positive and negative integers should be considered because of the minus sign. Then examine the sums.

$$12, 1 \qquad 12 + 1 = 13$$
$$6, 2 \qquad 6 + 2 = 8$$
$$3, 4 \qquad 3 + 4 = 7$$
$$-12, -1 \qquad -12 + (-1) = -13$$
$$-6, -2 \qquad -6 + (-2) = -8$$
$$-3, -4 \qquad -3 + (-4) = -7$$

None of the pairs of integers has a sum of -5. Because of this, the trinomial $x^2 - 5x + 12$ *cannot be factored using only integer factors,* showing that it is a **prime polynomial.**

(b) $k^2 - 8k + 11$

There is no pair of integers whose product is 11 and whose sum is -8, so $k^2 - 8k + 11$ is a prime polynomial.   ✦

Example 4   Factor $z^2 - 2bz - 3b^2$.

To factor $z^2 - 2bz - 3b^2$, look for two expressions whose product is $-3b^2$ and whose sum is $-2b$. The expressions are $-3b$ and b, with

$$z^2 - 2bz - 3b^2 = (z - 3b)(z + b).$$ ✦

2   The trinomial in the next example does not have a coefficient of 1 for the squared term. (In fact, there is no squared term.) A preliminary step must be taken before using the steps discussed above.

Example 5   Factor $4x^5 - 28x^4 + 40x^3$.

First, factor out the greatest common factor, $4x^3$.

$$4x^5 - 28x^4 + 40x^3 = 4x^3(x^2 - 7x + 10)$$

Now factor $x^2 - 7x + 10$. The integers -5 and -2 have a product of 10 and a sum of -7. The complete factored form is

$$4x^5 - 28x^4 + 40x^3 = 4x^3(x - 5)(x - 2).$$ ✦

When you factor, always remember to look for a common factor first. Do not forget to write the common factor as part of the answer. Multiplying out the factored form should always give the original polynomial.

4.2   Exercises

Complete the factoring.

1. $x^2 + 10x + 21 = (x + 7)($ $)$
2. $p^2 + 11p + 30 = (p + 5)($ $)$
3. $r^2 + 15r + 56 = (r + 7)($ $)$
4. $x^2 + 15x + 44 = (x + 4)($ $)$
5. $t^2 - 14t + 24 = (t - 2)($ $)$
6. $x^2 - 9x + 8 = (x - 1)($ $)$
7. $x^2 - 12x + 32 = (x - 4)($ $)$
8. $y^2 - 2y - 15 = (y + 3)($ $)$
9. $m^2 + 2m - 24 = (m - 4)($ $)$
10. $x^2 + 9x - 22 = (x - 2)($ $)$
11. $x^2 - 7xy + 10y^2 = (x - 2y)($ $)$
12. $k^2 - 3kh - 28h^2 = (k - 7h)($ $)$

Factor as completely as possible. If a polynomial cannot be factored, write prime. *See Examples 1–3.*

13. $x^2 + 6x + 5$
14. $y^2 + 9y + 8$
15. $a^2 + 9a + 20$
16. $b^2 + 8b + 15$
17. $x^2 - 8x + 7$
18. $m^2 + m - 20$
19. $p^2 + 4p + 5$
20. $n^2 + 4n + 12$
21. $y^2 - 6y + 8$

22. $r^2 - 11r + 30$

23. $s^2 + 2s - 35$

24. $h^2 + 11h + 12$

25. $n^2 - 12n - 35$

26. $a^2 - 2a - 99$

27. $b^2 - 11b + 24$

28. $x^2 - 9x + 20$

29. $k^2 - 10k + 25$

30. $z^2 - 14z + 49$

Factor completely in Exercises 31–48. See Examples 4 and 5.

31. $x^2 + 4ax + 3a^2$

32. $x^2 - mx - 6m^2$

33. $y^2 - by - 30b^2$

34. $z^2 + 2zx - 15x^2$

35. $x^2 + xy - 30y^2$

36. $a^2 - ay - 56y^2$

37. $r^2 - 2rs + s^2$

38. $m^2 - 2mn - 3n^2$

39. $p^2 - 3pq - 10q^2$

40. $c^2 - 5cd + 4d^2$

41. $3m^3 + 12m^2 + 9m$

42. $3y^5 - 18y^4 + 15y^3$

43. $6a^2 - 48a - 120$

44. $h^7 - 5h^6 - 14h^5$

45. $3r^3 - 30r^2 + 72r$

46. $2x^6 - 8x^5 - 42x^4$

47. $3x^4 - 3x^3 - 90x^2$

48. $2y^3 - 8y^2 - 10y$

49. Use the FOIL method from Section 3.6 to show that $(2x + 4)(x - 3) = 2x^2 - 2x - 12$. Why is it incorrect to completely factor $2x^2 - 2x - 12$ as $(2x + 4)(x - 3)$?

50. Why is it incorrect to completely factor $3x^2 + 9x - 12$ as the product $(x - 1)(3x + 12)$?

51. What polynomial can be factored to give $(y - 7)(y + 6)$?

52. What polynomial can be factored to give $(a + 9)(a + 6)$?

Factor completely.

53. $a^5 + 3a^4b - 4a^3b^2$

54. $m^3n - 2m^2n^2 - 3mn^3$

55. $y^3z + y^2z^2 - 6yz^3$

56. $k^7 - 2k^6m - 15k^5m^2$

57. $z^{10} - 4z^9y - 21z^8y^2$

58. $x^9 + 5x^8w - 24x^7w^2$

59. $(a + b)x^2 + (a + b)x - 12(a + b)$

60. $(x + y)n^2 + (x + y)n + 16(x + y)$

61. $(2p + q)r^2 - 12(2p + q)r + 27(2p + q)$

62. $(3m - n)k^2 - 13(3m - n)k + 40(3m - n)$

Review Exercises   Find each product. See Section 3.6.

63. $(2y - 7)(y + 4)$

64. $(3a + 2)(2a + 1)$

65. $(5z + 2)(3z - 2)$

66. $(4m - 3)(2m + 5)$

67. $(4p + 1)(2p - 3)$

68. $(6r - 2)(3r + 1)$

4.3   More on Factoring Trinomials

Objectives

Factor trinomials not having 1 as the coefficient of the squared term by

1 trying different combinations;

2 factoring by grouping.

Trinomials such as $2x^2 + 7x + 6$, in which the coefficient of the squared term is *not* 1, are factored with an extension of the method presented in the last section. Recall that a trinomial such as $m^2 + 3m + 2$ is factored by finding two numbers whose product is 2 and whose sum is 3.

1   Factor $2x^2 + 7x + 6$ by finding integers a, b, c, and d such that

$$2x^2 + 7x + 6 = (ax + b)(cx + d)$$

where, using FOIL, $ac = 2$, $bd = 6$, and $ad + bc = 7$. The possible factors of $2x^2$ are $2x$ and x, or $-2x$ and $-x$. Since the polynomial has only positive coefficients, use the factors with positive coefficients. Then the factored form of $2x^2 + 7x + 6$ can be set up as

$$2x^2 + 7x + 6 = (2x \quad)(x \quad).$$

The product 6 can be factored as $6 \cdot 1$, $1 \cdot 6$, $2 \cdot 3$, or $3 \cdot 2$. Try each pair to find the correct choices for b and d.

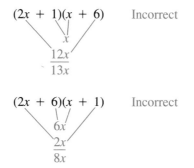

$(2x + 1)(x + 6)$ Incorrect

$\begin{array}{c} x \\ 12x \\ \hline 13x \end{array}$

$(2x + 6)(x + 1)$ Incorrect

$\begin{array}{c} 6x \\ 2x \\ \hline 8x \end{array}$

Since $2x + 6 = 2(x + 3)$, the binomial $2x + 6$ has a common factor of 2, while $2x^2 + 7x + 6$ does not have a common factor other than 1. The product $(2x + 6)(x + 1)$ cannot be correct.

> If the original polynomial has no common factor, then none of its binomial factors will either.

Now try the numbers 2 and 3 as factors of 6. Because of the common factor of 2 in $2x + 2$, $(2x + 2)(x + 3)$ will not work. Try $(2x + 3)(x + 2)$.

$(2x + 3)(x + 2) = 2x^2 + 7x + 6$

$\begin{array}{c} 3x \\ 4x \\ \hline 7x \end{array}$

Finally, $2x^2 + 7x + 6$ factors as

$$2x^2 + 7x + 6 = (2x + 3)(x + 2).$$

Check by multiplying $2x + 3$ and $x + 2$.

Example 1   Factor $8p^2 + 14p + 5$.

The number 8 has several possible pairs of factors, but 5 has only 1 and 5 or -1 and -5. For this reason, it is easier to begin by considering the factors of 5. Ignore the negative factors since all coefficients in the trinomial are positive. If $8p^2 + 14p + 5$ can be factored, it will be factored as

$$(\quad + 5)(\quad + 1).$$

The possible pairs of factors of $8p^2$ are $8p$ and p, or $4p$ and $2p$. Try various combinations.

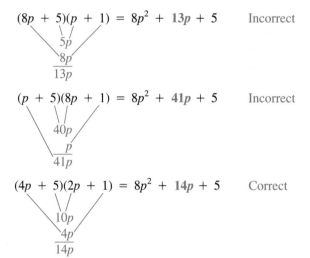

$$(8p + 5)(p + 1) = 8p^2 + 13p + 5 \qquad \text{Incorrect}$$

$$(p + 5)(8p + 1) = 8p^2 + 41p + 5 \qquad \text{Incorrect}$$

$$(4p + 5)(2p + 1) = 8p^2 + 14p + 5 \qquad \text{Correct}$$

Finally, $8p^2 + 14p + 5$ factors as $(4p + 5)(2p + 1)$. ❖

Example 2   Factor $6x^2 - 11x + 3$.

There are several possible pairs of factors for 6, but 3 has only 1 and 3 or -1 and -3, so it is better to begin by factoring 3. Since the last term of the trinomial $6x^2 - 11x + 3$ is positive and the middle term has a negative coefficient, only negative factors should be considered. Try -3 and -1 as factors of 3:

$$(\quad - 3)(\quad - 1).$$

The factors of $6x^2$ are either $6x$ and x, or $2x$ and $3x$. Try $2x$ and $3x$.

$$(2x - 3)(3x - 1) = 6x^2 - 11x + 3 \qquad \text{Correct}$$

Therefore, $6x^2 - 11x + 3 = (2x - 3)(3x - 1)$. ❖

Example 3   Factor $8x^2 + 6x - 9$.

The integer 8 has several possible pairs of factors, as does -9.

The last term is negative, so one positive factor and one negative factor of -9 is needed. Since the coefficient of the middle term is small, it is wise to avoid large factors such as 8 or 9. Try 4 and 2 as factors of 8, and 3 and -3 as factors of -9.

$$(4x + 3)(2x - 3) = 8x^2 - 6x - 9 \qquad \text{Incorrect}$$

$$\begin{array}{c} 6x \\ -12x \\ \hline -6x \end{array}$$

Try exchanging 3 and -3.

$$(4x - 3)(2x + 3) = 8x^2 + 6x - 9 \qquad \text{Correct}$$

$$\begin{array}{c} -6x \\ 12x \\ \hline 6x \end{array}$$

This last result is correct.   ✚

Example 4   Factor $12a^2 - ab - 20b^2$.

There are several possible factors of $12a^2$, including $12a$ and a, $6a$ and $2a$, and $3a$ and $4a$, just as there are many possible factors of $-20b^2$, including $-20b$ and b, $10b$ and $-2b$, $-10b$ and $2b$, $4b$ and $-5b$, and $-4b$ and $5b$. Once again, since the desired middle term is small, avoid the larger factors. Try as factors $6a$ and $2a$ and $4b$ and $-5b$.

$$(6a + 4b)(2a - 5b)$$

This cannot be correct, as mentioned before, since $6a + 4b$ has a common factor while the given trinomial has none. Try $3a$ and $4a$ with $4b$ and $-5b$.

$$(3a + 4b)(4a - 5b) = 12a^2 + ab - 20b^2 \qquad \text{Incorrect}$$

Here the middle term has the wrong sign, so change the sign on the factors.

$$(3a - 4b)(4a + 5b) = 12a^2 - ab - 20b^2 \qquad \text{Correct}   ✚$$

Example 5   Factor $28x^5 - 58x^4 - 30x^3$.

First factor out the greatest common factor, $2x^3$.

$$28x^5 - 58x^4 - 30x^3 = 2x^3(14x^2 - 29x - 15)$$

Now try to factor $14x^2 - 29x - 15$. Try $7x$ and $2x$ as factors of $14x^2$ and -3 and 5 as factors of -15.

$$(7x - 3)(2x + 5) = 14x^2 + 29x - 15 \qquad \text{Incorrect}$$

The middle term differs only in sign, so change the signs on the two factors.

$$(7x + 3)(2x - 5) = 14x^2 - 29x - 15 \qquad \text{Correct}$$

Finally, the factored form of $28x^5 - 58x^4 - 30x^3$ is

$$28x^5 - 58x^4 - 30x^3 = 2x^3(7x + 3)(2x - 5). \qquad \blacklozenge$$

Do not forget to include the common factor in the final result.

2 The rest of this section shows an alternative method of factoring trinomials in which the coefficient of the squared term is not 1. This method uses factoring by grouping, introduced in Section 4.1. In the next example, the alternative method is used to factor $2x^2 + 7x + 6$, the same trinomial factored at the beginning of this section.

Recall that a trinomial such as $m^2 + 3m + 2$ is factored by finding two numbers whose product is 2 and whose sum is 3. To factor $2x^2 + 7x + 6$ by the alternative method, look for two integers whose product is $2 \cdot 6 = 12$ and whose sum is 7.

$$\overbrace{2x^2 + 7x + 6}$$
Sum is 7

Product is $2 \cdot 6 = 12$

By considering the pairs of integers whose product is 12, the necessary integers are found to be 3 and 4. Use these integers to write the middle term, $7x$, as $7x = 3x + 4x$. With this, the trinomial $2x^2 + 7x + 6$ becomes

$$2x^2 + 7x + 6 = 2x^2 + \underbrace{3x + 4x}_{7x \; = \; 3x + 4x} + 6.$$

Factor the new polynomial by grouping.

$$2x^2 + 3x + 4x + 6 = x(2x + 3) + 2(2x + 3)$$
$$= (2x + 3)(x + 2)$$

The common factor of $2x + 3$ was factored out to get

$$2x^2 + 7x + 6 = (2x + 3)(x + 2).$$

Check by finding the product of $2x + 3$ and $x + 2$.

The middle term in the polynomial $2x^2 + 7x + 6$ could have been written as $7x = 4x + 3x$ to get

$$2x^2 + 7x + 6 = 2x^2 + 4x + 3x + 6$$
$$= 2x(x + 2) + 3(x + 2)$$
$$= (x + 2)(2x + 3).$$

Either result is correct.

Example 6   Factor each trinomial.

(a) $6r^2 + r - 1$

Find two integers with a product of $6(-1) = -6$ and a sum of 1.

$$\underset{\underset{\textstyle \text{Product is } 6(-1) = -6}{\uparrow \qquad\qquad \uparrow}}{\overset{\overset{\textstyle \text{Sum is 1}}{\downarrow}}{6r^2 + r - 1 = 6r^2 + 1r - 1}}$$

The integers are -2 and 3. Write the middle term, $+r$, as $-2r + 3r$, so that

$$6r^2 + r - 1 = 6r^2 \quad 2r + 3r - 1.$$

Factor by grouping on the right-hand side.

$$\begin{aligned}
6r^2 + r - 1 &= 6r^2 - 2r + 3r - 1 \\
&= 2r(3r - 1) + 1(3r - 1) \\
&= (3r - 1)(2r + 1)
\end{aligned}$$

(b) $12z^2 - 5z - 2$

Look for two integers whose product is $12(-2) = -24$ and whose sum is -5. The required integers are 3 and -8, and

$$\begin{aligned}
12z^2 - 5z - 2 &= 12z^2 + 3z - 8z - 2 \\
&= 3z(4z + 1) - 2(4z + 1) \\
&= (4z + 1)(3z - 2). \quad \clubsuit
\end{aligned}$$

4.3   Exercises

Complete the factoring.

1. $2x^2 - x - 1 = (2x + 1)(\qquad)$ **2.** $3a^2 + 5a + 2 = (3a + 2)(\qquad)$

3. $5b^2 - 16b + 3 = (5b - 1)(\qquad)$ **4.** $2x^2 + 11x + 12 = (2x + 3)(\qquad)$

5. $4y^2 + 17y - 15 = (y + 5)(\qquad)$ **6.** $7z^2 + 10z - 8 = (z + 2)(\qquad)$

7. $15x^2 + 7x - 4 = (3x - 1)(\qquad)$ **8.** $12c^2 - 7c - 12 = (4c + 3)(\qquad)$

9. $2m^2 + 19m - 10 = (2m - 1)(\qquad)$ **10.** $6x^2 + x - 12 = (2x + 3)(\qquad)$

11. $6a^2 + 7ab - 20b^2 = (2a + 5b)(\qquad)$

12. $9m^2 - 3mn - 2n^2 = (3m - 2n)(\qquad)$

13. $4k^2 + 13km + 3m^2 = (4k + m)(\qquad)$

14. $6x^2 - 13xy - 5y^2 = (3x + y)(\qquad)$

15. $4x^3 - 10x^2 - 6x = 2x(\qquad) = 2x(2x + 1)(\qquad)$

16. $15r^3 - 39r^2 - 18r = 3r(\qquad) = 3r(5r + 2)(\qquad)$

17. $6m^6 + 7m^5 - 20m^4 = m^4(\qquad) = m^4(3m - 4)(\qquad)$

18. $16y^5 - 4y^4 - 6y^3 = 2y^3(\qquad) = 2y^3(4y - 3)(\qquad)$

Factor as completely as possible. Remember to look for a greatest common factor first. Use either method. See Examples 1–6.

19. $2x^2 + 7x + 3$

20. $3y^2 + 13y + 4$

21. $3a^2 + 10a + 7$

22. $7r^2 + 8r + 1$

23. $4r^2 + r - 3$

24. $3p^2 + 2p - 8$

25. $15m^2 + m - 2$

26. $6x^2 + x - 1$

27. $8m^2 - 10m - 3$

28. $2a^2 + 30 - 17a$

29. $5a^2 - 6 - 7a$

30. $11s + 12s^2 - 5$

31. $3r^2 + r - 10$

32. $20x^2 - 28x - 3$

33. $4y^2 + 69y + 17$

34. $21m^2 + 13m + 2$

35. $38x^2 + 23x + 2$

36. $20y^2 + 39y - 11$

37. $10x^2 + 11x - 6$

38. $2 + 7b + 6b^2$

39. $10 + 19w + 6w^2$

40. $20q^2 - 41q + 20$

41. $6q^2 + 23q + 21$

42. $8x^2 + 47x - 6$

43. $10m^2 - 23m + 12$

44. $4t^2 - 5t - 6$

45. $8k^2 + 2k - 15$

46. $15p^2 - p - 6$

47. $10m^2 - m - 24$

48. $16a^2 + 30a + 9$

49. $8x^2 - 14x + 3$

50. $24b^2 - 37b - 5$

51. $40m^2 + m - 6$

52. $15a^2 + 22a + 8$

53. $12p^2 + 7pq - 12q^2$

54. $6m^2 - 5mn - 6n^2$

55. $25a^2 + 25ab + 6b^2$

56. $6x^2 - 5xy - y^2$

57. $6a^2 - 7ab - 5b^2$

58. $2m^3 + 2m^2 - 40m$

59. $15n^4 - 39n^3 + 18n^2$

60. $24a^4 + 10a^3 - 4a^2$

61. $32z^2w^4 - 20zw^4 - 12w^4$

62. $15x^2y^2 - 7xy^2 - 4y^2$

63. $4k^4 - 2k^3w - 6k^2w^2$

64. $4a^4 - a^3b - 3a^2b^2$

65. $6m^6n + 7m^5n^2 + 2m^4n^3$

66. $12k^3q^4 - 4k^2q^5 - kq^6$

67. $18z^3y - 3z^2y^2 - 105zy^3$

68. $18x^2(y - 3)^2 + 5x(y - 3)^2 - 75(y - 3)^2$

69. $25q^2(m + 1)^3 - 5q(m + 1)^3 - 2(m + 1)^3$

Review Exercises *Find each product. See Sections 3.5 and 3.6.*

70. $(3r - 1)(3r + 1)$

71. $(5p - 3q)(5p + 3q)$

72. $(2m - 3)^2$

73. $(x - 1)(x^2 + x + 1)$

74. $(2z - 3)(4z^2 + 6z + 9)$

75. $(3k - 1)(9k^2 + 3k + 1)$

4.4   Special Factorizations

Objectives

Factor:

1 the difference of two squares;

2 a perfect square trinomial;

3 the difference of two cubes;

4 the sum of two cubes.

1   Recall from the last chapter that

$$(x + y)(x - y) = x^2 - y^2.$$

Based on this product, a **difference of two squares** can be factored as

$$x^2 - y^2 = (x + y)(x - y).$$

Example 1   Factor each difference of two squares.

(a) $x^2 - 49 = x^2 - 7^2 = (x + 7)(x - 7)$

(b) $y^2 - m^2 = (y + m)(y - m)$

(c) $z^2 - \dfrac{9}{16} = z^2 - \left(\dfrac{3}{4}\right)^2 = \left(z + \dfrac{3}{4}\right)\left(z - \dfrac{3}{4}\right)$

(d) $p^2 + 16$

Since $p^2 + 16$ is the *sum* of two squares, it is not equal to $(p + 4)(p - 4)$. Also, using FOIL,

$$(p - 4)(p - 4) = p^2 - 8p + 16 \neq p^2 + 16,$$

and

$$(p + 4)(p + 4) = p^2 + 8p + 16 \neq p^2 + 16,$$

so $p^2 + 16$ is a prime polynomial.   ✛

As Example 1(d) suggests, assuming that the two squares have no common factor,

the sum of two squares usually cannot be factored.

Example 2   Factor $25m^2 - 16$.

This is the difference of two squares, since

$$25m^2 - 16 = (5m)^2 - 4^2.$$

Now factor $(5m)^2 - 4^2$ as

$$(5m + 4)(5m - 4).   ✛$$

Example 3   Factor completely.

(a) $9a^2 - 4b^2 = (3a)^2 - (2b)^2 = (3a + 2b)(3a - 2b)$

(b) $81y^2 - 36$

First factor out the common factor of 9.

$$81y^2 - 36 = 9(9y^2 - 4)$$
$$= 9(3y + 2)(3y - 2)$$

(c) $p^4 - 36 = (p^2)^2 - 6^2 = (p^2 + 6)(p^2 - 6)$

Neither $p^2 + 6$ nor $p^2 - 6$ can be factored further.

(d) $m^4 - 16 = (m^2)^2 - 4^2 = (m^2 + 4)(m^2 - 4)$
While $m^2 + 4$ cannot be factored further, $m^2 - 4$ can be factored, giving
$$m^4 - 16 = (m^2 + 4)(m + 2)(m - 2). \quad \clubsuit$$

2   The expressions 144, $4x^2$, and $81m^6$ are called *perfect squares,* since
$$144 = 12^2, \quad 4x^2 = (2x)^2, \quad \text{and} \quad 81m^6 = (9m^3)^2.$$

A **perfect square trinomial** is a trinomial that is the square of a binomial. For example, $x^2 + 8x + 16$ is a perfect square trinomial, since it is the square of the binomial $x + 4$:
$$x^2 + 8x + 16 = (x + 4)^2.$$

For a trinomial to be a perfect square, two of its terms must be perfect squares. For this reason, $16x^2 + 4x + 15$ is not a perfect square trinomial since only the term $16x^2$ is a perfect square.

On the other hand, just because two of the terms are perfect squares, the trinomial may not be a perfect square trinomial. For example, $x^2 + 6x + 36$ has two perfect square terms, but it is not a perfect square trinomial. (Try to find a binomial that can be squared to give $x^2 + 6x + 36$.)

Multiply to see that the square of a binomial gives the following perfect square trinomials.
$$x^2 + 2xy + y^2 = (x + y)^2$$
$$x^2 - 2xy + y^2 = (x - y)^2$$

The middle term of a perfect square trinomial is always twice the product of the two terms in the squared binomial. (This was shown in Section 3.6.) Use this to check any attempt to factor a trinomial that appears to be a perfect square.

Example 4   Factor $x^2 + 10x + 25$.
   The term x^2 is a perfect square, and so is 25. Try to factor the trinomial as
$$x^2 + 10x + 25 = (x + 5)^2.$$

To check, take twice the product of the two terms in the squared binomial.
$$2 \cdot x \cdot 5 = 10x$$

Since $10x$ is the middle term of the trinomial, the trinomial is a perfect square and can be factored as $(x + 5)^2$. $\clubsuit$

Example 5   Factor each perfect square trinomial.
   (a) $x^2 - 22x + 121$
   The first and last terms are perfect squares ($121 = 11^2$). Check to see whether the middle term of $x^2 - 22x + 121$ is twice the product of the first and last terms of the binomial $x - 11$.

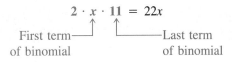

$$2 \cdot x \cdot 11 = 22x$$

First term——— ———Last term
of binomial   of binomial

Since twice the product of the first and last terms of the binomial is the middle term, $x^2 - 22x + 121$ is a perfect square trinomial and

$$x^2 - 22x + 121 = (x - 11)^2.$$

The middle sign in the binomial, a minus sign in this case, is always the same as the middle sign in the trinomial. Also, the first and last terms of a perfect square trinomial must be *positive*, since they are squares. The polynomial $x^2 - 2x - 1$ cannot be a perfect square because the last term is negative.

(b) $9m^2 - 24m + 16 = (3m)^2 - 2(3m)(4) + 4^2 = (3m - 4)^2$

(c) $25y^2 + 20y + 16$

The first and last terms are perfect squares.

$$25y^2 = (5y)^2 \quad \text{and} \quad 16 = 4^2$$

Twice the product of the first and last terms of the binomial $5y + 4$ is

$$2 \cdot 5y \cdot 4 = 40y,$$

which is not the middle term of $25y^2 + 20y + 16$. This polynomial is not a perfect square. In fact, the polynomial cannot be factored even with the methods of Section 4.3; it is a prime polynomial. ✦

3 The difference of two squares was factored above; it is possible also to factor the **difference of two cubes.** Use the pattern

$$x^3 - y^3 = (x - y)(x^2 + xy + y^2).$$

Multiply on the right to see that the pattern gives the correct factors.

$$
\begin{array}{r}
x^2 + xy + y^2 \\
x - y \\
\hline
-x^2y - xy^2 - y^3 \\
x^3 + x^2y + xy^2 \\
\hline
x^3 - y^3
\end{array}
$$

Memorize this pattern.

Example 6   Factor the following.

(a) $m^3 - 125$

Let $x = m$ and $y = 5$ in the pattern for the difference of two cubes.

$$
\begin{aligned}
m^3 - 125 &= (m - 5)(m^2 + 5m + 5^2) \\
&= (m - 5)(m^2 + 5m + 25)
\end{aligned}
$$

(b) $8p^3 - 27$

Substitute into the rule using $2p$ for x and 3 for y.

$$8p^3 - 27 = 2p^{\ 3} - 3^3$$
$$= (2p - 3)[(2p)^2 + (2p)3 + 3^2]$$
$$= (2p - 3)(4p^2 + 6p + 9)$$

(c) $4m^3 - 32 = 4(m^3 - 8)$
$$= 4(m - 2)(m^2 + 2m + 4)$$

(d) $125t^3 - 216s^6 = (5t)^3 - (6s^2)^3$
$$= (5t - 6s^2)[(5t)^2 + (5t)(6s^2) + (6s^2)^2]$$
$$= (5t - 6s^2)(25t^2 + 30ts^2 + 36s^4) \quad ✣$$

A common error in factoring the difference of two cubes, $x^3 - y^3 = (x - y)(x^2 + xy + y^2)$, is to try to factor $x^2 + xy + y^2$. It is easy to confuse this factor with a perfect square trinomial, $x^2 + 2xy + y^2$. Because of the lack of a 2 in $x^2 + xy + y^2$, it is very unusual to be able to further factor an expression of the form $x^2 + xy + y^2$.

4 A sum of two squares, such as $m^2 + 25$, cannot be factored, but the **sum of two cubes** can be factored by the following pattern, *which should be memorized.*

$$x^3 + y^3 = (x + y)(x^2 - xy + y^2)$$

Example 7   Factor.

(a) $k^3 + 27 = k^3 + 3^3$
$$= (k + 3)(k^2 - 3k + 3^2)$$
$$= (k + 3)(k^2 - 3k + 9)$$

(b) $8m^3 + 125 = (2m)^3 + 5^3$
$$= (2m + 5)[(2m)^2 - (2m)(5) + 5^2]$$
$$= (2m + 5)(4m^2 - 10m + 25)$$

(c) $1000a^6 + 27b^3 = (10a^2)^3 + (3b)^3$
$$= (10a^2 + 3b)[(10a^2)^2 - (10a^2)(3b) + (3b)^2]$$
$$= (10a^2 + 3b)(100a^4 - 30a^2b + 9b^2) \quad ✣$$

The methods of factoring discussed in this section are summarized here. All these rules should be memorized.

Special Factorizations		
Difference of two squares	$x^2 - y^2 = (x + y)(x - y)$	
Perfect square trinomials	$x^2 + 2xy + y^2 = (x + y)^2$	
	$x^2 - 2xy + y^2 = (x - y)^2$	
Difference of two cubes	$x^3 - y^3 = (x - y)(x^2 + xy + y^2)$	
Sum of two cubes	$x^3 + y^3 = (x + y)(x^2 - xy + y^2)$	

4.4　Exercises

Factor each binomial completely. See Examples 1–3.

1. $x^2 - 16$　　　**2.** $m^2 - 25$　　　**3.** $a^2 - b^2$　　　**4.** $r^2 - t^2$

5. $9m^2 - 1$　　　**6.** $16y^2 - 9$　　　**7.** $25m^2 - 16$　　　**8.** $144y^2 - 25$

9. $36t^2 - 16$　　　**10.** $9 - 36a^2$　　　**11.** $25a^2 - 16r^2$　　　**12.** $100k^2 - 49m^2$

13. $x^2 + 16$　　　**14.** $m^2 + 100$　　　**15.** $p^4 - 36$　　　**16.** $r^4 - 9$

17. $a^4 - 1$　　　**18.** $x^4 - 16$　　　**19.** $m^4 - 81$　　　**20.** $p^4 - 256$

Factor any perfect square trinomials. See Examples 4 and 5.

21. $a^2 + 4a + 4$　　　**22.** $p^2 + 2p + 1$　　　**23.** $x^2 - 10x + 25$

24. $y^2 - 8y + 16$　　　**25.** $49 + 14a + a^2$　　　**26.** $100 - 20m + m^2$

27. $k^2 + 121 + 22k$　　　**28.** $r^2 + 144 + 24r$　　　**29.** $y^2 - 10y + 100$

30. $100a^2 - 140ab + 49b^2$　　　**31.** $49x^2 + 28xy + 4y^2$　　　**32.** $64y^2 - 48ya + 9a^2$

33. $4c^2 + 12cd + 9d^2$　　　**34.** $16t^2 - 40tr + 25r^2$　　　**35.** $25h^2 - 20hy + 4y^2$

36. $9x^2 + 24xy + 16y^2$　　　**37.** $x^3y + 6x^2y^2 + 9xy^3$　　　**38.** $4k^3w + 20k^2w^2 + 25kw^3$

Factor each sum or difference of cubes. See Examples 6 and 7.

39. $8a^3 + 1$　　　**40.** $8a^3 - 1$　　　**41.** $27x^3 - 125$　　　**42.** $64p^3 + 27$

43. $8p^3 + q^3$　　　**44.** $y^3 - 8x^3$　　　**45.** $27a^3 - 64b^3$　　　**46.** $125t^3 + 8s^3$

47. $64x^3 + 125y^3$　　　**48.** $216z^3 - w^3$　　　**49.** $125m^3 - 8p^3$　　　**50.** $343r^3 + 1000s^3$

51. $1000z^3 + 27x^3$　　　**52.** $64y^3 - 1331w^3$　　　**53.** $64y^6 + 1$　　　**54.** $m^6 - 8$

55. $8k^6 - 27q^3$　　　**56.** $125z^3 + 64r^6$　　　**57.** $1000a^3 - 343b^9$　　　**58.** $27r^9 + 125s^3$

Factor completely in Exercises 59–68.

59. $(m + n)^2 - (m - n)^2$　　　　　　**60.** $(x + 2y)^2 - (2x + y)^2$

61. $(a - b)^3 - (a + b)^3$　　　　　　**62.** $(r + t)^3 + (r - t)^3$

63. $(2x + 1)^2 + 2(2x + 1) + 1$　　　　**64.** $4(p - q)^2 - 4(p - q) + 1$

65. $(x^2 + 2x + 1) - 4$　　　　　　**66.** $(a + 1)^2 - 4$

67. $m^2 - p^2 + 2m + 2p$　　　　　　**68.** $3r - 3k + 3r^2 - 3k^2$

69. Find a value of b so that $x^2 + bx + 25 = (x + 5)^2$.

70. For what value of c is $4m^2 - 12m + c = (2m - 3)^2$?

71. Find a so that $ay^2 - 12y + 4 = (3y - 2)^2$.

72. Find b so that $100a^2 + ba + 9 = (10a + 3)^2$.

Review Exercises　*Solve each equation. See Section 2.4.*

73. $m - 2 = 0$　　　**74.** $r + 1 = 0$　　　**75.** $3k - 2 = 0$　　　**76.** $4z + 5 = 0$

77. $7a + 9 = 0$　　　**78.** $3x + 7 = 0$　　　**79.** $8y + 5 = 0$　　　**80.** $12k - 11 = 0$

Supplementary Factoring Exercises

Factoring a Polynomial	*Step 1* Is there a common factor?
	Step 2 How many terms are in the polynomial?
	Two terms Check to see whether it is either the difference of two squares or the sum or difference of two cubes.
	Three terms Is it a perfect square trinomial? If the trinomial is not a perfect square, check to see whether the coefficient of the squared term is 1. If so, use the method of Section 4.2. If the coefficient of the squared term of the trinomial is not 1, use the general factoring methods of Section 4.3.
	Four terms Can the polynomial be factored by grouping?
	Step 3 Can any factors be factored further?

Factor as completely as possible.

1. $a^2 - 4a - 12$

2. $a^2 + 17a + 72$

3. $6y^2 - 6y - 12$

4. $7y^6 + 14y^5 - 168y^4$

5. $6a + 12b + 18c$

6. $m^2 - 3mn - 4n^2$

7. $p^2 - 17p + 66$

8. $z^2 - 6z + 7z - 42$

9. $10z^2 - 7z - 6$

10. $2m^2 - 10m - 48$

11. $m^2 - n^2 + 5m - 5n$

12. $15y + 5$

13. $8a^5 - 8a^4 - 48a^3$

14. $8k^2 - 10k - 3$

15. $z^2 - 3za - 10a^2$

16. $50z^2 - 100$

17. $x^2 - 4x - 5x + 20$

18. $100n^2r^2 + 30nr^3 - 50n^2r$

19. $6n^2 - 19n + 10$

20. $9y^2 + 12y - 5$

21. $16x + 20$

22. $m^2 + 2m - 15$

23. $6y^2 - 5y - 4$

24. $m^2 - 81$

25. $6z^2 + 31z + 5$

26. $5z^2 + 24z - 5 + 3z + 15$

27. $4k^2 - 12k + 9$

28. $8p^2 + 23p - 3$

29. $54m^2 - 24z^2$

30. $8m^2 - 2m - 3$

31. $3k^2 + 4k - 4$

32. $45a^3b^5 - 60a^4b^2 + 75a^6b^4$

33. $14k^3 + 7k^2 - 70k$

34. $5 + r - 5s - rs$

35. $y^4 - 16$

36. $20y^5 - 30y^4$

37. $8m - 16m^2$

38. $k^2 - 16$

39. $z^3 - 8$

40. $y^2 - y - 56$

41. $k^2 + 9$

42. $27p^{10} - 45p^9 - 252p^8$

43. $32m^9 + 16m^5 + 24m^3$

44. $8m^3 + 125$

45. $16r^2 + 24rm + 9m^2$

46. $z^2 - 12z + 36$

47. $15h^2 + 11hg - 14g^2$

48. $5z^3 - 45z^2 + 70z$

49. $k^2 - 11k + 30$

50. $64p^2 - 100m^2$

51. $3k^3 - 12k^2 - 15k$

52. $y^2 - 4yk - 12k^2$

53. $1000p^3 + 27$

54. $64r^3 - 343$

55. $6 + 3m + 2p + mp$

56. $2m^2 + 7mn - 15n^2$

57. $16z^2 - 8z + 1$

58. $125m^4 - 400m^3n + 195m^2n^2$

59. $108m^2 - 36m + 3$

60. $100a^2 - 81y^2$

61. $64m^2 - 40mn + 25n^2$

62. $4y^2 - 25$

63. $32z^3 + 56z^2 - 16z$

64. $10m^2 + 25m - 60$

65. $20 + 5m + 12n + 3mn$

66. $4 - 2q - 6p + 3pq$

67. $6a^2 + 10a - 4$

68. $36y^6 - 42y^5 - 120y^4$

69. $a^3 - b^3 + 2a - 2b$

70. $16k^2 - 48k + 36$

71. $64m^2 - 80mn + 25n^2$

72. $72y^3z^2 + 12y^2 - 24y^4z^2$

73. $8k^2 - 2kh - 3h^2$

74. $2a^2 - 7a - 30$

75. $(m + 1)^3 + 1$

76. $8a^3 - 27$

77. $10y^2 - 7yz - 6z^2$

78. $m^2 - 4m + 4$

79. $8a^2 + 23ab - 3b^2$

80. $a^4 - 625$

81. $9m^2 - 64$

82. $24k^4p + 60k^3p^2 + 150k^2p^3$

83. $9z^2 + 64$

84. $15t - 15 - t^2 + 1$

85. $a^2 + 8a + 16$

86. $z^2 - z^3 + m - mz$

4.5 Solving Quadratic Equations

Objectives

1 Solve quadratic equations by factoring.

2 Solve other equations by factoring.

This section introduces *quadratic equations*, equations that contain a squared term and no terms of higher degree.

Quadratic Equations	An equation that can be put in the form $$ax^2 + bx + c = 0,$$ where a, b, and c are real numbers, with $a \neq 0$, is a **quadratic equation.**

The form $ax^2 + bx + c = 0$ is the **standard form** of a quadratic equation.

For example,

$$x^2 + 5x + 6 = 0, \qquad 2a^2 - 5a = 3, \qquad \text{and} \qquad y^2 = 4$$

are all quadratic equations but only $x^2 + 5x + 6 = 0$ is in standard form.

1 Some quadratic equations can be solved by factoring. A more general method for solving those equations that cannot be solved by factoring is given in Chapter 9. Use the **zero-factor property** to solve a quadratic equation by factoring.

Zero-Factor Property	If a and b are real numbers and if $ab = 0$, then $a = 0$ or $b = 0$.

In other words, if the product of two numbers is zero, then at least one of the numbers must be zero.

Example 1   Solve the equation $(x + 3)(2x - 1) = 0$.

The product $(x + 3)(2x - 1)$ is equal to zero. By the zero-factor property, the only way that the product of these two factors can be zero is if at least one of the factors is zero. Therefore, either $x + 3 = 0$ or $2x - 1 = 0$. Solve each of these two linear equations as in Chapter 2.

$$x + 3 = 0 \qquad \text{or} \qquad 2x - 1 = 0$$
$$x = -3 \qquad \text{or} \qquad 2x = 1$$
$$x = \frac{1}{2}$$

The given equation $(x + 3)(2x - 1) = 0$ has two solutions, -3 and $1/2$. Check these answers by substituting -3 for x in the original equation. Then start over and substitute $1/2$ for x.

If $x = -3$, then
$$(-3 + 3)[2(-3) - 1] = 0$$
$$0(-7) = 0.$$

If $x = 1/2$, then
$$\left(\frac{1}{2} + 3\right)\left(2 \cdot \frac{1}{2} - 1\right) = 0$$
$$\frac{7}{2}(1 - 1) = 0$$
$$\frac{7}{2} \cdot 0 = 0.$$

Both -3 and $1/2$ result in true equations, so they are solutions to the original equation. ✦

In Example 1 the equation to be solved was presented in factored form. If the equation is not already factored, first make sure that the equation is in standard form. Then factor the polynomial.

Example 2   Solve the equation $x^2 - 5x = -6$.

First, rewrite the equation with all terms on one side by adding 6 on both sides.

$$x^2 - 5x = -6$$
$$x^2 - 5x + 6 = 0 \qquad \text{Add 6}$$

Now factor $x^2 - 5x + 6$. Find two numbers whose product is 6 and whose sum is -5. These two numbers are -2 and -3, so the equation becomes

$$(x - 2)(x - 3) = 0.$$

Proceed as in Example 1. Set each factor equal to 0.

$$x - 2 = 0 \qquad \text{or} \qquad x - 3 = 0$$

Solve the equation on the left by adding 2 on both sides. In the equation on the right, add 3 on both sides. Doing this gives

$$x = 2 \qquad \text{or} \qquad x = 3.$$

Check both solutions by substituting first 2 and then 3 for x in the original equation. ✤

Example 3   Solve the equation $4p^2 + 40 = 26p$.

Subtract $26p$ from each side and write in descending powers to get

$$4p^2 - 26p + 40 = 0.$$

Factor out the common factor of 2:

$$2(2p^2 - 13p + 20) = 0.$$

Divide both sides by 2 to get

$$2p^2 - 13p + 20 = 0.$$

Factor $2p^2 - 13p + 20$ as $(2p - 5)(p - 4)$, giving

$$(2p - 5)(p - 4) = 0.$$

Set each of these two factors equal to 0.

$$2p - 5 = 0 \qquad \text{or} \qquad p - 4 = 0$$

Solve the equation on the left by first adding 5 on both sides of the equation. Then divide both sides by 2. Solve the equation on the right by adding 4 to both sides.

$$2p - 5 = 0 \qquad \text{or} \qquad p - 4 = 0$$
$$2p = 5 \qquad \text{or} \qquad p = 4$$
$$p = \frac{5}{2}$$

The solutions of $4p^2 + 40 = 26p$ are 5/2 and 4; check them by substituting in the original equation. ✤

Example 4 Solve each equation.

(a) $16m^2 - 25 = 0$

Factor the left-hand side of the equation as the difference of two squares.

$$(4m + 5)(4m - 5) = 0$$

Set each factor equal to 0.

$$4m + 5 = 0 \qquad \text{or} \qquad 4m - 5 = 0$$

Solve each equation.

$$4m = -5 \qquad \text{or} \qquad 4m = 5$$
$$m = -\frac{5}{4} \qquad \text{or} \qquad m = \frac{5}{4}$$

The two solutions, $-5/4$ and $5/4$, should be checked in the original equation.

(b) $k(2k + 5) = 3$

Multiply on the left-hand side and then get all terms on one side.

$$k(2k + 5) = 3$$
$$2k^2 + 5k = 3$$
$$2k^2 + 5k - 3 = 0$$

Now factor.

$$(2k - 1)(k + 3) = 0$$

Set each factor equal to 0 and solve the equations.

$$2k - 1 = 0 \qquad \text{or} \qquad k + 3 = 0$$
$$2k = 1 \qquad \text{or} \qquad k = -3$$
$$k = \frac{1}{2}$$

The two solutions are $1/2$ and -3. ✦

In Example 4(b) the zero-factor property could not be used to solve the original equation because of the 3 on the right. The equation must have zero on one side to use the property.

In summary, go through the following steps to solve quadratic equations by factoring.

Solving a Quadratic Equation by Factoring	*Step 1* Write the equation in standard form: all terms on one side of the equals sign, with 0 on the other side.
	Step 2 Factor completely.
	Step 3 Set each factor with a variable equal to 0, and solve the resulting equations.
	Step 4 Check each solution in the original equation.
	Remember: Not all quadratic equations can be solved by factoring.

2 The zero-factor property also can be used to solve equations that result in more than two factors, as shown in Example 5. (These equations are *not* quadratic equations. Why not?)

Example 5 Solve the equation $6z^3 - 6z = 0$.

First, factor out the greatest common factor in $6z^3 - 6z$.

$$6z^3 - 6z = 0$$
$$6z(z^2 - 1) = 0$$

Now factor $z^2 - 1$ as $(z + 1)(z - 1)$ to get

$$6z(z + 1)(z - 1) = 0.$$

By an extension of the zero-factor property, this product can equal zero only if at least one of the factors is zero. Write three equations, one for each factor with a variable.

$$6z = 0 \qquad \text{or} \qquad z + 1 = 0 \qquad \text{or} \qquad z - 1 = 0$$

Solving these three equations gives three solutions,

$$z = 0 \qquad \text{or} \qquad z = -1 \qquad \text{or} \qquad z = 1.$$

Check by substituting, in turn, 0, -1, and 1 in the original equation. ✤

Example 6 Solve the equation $(2x - 1)(x^2 - 9x + 20) = 0$.

Factor $x^2 - 9x + 20$ as $(x - 5)(x - 4)$. Then rewrite the original equation as

$$(2x - 1)(x - 5)(x - 4) = 0.$$

Set each of these three factors equal to 0.

$$2x - 1 = 0 \qquad \text{or} \qquad x - 5 = 0 \qquad \text{or} \qquad x - 4 = 0$$

Solving these three equations gives

$$x = \frac{1}{2} \qquad \text{or} \qquad x = 5 \qquad \text{or} \qquad x = 4$$

as the solutions of the original equation. Check each solution. ✤

4.5 Exercises

Solve each equation. See Example 1.

1. $(x - 2)(x + 4) = 0$ **2.** $(y - 3)(y + 5) = 0$ **3.** $(3x + 5)(2x - 1) = 0$

4. $(2a + 3)(a - 2) = 0$ **5.** $(5p + 1)(2p - 1) = 0$ **6.** $(3k - 8)(k + 7) = 0$

7. $(2m + 9)(3m - 1) = 0$ **8.** $(9a - 2)(3a + 1) = 0$ **9.** $(x - 1)(3x + 5) = 0$

10. $(k - 3)(k + 5) = 0$ **11.** $(3r - 7)(2r + 8) = 0$ **12.** $(5a + 2)(3a - 1) = 0$

Solve each equation. See Examples 2–4.

13. $x^2 + 5x + 6 = 0$

14. $y^2 - 3y + 2 = 0$

15. $r^2 - 5r - 6 = 0$

16. $y^2 - y - 12 = 0$

17. $m^2 + 3m - 28 = 0$

18. $p^2 - p - 6 = 0$

19. $a^2 = 24 - 5a$

20. $m^2 = 3m + 4$

21. $z^2 = -2 - 3z$

22. $p^2 = 2p + 3$

23. $3a^2 + 5a - 2 = 0$

24. $6r^2 - r - 2 = 0$

25. $2k^2 - k - 10 = 0$

26. $6x^2 - 7x - 5 = 0$

27. $18a^2 = 15 - 39a$

28. $18s^2 + 24s = -8$

29. $2z^2 + 3z = 20$

30. $25p^2 + 20p + 4 = 0$

31. $3a^2 + 7a = 20$

32. $6z^2 + 11z + 3 = 0$

33. $15r^2 = r + 2$

34. $3m^2 = 5m + 28$

35. $10b^2 + 15b - 45 = 0$

36. $20b^2 = 32b + 16$

37. $16r^2 - 25 = 0$

38. $4k^2 - 9 = 0$

39. $9m^2 - 36 = 0$

40. $16x^2 - 64 = 0$

41. $m(m - 7) = -10$

42. $z(2z + 7) = 4$

43. $2(x^2 - 66) = -13x$

44. $3(m^2 + 4) = 20m$

45. $3r(r + 1) = (2r + 3)(r + 1)$

46. $(3k + 1)(k + 1) = 2k(k + 3)$

47. $12k(k - 4) = 3(k - 4)$

48. $y^2 = 4(y - 1)$

Solve each equation. See Examples 5 and 6.

49. $(2r - 5)(3r^2 - 16r + 5) = 0$

50. $(3m - 4)(6m^2 + m - 2) = 0$

51. $(2x + 7)(x^2 - 2x - 3) = 0$

52. $(x - 1)(6x^2 + x - 12) = 0$

53. $x^3 - 25x = 0$

54. $m^3 - 4m = 0$

55. $9y^3 - 49y = 0$

56. $16r^3 - 9r = 0$

57. $r^3 - 2r^2 - 8r = 0$

58. $x^3 - x^2 - 6x = 0$

59. $a^3 + a^2 - 20a = 0$

60. $y^3 - 6y^2 + 8y = 0$

61. $r^4 = 2r^3 + 15r^2$

62. $x^3 = 3x + 2x^2$

63. $6p^2(p + 1) = 4(p + 1) - 5p(p + 1)$

64. $6x^2(2x + 3) - 5x(2x + 3) = 4(2x + 3)$

65. $(k + 3)^2 - (2k - 1)^2 = 0$

66. $(4y - 3)^3 - 9(4y - 3) = 0$

Solve each problem.

67. When three times a number is added to the square of a number, the result is 10. Find the number.

68. When a number is subtracted from its square, the result is 6. Find the number.

Review Exercises *Solve each word problem. See Section 2.5.*

69. The length of a rectangle is 3 meters more than the width. The perimeter of the rectangle is 34 meters. Find the width of the rectangle.

70. A rectangle has a length 4 meters less than twice the width. The perimeter of the rectangle is 4 meters more than five times the width. Find the width of the rectangle.

71. When two consecutive integers are added, the result is 17 less than three times the smaller integer. Find the smaller integer.

72. If three consecutive even integers are added, the result is 44 less than four times the smallest integer. Find the smallest integer.

4.6 Applications of Quadratic Equations

Objectives

Solve word problems about

1️⃣ geometric figures;

2️⃣ the Pythagorean formula;

3️⃣ consecutive integers.

1️⃣ In this section we look at problems whose solutions require quadratic equations, starting with a problem about area.

Example 1 The width of a rectangular box is 4 centimeters less than the length. The area is 96 square centimeters. Find the length and width of the box.

Let L = the length of the box;

$L - 4$ = the width (The width is 4 less than the length).

See Figure 4.1. The area of a rectangle is given by the formula

$$\text{area} = LW = \text{length} \times \text{width}.$$

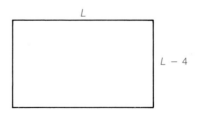

Figure 4.1

Substitute 96 for the area, L for the length, and $L - 4$ for the width into the formula for area getting

$$96 = L(L - 4).$$

Multiply on the right.

$$96 = L(L - 4)$$
$$96 = L^2 - 4L$$
$$0 = L^2 - 4L - 96. \qquad \text{Subtract 96 from both sides}$$
$$0 = (L - 12)(L + 8) \qquad \text{Factor}$$

Place each factor equal to 0.

$$L - 12 = 0 \qquad \text{or} \qquad L + 8 = 0$$

Solve each equation.

$$L = 12 \qquad \text{or} \qquad L = -8$$

The solutions of the equations are 12 and -8. Always be careful, however, to check solutions against physical facts. Since a rectangle cannot have a negative length, discard the solution -8. Then 12 centimeters is the length of the box, and $12 - 4 = 8$ centimeters is the width. As a check, note that the width is 4 less than the length and the area is $8 \cdot 12 = 96$ square centimeters as required. ✦

Example 2 The length of a rectangular rug is 4 feet more than the width. (See Figure 4.2.) The length of the rug is 2 feet less than the square of the width. Find the length and width of the rug.

Let $\qquad W =$ the width of the rug,
$\qquad W + 4 =$ the length.

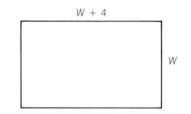

W + 4

W

Figure 4.2

From the information given in the problem, the length, $W + 4$, is 2 feet less than the square of the width, W^2.

length	is	square of the width	less 2
↓	↓	↓	↓
$W + 4$	$=$	W^2	$- 2$

Solve this equation by first getting 0 on one side.

$$0 = W^2 - W - 6 \qquad \text{Subtract } W \text{ and 4 on both sides}$$
$$0 = (W - 3)(W + 2) \qquad \text{Factor}$$
$$W - 3 = 0 \quad \text{or} \quad W + 2 = 0$$
$$W = 3 \quad \text{or} \quad W = -2$$

A rug cannot have a negative width, so ignore -2. The only useful solution is 3: the width is 3 feet and the length is $3 + 4 = 7$ feet. (Check to see that the length is 2 feet less than the square of the width.) The rug is 3 feet by 7 feet. ✤

2 The next example requires the **Pythagorean formula** from geometry.

<table>
<tr>
<td>

Pythagorean Formula

</td>
<td>

If a right triangle (a triangle with a 90° angle) has longest side of length c and two other sides of length a and b, then

$$a^2 + b^2 = c^2.$$

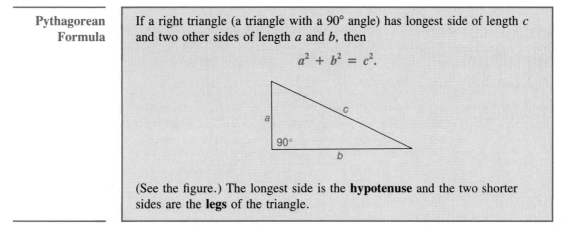

(See the figure.) The longest side is the **hypotenuse** and the two shorter sides are the **legs** of the triangle.

</td>
</tr>
</table>

Example 3 The hypotenuse of a right triangle is 2 feet more than the shorter leg. The longer leg is 1 foot more than the shorter leg. Find the lengths of the sides of the triangle.
Let x be the length of the shorter leg. Then

$$x = \text{shorter leg},$$
$$x + 1 = \text{longer leg},$$
$$x + 2 = \text{hypotenuse}.$$

Place these on a right triangle, as in Figure 4.3.

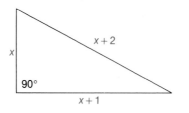

Figure 4.3

By the Pythagorean formula,

$$x^2 + (x + 1)^2 = (x + 2)^2.$$

Since $(x + 1)^2 = x^2 + 2x + 1$, and since $(x + 2)^2 = x^2 + 4x + 4$, the equation becomes

$$x^2 + x^2 + 2x + 1 = x^2 + 4x + 4$$

$$x^2 - 2x - 3 = 0 \qquad \text{Get 0 on one side}$$

$$(x - 3)(x + 1) = 0. \qquad \text{Factor}$$

Set each factor equal to 0.

$$x - 3 = 0 \qquad \text{or} \qquad x + 1 = 0$$

$$x = 3 \qquad \text{or} \qquad x = -1$$

Since -1 cannot be the length of a side of a triangle, 3 is the only possible answer. The triangle has a shorter leg of length 3, a longer leg of length $3 + 1 = 4$, and a hypotenuse of length $3 + 2 = 5$. Check that $3^2 + 4^2 = 5^2$. ✤

3   As mentioned earlier, **consecutive integers** are integers that are next to each other, such as 5 and 6, or -11 and -10. **Consecutive odd integers** are odd integers that are next to each other, such as 21 and 23, or -17 and -15. The next example shows how quadratic equations can occur in work with consecutive integers. When working with consecutive integers, the list in Section 2.5 may be helpful.

Example 4   The product of two consecutive odd integers is 1 less than five times their sum. Find the integers.

Let $s =$ the smaller of the two integers,

$s + 2 =$ the larger of the two integers (since they are consecutive *odd* integers).

According to the problem, the product is 1 less than five times the sum.

	product	is	five times sum	less 1
	↓	↓	↓	↓
	$s(s + 2)$	$=$	$5(s + s + 2)$	$- 1$

Simplify this equation and solve it.

$$s^2 + 2s = 5s + 5s + 10 - 1$$

$$s^2 + 2s = 10s + 9$$

$$s^2 - 8s - 9 = 0 \qquad \text{Subtract } 10s \text{ and } 9$$

$$(s - 9)(s + 1) = 0 \qquad \text{Factor}$$

$$s - 9 = 0 \qquad \text{or} \qquad s + 1 = 0$$

$$s = 9 \qquad \text{or} \qquad s = -1$$

Since two consecutive odd integers are required,

if $s = 9$ is the first, then $s + 2 = 11$ is the second;

if $s = -1$ is the first, then $s + 2 = 1$ is the second.

Check that both pairs of integers satisfy the problem: 9 and 11 or -1 and 1.   ✚

4.6   Exercises

Solve each problem. See Examples 1 and 2.

1. The length of a rectangle is 5 centimeters more than the width. The area is 66 square centimeters. Find the width of the rectangle.

2. The length of a rectangle is 1 foot more than the width. The area is 56 square feet. Find the width of the rectangle.

3. The width of a rectangle is 3 meters less than its length. The area of the rectangle is 70 square meters. Find the length of the rectangle.

4. The width of a rectangle is 7 meters less than its length. The area is 8 square meters. Find the length of the rectangle.

5. The length of a rectangle is 3 more than the width. The area is numerically 4 less than the perimeter. Find the width of the rectangle.

6. The width of a rectangle is 5 less than the length. The area is numerically 10 more than the perimeter. Find the length of the rectangle.

Exercises 7 and 8 require the formula for the area of a triangle.

$$\text{area} = \frac{1}{2} bh$$

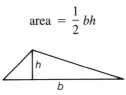

7. The area of a triangle is 25 square centimeters. The base is twice the height. Find the height of the triangle.

8. The height of a triangle is 3 inches more than the base. The area of the triangle is 27 square inches. Find the base of the triangle.

Work the following problems.

9. One square has sides 1 foot less than the length of the sides of a second square. If the difference of the areas of the two squares is 37 square feet, find the length of the side of the second square.

10. The sides of one square have a length 2 meters more than the sides of a second square. If the area of the larger square is subtracted from three times the area of the smaller square, the answer is 12 square meters. Find the length of the side of the second square.

11. John wishes to build a box to hold his tools. The box is to be 4 feet high, and the width of the box is to be 1 foot less than the length. The volume of the box will be 120 cubic feet. Find the length of the box. (Hint: The formula for the volume of a box is $V = LWH$.)

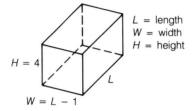

L = length
W = width
H = height

12. The volume of a box must be 315 cubic meters. The length of the box is to be 7 meters, and the height is to be 4 meters more than the width. Find the width of the box.

The following exercises require the Pythagorean formula. See Example 3.

13. The hypotenuse of a right triangle is 1 centimeter longer than the longer leg. The shorter leg is 7 centimeters shorter than the longer leg. Find the length of the longer leg of the triangle.

14. The longer leg of a right triangle is 1 meter longer than the shorter leg. The hypotenuse is 1 meter shorter than twice the shorter leg. Find the length of the shorter leg of the triangle.

15. A ladder is resting against a wall. The top of the ladder touches the wall at a height of 15 feet. Find the distance from the wall to the bottom of the ladder if the length of the ladder is one foot more than twice its distance from the wall.

16. Two cars leave an intersection. One car travels north; the other travels east. When the car traveling north had gone 24 miles, the distance between the cars was four miles more than three times the distance traveled by the car heading east. Find the distance between the cars at that time.

Work the following problems. See Example 4.

17. The product of two consecutive integers is 2 more than twice their sum. Find the first integer.

18. The product of two consecutive even integers is 60 more than twice the larger. Find the first integer.

19. Find three consecutive even integers such that four times the sum of all three equals the product of the smaller two.

20. If the square of the sum of two consecutive integers is reduced by three times their product, the result is 31. Find the integers.

21. One number is 4 more than another. The square of the smaller increased by three times the larger is 66. Find the numbers.

22. If the square of the larger of two numbers is reduced by six times the smaller, the result is five times the larger. The larger is twice the smaller. Find the numbers.

Work the following problems.

23. The length of a rectangle is twice its width. If the width were increased by 2 inches while the length remained the same, the resulting rectangle would have an area of 48 square inches. Find the width of the original rectangle.

24. When four times an integer is subtracted from twice the square of the integer, the result is 16. Find the integer.

25. The hypotenuse of a right triangle is 1 foot more than twice the length of the shorter leg. The longer leg is 1 foot less than twice the length of the shorter leg. Find the length of the shorter leg.

26. If the length of the shorter leg of a right triangle is tripled, with 4 centimeters added to the result, the result is the length of the hypotenuse. The length of the longer leg is 10 centimeters more than twice the length of the shorter leg. Find the length of the shorter leg of the triangle.

27. The sum of three times the square of an integer and twice the integer is 8. Find the integer.

28. The length of a rectangle is three times its width. If the length were decreased by 1 centimeter while the width stayed the same, the area of the new figure would be 44 square centimeters. Find the width of the original rectangle.

Exercises 29 and 30 require the formula for the volume of a pyramid.

$$V = \frac{1}{3}Bh, \text{ where } B \text{ is the area of the base}$$

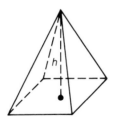

29. The volume of a pyramid is 32 cubic meters. Suppose the numerical value of the height is 10 less than the numerical value of the area of the base. Find the area of the base.

30. Suppose a pyramid has a rectangular base whose width is 3 centimeters less than the length. If the height is 8 centimeters and the volume is 144 cubic centimeters, find the length of the base.

Work the following problems involving formulas.

If an object is dropped, the distance d it falls in t seconds (disregarding air resistance) is given by

$$d = \frac{1}{2}gt^2,$$

where g is approximately 32 feet per second per second. Find the distance an object would fall in the following times.

31. 4 seconds   **32.** 8 seconds

How long would it take an object to fall the following distances?

33. 1600 feet   **34.** 2304 feet

If an object is projected straight up with an initial velocity of v_0 feet per second, its height h after t seconds is given by

$$h = v_0 t - 16t^2.$$

Suppose an object is thrown upward with an initial velocity of 64 feet per second. Find its height after the following times.

35. 1 second   **36.** 2 seconds   **37.** 3 seconds

38. When will the object hit the ground? (Hint: It will hit the ground when $h = 0$.)

Find a polynomial representing the area of each shaded region.

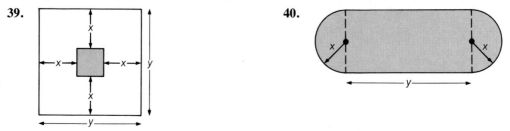

39.   **40.**

Review Exercises   *Decide whether the following inequalities are true or false when x is replaced with -2. See Sections 1.5–1.8.*

41. $x^2 - x + 4 \leq 0$

42. $x^2 + 2x - 3 < 0$

43. $2x^2 + 3x + 2 > 0$

44. $3x^2 - 5x + 1 < 0$

45. $(x - 1)(x^2 + 3x + 7) \leq 0$

46. $(x + 3)(x^2 - 5x + 1) \geq 0$

4.7 Solving Quadratic Inequalities

Objective

1 Solve quadratic inequalities and graph their solutions.

1 A **quadratic inequality** is an inequality that involves a second-degree polynomial. Examples of quadratic inequalities include

$$2x^2 + 3x - 5 < 0, \qquad x^2 \le 4, \qquad \text{and} \qquad x^2 + 5x + 6 > 0.$$

Examples 1 and 2 show how to solve such inequalities.

Example 1 Solve $x^2 + 5x + 6 > 0$.

To begin, find the solution of the corresponding quadratic equation,

$$x^2 + 5x + 6 = 0.$$

Factor to get

$$(x + 2)(x + 3) = 0,$$

from which

$$x + 2 = 0 \qquad \text{or} \qquad x + 3 = 0$$
$$x = -2 \qquad \text{or} \qquad x = -3.$$

Since -2 and -3 are the only values that satisfy $x^2 + 5x + 6 = 0$, all other values of x will make $x^2 + 5x + 6$ either less than 0 or greater than 0. The values -2 and -3 determine three regions on the number line, as shown in Figure 4.4. Region A includes all numbers less than -3. Region B includes the numbers between -3 and -2, and Region C includes all numbers greater than -2.

Figure 4.4

All the values of x in a given region will make $x^2 + 5x + 6$ positive, or else all the values of x in the given region will make $x^2 + 5x + 6$ negative. Test one value of x from each region to see which regions satisfy $x^2 + 5x + 6 > 0$.

First, are the points in Region A part of the solution? As a trial value, choose any number less than -3, say -4.

$$x^2 + 5x + 6 > 0$$
$$(-4)^2 + 5(-4) + 6 > 0 \qquad \text{Let } x = -4$$
$$16 - 20 + 6 > 0 \qquad \text{Simplify}$$
$$2 > 0 \qquad \text{True}$$

Since $2 > 0$ is true, all the points in Region A belong to the solution of the given inequality.

For Region B, choose a value between -3 and -2, say $-2\ 1/2$, or $-5/2$.

$$\left(-\frac{5}{2}\right)^2 + 5\left(-\frac{5}{2}\right) + 6 > 0 \qquad \text{Let } x = -\frac{5}{2}$$

$$\frac{25}{4} + \left(-\frac{25}{2}\right) + 6 > 0 \qquad \text{Simplify}$$

$$-\frac{1}{4} > 0 \qquad \text{False}$$

Since $-1/4 > 0$ is false, no point in Region B belongs to the solution.

For Region C, try the number 0.

$$0^2 + 5(0) + 6 > 0 \qquad \text{Let } x = 0$$

$$6 > 0 \qquad \text{True}$$

Since $6 > 0$ is true, the points in Region C belong to the solution.

The solution is shown in Figure 4.5. The graph of the quadratic inequality $x^2 + 5x + 6 > 0$ includes all values of x less than -3, and all values of x greater than -2. Write the solution as

$$x < -3 \qquad \text{or} \qquad x > -2. \quad \clubsuit$$

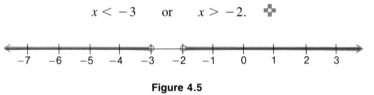

Figure 4.5

There is no shortcut way to write the solution $x < -3$ or $x > -2$.

Example 2   Solve $x^2 - 3x - 10 \le 0$.

Begin by factoring the corresponding equation $x^2 - 3x - 10 = 0$, to get $(x - 5)(x + 2) = 0$. The solutions of the equation are -2 and 5. These points determine three regions on the number line. See Figure 4.6. This time, these points also will belong to the solution, because all values of x that make $x^2 - 3x - 10 < 0$ or that make $x^2 - 3x - 10 = 0$ are solutions.

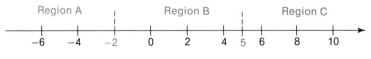

Figure 4.6

Do the points in Region A belong to the solution? Decide by selecting any point in Region A, such as -6. Does -6 satisfy the original inequality?

$$x^2 - 3x - 10 \le 0$$
$$(-6)^2 - 3(-6) - 10 \le 0 \qquad \text{Let } x = -6$$
$$36 + 18 - 10 \le 0 \qquad \text{Simplify}$$
$$44 \le 0 \qquad \text{False}$$

Since $44 \le 0$ is false, the points in Region A do not belong to the solution. What about Region B? Try the value $x = 0$.

$$0^2 - 3(0) - 10 \le 0 \qquad \text{Let } x = 0$$
$$-10 \le 0 \qquad \text{True}$$

Since $-10 \le 0$ is true, the points in Region B do belong to the solution. Try $x = 6$ to check Region C.

$$6^2 - 3(6) - 10 \le 0 \qquad \text{Let } x = 6$$
$$36 - 18 - 10 \le 0 \qquad \text{Simplify}$$
$$8 \le 0 \qquad \text{False}$$

Since $8 \le 0$ is false, the points in Region C do not belong to the solution.

The points in Region B are the only ones that satisfy $x^2 - 3x - 10 \le 0$. As shown in Figure 4.7, the solution includes the points in Region B together with the endpoints -2 and 5. The solution is written

$$-2 \le x \le 5. \quad \clubsuit$$

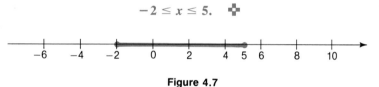

Figure 4.7

4.7 Exercises

Solve each inequality and graph the solution. See Examples 1 and 2.

1. $(m + 2)(m - 5) < 0$

2. $(k - 1)(k + 3) > 0$

3. $(t + 6)(t + 5) \ge 0$

4. $(g - 2)(g - 4) \le 0$

5. $(a + 3)(a - 3) < 0$

6. $(b - 2)(b + 2) > 0$

7. $(a + 6)(a - 7) \ge 0$

8. $(z - 5)(z - 4) \le 0$

9. $m^2 + 5m + 6 > 0$

10. $y^2 - 3y + 2 < 0$

11. $z^2 - 4z - 5 \le 0$

12. $3p^2 - 5p - 2 \le 0$

13. $5m^2 + 3m - 2 < 0$

14. $2k^2 + 7k - 4 > 0$

15. $6r^2 - 5r - 4 < 0$

16. $6r^2 + 7r - 3 > 0$

17. $q^2 - 7q + 6 < 0$

18. $2k^2 - 7k - 15 \le 0$

19. $6m^2 + m - 1 > 0$

20. $30r^2 + 3r - 6 \le 0$

21. $12p^2 + 11p + 2 < 0$

22. $a^2 - 16 < 0$

23. $9m^2 - 36 > 0$

24. $r^2 - 100 \ge 0$

25. $r^2 > 16$

26. $m^2 \ge 25$

The following inequalities are not quadratic inequalities, but they may be solved in a similar manner.

27. $(a + 2)(3a - 1)(a - 4) \geq 0$

28. $(2p - 7)(p - 1)(p + 3) \leq 0$

29. $(r - 2)(r^2 - 3r - 4) < 0$

30. $(m + 5)(m^2 - m - 6) > 0$

Review Exercises Write in lowest terms. See Section 1.1.

31. $\dfrac{8}{12}$ **32.** $\dfrac{5}{10}$ **33.** $\dfrac{14}{42}$ **34.** $\dfrac{18}{32}$

35. $\dfrac{50}{72}$ **36.** $\dfrac{25}{60}$ **37.** $\dfrac{26}{156}$ **38.** $\dfrac{34}{136}$

Chapter 4 *Summary*

Special Factorizations	Difference of two squares	$x^2 - y^2 = (x + y)(x - y)$
	Perfect square trinomials	$x^2 + 2xy + y^2 = (x + y)^2$
		$x^2 - 2xy + y^2 = (x - y)^2$
	Difference of two cubes	$x^3 - y^3 = (x - y)(x^2 + xy + y^2)$
	Sum of two cubes	$x^3 + y^3 = (x + y)(x^2 - xy + y^2)$

Factoring a Polynomial

Step 1 Is there a common factor?

Step 2 How many terms are in the polynomial?

Two terms Check to see whether it is either the difference of two squares or the sum or difference of two cubes.

Three terms Is it a perfect square trinomial? If the trinomial is not a perfect square, check to see whether the coefficient of the squared term is 1. If so, use the method of Section 4.2. If the coefficient of the squared term of the trinomial is not 1, use the general factoring methods of Section 4.3.

Four terms Can the polynomial be factored by grouping?

Step 3 Can any factors be factored further?

Zero-Factor Property

If a and b are real numbers and if $ab = 0$, then $a = 0$ or $b = 0$.

In other words, if the product of two numbers is zero, then at least one of the numbers must be zero.

Solving a Quadratic Equation by Factoring

Step 1 Write the equation in standard form: all terms on one side of the equals sign, with 0 on the other side.

Step 2 Factor completely.

Step 3 Set each factor with a variable equal to 0, and solve the resulting equations.

Step 4 Check each solution in the original equation.

Remember: Not all quadratic equations can be solved by factoring.

Chapter 4   *Review Exercises*

❶ **[4.1]** *Factor out the greatest common factor.*

1. $6 - 18r^5 + 12r^3$

2. $32y^4r^3 - 48y^5r^2 + 24y^7r^5$

Factor by grouping.

3. $6p^2 + 9p + 4p + 6$

4. $12r^2 + 18rq - 10rq - 15q^2$

[4.2] *Factor completely.*

5. $r^2 - 6r - 27$

6. $p^2 + p - 30$

7. $z^2 - 7z - 44$

8. $z^2 - 11zx + 10x^2$

9. $p^7 - p^6q - 2p^5q^2$

10. $5p^6 - 45p^5 + 70p^4$

[4.3] *Factor completely.*

11. $2k^2 - 5k + 2$

12. $3z^2 + 11z - 4$

13. $6r^2 - 5r - 6$

14. $3p^2 - 2pq - 8q^2$

15. $7m^2 + 19mn - 6n^2$

16. $10r^3s + 17r^2s^2 + 6rs^3$

[4.4] *Factor completely.*

17. $100a^2 - 9$

18. $49y^2 - 25z^2$

19. $144p^2 - 36q^2$

20. $y^4 - 625$

21. $16m^2 + 40mn + 25n^2$

22. $25a^2 + 15ab + 9b^2$

23. $54x^3 - 72x^2 + 24x$

24. $125r^3 - 216s^3$

25. $343x^3 + 64$

[4.5] *Solve each equation.*

26. $r^2 + 3r = 10$

27. $m^2 - 5m + 4 = 0$

28. $k^2 = 8k - 15$

29. $p^2 = 12(p - 3)$

30. $(2p+3)(p^2-4p+3) = 0$

31. $x^3 - 9x = 0$

[4.6] *Solve each word problem.*

32. The length of a rectangle is 6 meters more than the width. The area is 40 square meters. Find the width of the rectangle.

33. The product of two consecutive even integers is 4 more than twice their sum. Find the integers.

34. The hypotenuse of a right triangle is 4 meters more than three times the length of the shorter leg. The length of the longer leg is 1 meter less than that of the hypotenuse. Find the length of the shorter leg of the triangle.

[4.7] *Solve each inequality.*

35. $(q + 5)(q - 3) > 0$

36. $(2r - 1)(r + 4) \geq 0$

37. $m^2 - 5m + 6 \leq 0$

38. $z^2 - 8z + 15 < 0$

39. $2p^2 + 5p - 12 \geq 0$

40. $(m + 3)(m - 1)(2m + 5) \geq 0$

❷ *Factor completely.*

41. $r^2 - 4rs - 96s^2$

42. $24k^5 - 20k^4 + 4k^3$

43. $15m^3n^4 - 20m^2n^5 + 50m^3n^6$

44. $1000r^3 - 27$

45. $m^2 + 9$

46. $4y^2 + 3y + 8y + 6$

47. $6m^3 - 21m^2 - 45m$

48. $100y^6 - 50y^3 + 300y^4$

49. $y^2 - 8yz + 15z^2$

50. $9r^2 - 42r + 49$

51. $2z^2 + 9zy - 5y^2$

52. $p^2 + 2pq - 120q^2$

53. $15m^2 + 20mp - 12mp - 16p^2$

Solve the following.

54. $3k^2 - 11k - 20 = 0$

55. $(3z - 1)(z^2 + 3z + 2) = 0$

56. $100b^2 - 49 = 0$

57. The area of a triangle is 12 square meters. The base is 2 meters longer than the height. Find the height of the triangle.

58. A pyramid has a rectangular base with a length that is 2 meters more than the width. The height of the pyramid is 6 meters, and the volume is 48 cubic meters. Find the width of the base.

59. A lot is shaped like a right triangle. The hypotenuse is 3 meters longer than the longer leg. The longer leg is 6 meters more than twice the length of the shorter leg. Find the lengths of the sides of the lot.

60. A bicyclist heading east and a motorist traveling south left an intersection at the same time. When the motorist had gone 17 miles further than the bicyclist, the distance between them was 1 mile more than the distance traveled by the motorist. How far apart were they then? (*Hint:* Draw a sketch.)

Chapter 4   *Test*

Factor completely.

1. $6ab - 36ab^2$

2. $15k^2t + 25kt^2$

3. $16m^2 - 24m^3 + 32m^4$

4. $28pq + 14p + 56p^2q^3$

5. $m^2 - 9m - 3m + 27$

6. $12 - 6a + 2b - ab$

7. $x^2 - 4x - 45$

8. $3p^4q + 18p^3q - 21p^2q$

9. $3a^2 + 13a - 10$

10. $30z^5 - 69z^4 - 15z^3$

11. $12r^2p^2 + 19rp^2 + 5p^2$

12. $6t^2 - tx - 2x^2$

13. $50m^2 - 98$

14. $144a^2 - 169b^2$

15. $a^4 - 625$

16. $4p^2 + 12p + 9$

17. $25z^2 - 10z + 1$

18. $4y^3 + 16y^2 + 16y$

19. $8p^3 - 125$

20. $27r^3 + 64t^6$

21. $m^2 - n^2 - 4m - 4n$

Solve each equation.

22. $3x^2 + 5x = 2$

23. $p(2p + 3) = 20$

24. $(m - 3)(6m^2 - 11m - 10) = 0$

25. $z^3 = 16z$

Solve each word problem.

26. The length of a rectangle is 1 inch less than twice its width. The area is 15 square inches. Find the width of the rectangle.

27. One number is 9 more than another. Their product is 11 more than five times their sum. Find the numbers.

28. The hypotenuse of a right triangle is 7 less than three times the shorter leg. The longer leg is 1 less than twice the shorter leg. Find the length of the shorter leg.

Graph the solution of each inequality.

29. $2p^2 + 5p - 3 \leq 0$

30. $m^2 + 2m - 24 > 0$

Rational Expressions

Objectives

1. Find the values for which a rational expression is undefined.
2. Find the numerical value of a rational expression.
3. Write rational expressions in lowest terms.

The quotient of two integers (with divisor not zero) is called a rational number. In the same way, the quotient of two polynomials with divisor not equal to zero is called a *rational expression*.

A **rational expression** is an expression of the form

$$\frac{P}{Q},$$

where P and Q are polynomials, with $Q \neq 0$.

Examples of rational expressions include

$$\frac{-6x}{x^3 + 8}, \qquad \frac{9x}{y + 3}, \qquad \text{and} \qquad \frac{2m^3}{9}.$$

1   A fraction with a zero denominator is *not* a rational expression, since division by zero is not possible. For that reason, be careful when substituting a number for a variable in the denominator of a rational expression. For example, in

$$\frac{8x^2}{x - 3}$$

x can take on any value except 3. When $x = 3$, the denominator becomes $3 - 3 = 0$, making the expression undefined.

Example 1   Find any values for which the following rational expressions are undefined.

(a) $\dfrac{p + 5}{3p + 2}$

This rational expression is undefined for any value of p that makes the denominator equal to zero. Find these values by solving the equation

$$3p + 2 = 0$$
$$3p = -2$$
$$p = -\frac{2}{3}.$$

Since $p = -2/3$ will make the denominator zero, the given expression is undefined for $-2/3$.

(b) $\dfrac{9m^2}{m^2 - 5m + 6}$

Find the numbers that make the denominator zero by solving the equation

$$m^2 - 5m + 6 = 0.$$

Factor the polynomial and set each factor equal to zero.

$$(m - 2)(m - 3) = 0$$
$$m - 2 = 0 \qquad \text{or} \qquad m - 3 = 0$$
$$m = 2 \qquad \text{or} \qquad m = 3$$

The original expression is undefined for 2 and for 3.

(c) $\dfrac{2r}{r^2 + 1}$

This denominator is never equal to zero, so there are no values for which the rational expression is undefined.   ✚

2   The following example shows how to find the numerical value of a rational expression for a given value of the variable.

Example 2   Find the numerical value of $\dfrac{3x + 6}{2x - 4}$ for the given values of x.

(a) $x = 1$

Find the value of the rational expression by substituting 1 for x.

$$\frac{3x + 6}{2x - 4} = \frac{3(1) + 6}{2(1) - 4} \qquad \text{Let } x = 1$$

$$= \frac{9}{-2}$$

$$= -\frac{9}{2}$$

(b) $x = 2$

Substituting 2 for x makes the denominator zero, so there is no value for the rational expression when $x = 2$.   ✦

3   A rational expression represents a number for each value of the variable that does not make the denominator zero. For this reason, the properties of rational numbers also apply to rational expressions. For example, the fundamental property of rational expressions permits rational expressions to be written in lowest terms.

Fundamental Property of Rational Expressions

If P/Q is a rational expression and if K represents any rational expression, where $K \neq 0$, then

$$\frac{PK}{QK} = \frac{P}{Q}.$$

This property is based on the identity property of multiplication, since

$$\frac{PK}{QK} = \frac{P}{Q} \cdot \frac{K}{K} = \frac{P}{Q} \cdot 1 = \frac{P}{Q}.$$

The next example shows how to write both a rational number and a rational expression in lowest terms. Notice the similarity in the procedures.

Example 3   Write each expression in lowest terms.

(a) $\dfrac{30}{72}$

(b) $\dfrac{14k^2}{2k^3}$

Begin by factoring.

$$\frac{30}{72} = \frac{2 \cdot 3 \cdot 5}{2 \cdot 2 \cdot 2 \cdot 3 \cdot 3}$$

$$\frac{14k^2}{2k^3} = \frac{2 \cdot 7 \cdot k \cdot k}{2 \cdot k \cdot k \cdot k}$$

Group any factors common to the numerator and denominator.

$$\frac{30}{72} = \frac{5(2 \cdot 3)}{2 \cdot 2 \cdot 3(2 \cdot 3)}$$

$$\frac{14k^2}{2k^3} = \frac{7(2 \cdot k \cdot k)}{k(2 \cdot k \cdot k)}$$

Use the fundamental property.

$$\frac{30}{72} = \frac{5}{2 \cdot 2 \cdot 3} = \frac{5}{12}$$

$$\frac{14k^2}{2k^3} = \frac{7}{k} \quad \clubsuit$$

Example 4   Write each rational expression in lowest terms.

(a) $\dfrac{3x - 12}{5x - 20}$

Begin by factoring both numerator and denominator. Then use the fundamental property.

$$\frac{3x - 12}{5x - 20} = \frac{3(x - 4)}{5(x - 4)} = \frac{3}{5}$$

(b) $\dfrac{m^2 + 2m - 8}{2m^2 - m - 6}$

Always begin by factoring both numerator and denominator, if possible. Then use the fundamental property.

$$\frac{m^2 + 2m - 8}{2m^2 - m - 6} = \frac{(m + 4)(m - 2)}{(2m + 3)(m - 2)} = \frac{m + 4}{2m + 3} \quad \clubsuit$$

In Example 4(a), the rational expression $\dfrac{3x - 12}{5x - 20}$ is restricted to values of x not equal to 4. For this reason,

$$\frac{3x - 12}{5x - 20} = \frac{3}{5} \qquad \text{for} \quad x \neq 4.$$

Similarly, in Example 4(b),

$$\frac{m^2 + 2m - 8}{2m^2 - m - 6} = \frac{m + 4}{2m + 3} \qquad \text{for} \quad x \neq -\frac{3}{2} \quad \text{or} \quad 2.$$

From now on such statements of equality will be written with the understanding that they apply only to those real numbers that make neither denominator equal to zero.

Example 5   Write $\dfrac{x - y}{y - x}$ in lowest terms.

At first glance, there does not seem to be any way in which $x - y$ and $y - x$ can be factored to get a common factor. However,

$$y - x = -1(-y + x) = -1(x - y).$$

With these factors, use the fundamental property to simplify the rational expression.

$$\frac{x - y}{y - x} = \frac{1(x - y)}{-1(x - y)} = \frac{1}{-1} = -1 \quad \clubsuit$$

As suggested by Example 5, the quotient of two nonzero expressions that differ only in sign is -1.

Example 6   Write each rational expression in lowest terms.

(a) $\dfrac{2 - m}{m - 2}$

Since $2 - m$ and $m - 2$ (or $-2 + m$) differ only in sign,

$$\frac{2 - m}{m - 2} = -1.$$

(b) $\dfrac{3 + r}{3 - r}$

The quantities $3 + r$ and $3 - r$ do *not* differ only in sign. This rational expression cannot be written in simpler form. $\clubsuit$

5.1   Exercises

Find any values for which the following expressions are undefined. See Example 1.

1. $\dfrac{3}{4x}$ **2.** $\dfrac{5}{2x}$ **3.** $\dfrac{x^2}{x + 5}$ **4.** $\dfrac{3x^2}{2x - 1}$

5. $\dfrac{a + 4}{a^2 - 8a + 15}$ **6.** $\dfrac{p + 6}{p^2 - p - 12}$ **7.** $\dfrac{9y}{y^2 + 16}$ **8.** $\dfrac{12z}{z^2 + 100}$

Find the numerical value of each expression when (a) $x = 2$ and (b) $x = -3$. See Example 2.

9. $\dfrac{4x - 2}{3x}$ **10.** $\dfrac{-5x + 1}{2x}$ **11.** $\dfrac{4x^2 - 2x}{3x}$ **12.** $\dfrac{x^2 - 1}{x}$

13. $\dfrac{(-8x)^2}{3x + 9}$ **14.** $\dfrac{2x^2 + 5}{3 + x}$ **15.** $\dfrac{x + 8}{x^2 - 4x + 2}$ **16.** $\dfrac{2x - 1}{x^2 - 7x + 3}$

17. $\dfrac{5x^2}{6 - 3x - x^2}$ **18.** $\dfrac{-2x^2}{8 + x - x^2}$ **19.** $\dfrac{2x + 5}{x^2 + 3x - 10}$ **20.** $\dfrac{3x - 7}{2x^2 - 3x - 2}$

Write each expression in lowest terms. See Examples 3 and 4.

21. $\dfrac{12k^2}{6k}$ **22.** $\dfrac{9m^3}{3m}$ **23.** $\dfrac{-12m^2p}{9mp^2}$ **24.** $\dfrac{6a^2b^3}{-24a^3b^2}$

25. $\dfrac{6(y + 2)}{8(y + 2)}$ **26.** $\dfrac{9(m + 2)}{5(m + 2)}$ **27.** $\dfrac{(x + 1)(x - 1)}{(x + 1)^2}$ **28.** $\dfrac{3(t + 5)}{(t + 5)(t - 1)}$

29. $\dfrac{12m^2 - 9}{3}$ **30.** $\dfrac{15p^2 - 10}{5}$ **31.** $\dfrac{32y + 20}{24}$ **32.** $\dfrac{40q - 25}{20}$

33. $\dfrac{2q - 6}{5q - 10}$ **34.** $\dfrac{9p + 12}{21p + 28}$ **35.** $\dfrac{m^2 - n^2}{m + n}$ **36.** $\dfrac{a^2 - b^2}{a - b}$

37. $\dfrac{5m^2 - 5m}{10m - 10}$ **38.** $\dfrac{3y^2 - 3y}{2(y - 1)}$ **39.** $\dfrac{16r^2 - 4s^2}{4r - 2s}$ **40.** $\dfrac{11s^2 - 22s^3}{6 - 12s}$

41. $\dfrac{m^2 - 4m + 4}{m^2 + m - 6}$ **42.** $\dfrac{a^2 - a - 6}{a^2 + a - 12}$ **43.** $\dfrac{x^2 + 3x - 4}{x^2 - 1}$ **44.** $\dfrac{8m^2 + 6m - 9}{16m^2 - 9}$

Write each expression in lowest terms. See Examples 5 and 6.

45. $\dfrac{m - 5}{5 - m}$ **46.** $\dfrac{3 - p}{p - 3}$ **47.** $\dfrac{-a + b}{b - a}$ **48.** $\dfrac{b - a}{a - b}$

49. $\dfrac{x^2 - 1}{1 - x}$ **50.** $\dfrac{p^2 - q^2}{q - p}$ **51.** $\dfrac{m^2 - 4m}{4m - m^2}$ **52.** $\dfrac{s^2 - r^2}{r^2 - s^2}$

Write each expression in lowest terms.

53. $\dfrac{a + b + a^2 + ba}{ab + b^2}$ **54.** $\dfrac{2p + pq - 8 - 4q}{8 + 4q}$

55. $\dfrac{m^2 - n^2 - 4m - 4n}{2m - 2n - 8}$ **56.** $\dfrac{x^2y + y + x^2z + z}{xy + xz}$

57. $\dfrac{b^3 - a^3}{a^2 - b^2}$ **58.** $\dfrac{k^3 + 8}{k^2 - 4}$

59. $\dfrac{z^3 + 27}{z^3 - 3z^2 + 9z}$ **60.** $\dfrac{1 - 8r^3}{8r^2 + 4r + 2}$

Review Exercises   Multiply or divide as indicated. See Section 1.1.

61. $\dfrac{3}{4} \cdot \dfrac{5}{8}$ **62.** $\dfrac{7}{10} \cdot \dfrac{3}{5}$ **63.** $\dfrac{8}{15} \cdot \dfrac{20}{3}$ **64.** $\dfrac{15}{4} \cdot \dfrac{12}{5}$

65. $\dfrac{6}{5} \div \dfrac{3}{10}$ **66.** $\dfrac{21}{8} \div \dfrac{7}{4}$ **67.** $\dfrac{27}{8} \div \dfrac{5}{12}$ **68.** $\dfrac{2}{3} \div \dfrac{4}{9}$

5.2   Multiplication and Division of Rational Expressions

Objectives

1 Multiply rational expressions.

2 Divide rational expressions.

1 The product of two fractions is found by multiplying the numerators and multiplying the denominators. Rational expressions are multiplied in the same way.

Multiplying Rational Expressions	The product of the rational expressions P/Q and R/S is $$\frac{P}{Q} \cdot \frac{R}{S} = \frac{PR}{QS}.$$

The next example shows the multiplication of both two rational numbers and two rational expressions. This parallel discussion lets you compare the steps.

Example 1   Multiply. Write answers in lowest terms.

(a) $\dfrac{3}{10} \cdot \dfrac{5}{9}$

(b) $\dfrac{6}{x} \cdot \dfrac{x^2}{12}$

Find the product of the numerators and the product of the denominators.

$$\frac{3}{10} \cdot \frac{5}{9} = \frac{3 \cdot 5}{10 \cdot 9} \qquad\qquad \frac{6}{x} \cdot \frac{x^2}{12} = \frac{6 \cdot x^2}{x \cdot 12}$$

Use the fundamental property to write each product in lowest terms.

$$\frac{3}{10} \cdot \frac{5}{9} = \frac{3 \cdot 5}{2 \cdot 5 \cdot 3 \cdot 3} = \frac{1}{6} \qquad\qquad \frac{6}{x} \cdot \frac{x^2}{12} = \frac{6 \cdot x \cdot x}{2 \cdot 6 \cdot x} = \frac{x}{2}$$

Notice in the second step above that the products were left in factored form since common factors must be identified to write the product in lowest terms. ❖

Example 2   Find the product of $\dfrac{x + y}{2x}$ and $\dfrac{x^2}{(x + y)^2}$.

Use the definition of multiplication.

$$\frac{x + y}{2x} \cdot \frac{x^2}{(x + y)^2} = \frac{(x + y)x^2}{2x(x + y)^2}$$

$$= \frac{(x + y)x \cdot x}{2x(x + y)(x + y)}$$

$$= \frac{x}{2(x + y)} \cdot \frac{x(x + y)}{x(x + y)} = \frac{x}{2(x + y)}$$

In the last steps, the product was written in lowest terms by factoring and using the fundamental property of rational expressions. ✛

Example 3 Find the product of $\dfrac{x^2 + 3x}{x^2 - 3x - 4}$ and $\dfrac{x^2 - 5x + 4}{x^2 + 2x - 3}$.

First factor the numerators and denominators whenever possible. Then use the fundamental property to write the product in lowest terms.

$$\frac{x^2 + 3x}{x^2 - 3x - 4} \cdot \frac{x^2 - 5x + 4}{x^2 + 2x - 3} = \frac{x(x + 3)}{(x - 4)(x + 1)} \cdot \frac{(x - 4)(x - 1)}{(x + 3)(x - 1)}$$

$$= \frac{x(x + 3)(x - 4)(x - 1)}{(x - 4)(x + 1)(x + 3)(x - 1)}$$

$$= \frac{x}{x + 1} \quad ✛$$

2 The fraction a/b is divided by the nonzero fraction c/d by multiplying a/b and the reciprocal of c/d, which is d/c. Division of rational expressions is defined in the same way.

Dividing Rational Expressions

If P/Q and R/S are any two rational expressions, with $R/S \neq 0$, then

$$\frac{P}{Q} \div \frac{R}{S} = \frac{P}{Q} \cdot \frac{S}{R} = \frac{PS}{QR}.$$

The next example shows the division of two rational numbers and the division of two rational expressions.

Example 4 Divide the following fractions. Write answers in lowest terms.

(a) $\dfrac{5}{8} \div \dfrac{7}{16}$

(b) $\dfrac{y}{y + 3} \div \dfrac{4y}{y + 5}$

Multiply the first expression and the reciprocal of the second.

$$\frac{5}{8} \div \frac{7}{16} = \frac{5}{8} \cdot \frac{16}{7} \quad \text{Reciprocal of } \frac{7}{16}$$

$$= \frac{5 \cdot 16}{8 \cdot 7}$$

$$= \frac{5 \cdot 8 \cdot 2}{8 \cdot 7}$$

$$= \frac{5 \cdot 2}{7}$$

$$= \frac{10}{7}$$

$$\frac{y}{y + 3} \div \frac{4y}{y + 5}$$

$$= \frac{y}{y + 3} \cdot \frac{y + 5}{4y} \quad \text{Reciprocal of } \frac{4y}{y + 5}$$

$$= \frac{y(y + 5)}{(y + 3)(4y)}$$

$$= \frac{y + 5}{4(y + 3)} \quad ✛$$

Example 5 Divide $\dfrac{(3m)^2}{(2p)^3} \div \dfrac{6m^3}{16p^2}.$

Use the properties of exponents as necessary.

$$\frac{(3m)^2}{(2p)^3} \div \frac{6m^3}{16p^2} = \frac{9m^2}{8p^3} \div \frac{6m^3}{16p^2}$$

$$= \frac{9m^2}{8p^3} \cdot \frac{16p^2}{6m^3} \qquad \text{Multiply by the reciprocal}$$

$$= \frac{9 \cdot 16m^2p^2}{8 \cdot 6p^3m^3} = \frac{3}{mp} \qquad \text{Fundamental property} \quad \clubsuit$$

Example 6 Divide: $\dfrac{x^2 - 4}{(x + 3)(x - 2)} \div \dfrac{(x + 2)(x + 3)}{2x}.$

First, use the definition of division.

$$\frac{x^2 - 4}{(x + 3)(x - 2)} \div \frac{(x + 2)(x + 3)}{2x} = \frac{x^2 - 4}{(x + 3)(x - 2)} \cdot \frac{2x}{(x + 2)(x + 3)}$$

Next, be sure all numerators and all denominators are factored.

$$= \frac{(x + 2)(x - 2)}{(x + 3)(x - 2)} \cdot \frac{2x}{(x + 2)(x + 3)}$$

Now multiply numerators and denominators and simplify.

$$= \frac{(x + 2)(x - 2)(2x)}{(x + 3)(x - 2)(x + 2)(x + 3)}$$

$$= \frac{2x}{(x + 3)^2} \quad \clubsuit$$

Example 7 Divide: $\dfrac{m^2 - 4}{m^2 - 1} \div \dfrac{2m^2 + 4m}{1 - m}.$

Use the definition of division; then factor.

$$\frac{m^2 - 4}{m^2 - 1} \div \frac{2m^2 + 4m}{1 - m} = \frac{m^2 - 4}{m^2 - 1} \cdot \frac{1 - m}{2m^2 + 4m}$$

$$= \frac{(m + 2)(m - 2)}{(m + 1)(m - 1)} \cdot \frac{1 - m}{2m(m + 2)}$$

As shown in Section 5.1, $\dfrac{1 - m}{m - 1} = -1$, so

$$\frac{(m + 2)(m - 2)}{(m + 1)(m - 1)} \cdot \frac{1 - m}{2m(m + 2)} = \frac{-1(m - 2)}{2m(m + 1)} = \frac{2 - m}{2m(m + 1)}. \quad \clubsuit$$

5.2   Exercises

Multiply or divide. Write each answer in lowest terms. See Examples 1 and 5.

1. $\dfrac{9m^2}{16} \cdot \dfrac{4}{3m}$

2. $\dfrac{21z^4}{8} \cdot \dfrac{12}{7z^3}$

3. $\dfrac{4p^2}{8p} \cdot \dfrac{3p^3}{16p^4}$

4. $\dfrac{6x^3}{9x} \cdot \dfrac{12x}{x^2}$

5. $\dfrac{8a^4}{12a^3} \cdot \dfrac{9a^5}{3a^2}$

6. $\dfrac{14p^5}{2p^2} \cdot \dfrac{8p^6}{28p^3}$

7. $\dfrac{3r^2}{9r^3} \div \dfrac{8r^4}{6r^5}$

8. $\dfrac{25m^{10}}{9m^5} \div \dfrac{15m^6}{10m^4}$

9. $\dfrac{3m^2}{(4m)^3} \div \dfrac{9m^3}{32m^4}$

10. $\dfrac{5x^3}{(4x)^2} \div \dfrac{15x^2}{8x^4}$

11. $\dfrac{-6r^4}{3r^5} \div \dfrac{(2r^2)^2}{-4}$

12. $\dfrac{-10a^6}{3a^2} \div \dfrac{(3a)^3}{81a}$

Multiply or divide. Write each answer in lowest terms. See Examples 2–4, 6, and 7.

13. $\dfrac{a + b}{2} \cdot \dfrac{12}{(a + b)^2}$

14. $\dfrac{3(x - 1)}{y} \cdot \dfrac{2y}{5(x - 1)^2}$

15. $\dfrac{2k + 8}{6} \div \dfrac{3k + 12}{2}$

16. $\dfrac{5m + 25}{10} \cdot \dfrac{12}{6m + 30}$

17. $\dfrac{9y - 18}{6y + 12} \cdot \dfrac{3y + 6}{15y - 30}$

18. $\dfrac{12p + 24}{36p - 36} \div \dfrac{6p + 12}{8p - 8}$

19. $\dfrac{3r + 12}{8} \cdot \dfrac{16r}{9r + 36}$

20. $\dfrac{2r + 2p}{8z} \div \dfrac{r^2 + rp}{72}$

21. $\dfrac{y^2 - 16}{y + 3} \div \dfrac{y - 4}{y^2 - 9}$

22. $\dfrac{9(y - 4)^2}{8(z + 3)^2} \cdot \dfrac{16(z + 3)}{3(y - 4)}$

23. $\dfrac{6(m + 2)}{3(m - 1)^2} \div \dfrac{(m + 2)^2}{9(m - 1)}$

24. $\dfrac{4y + 12}{2y - 10} \div \dfrac{y^2 - 9}{y^2 - y - 20}$

25. $\dfrac{2 - y}{8} \cdot \dfrac{7}{y - 2}$

26. $\dfrac{9 - 2z}{3} \cdot \dfrac{9}{2z - 9}$

27. $\dfrac{8 - r}{8 + r} \div \dfrac{r - 8}{r + 8}$

28. $\dfrac{6 - y}{6 + 2y} \div \dfrac{6 + y}{3 + y}$

29. $\dfrac{m^2 - 16}{4 - m} \cdot \dfrac{-4 + m}{-4 - m}$

30. $\dfrac{6r - 18}{3r^2 + 2r - 8} \cdot \dfrac{12r - 16}{4r - 12}$

31. $\dfrac{k^2 - k - 6}{k^2 + k - 12} \div \dfrac{k^2 + 2k - 3}{k^2 + 3k - 4}$

32. $\dfrac{m^2 + 3m + 2}{m^2 + 5m + 4} \cdot \dfrac{m^2 + 10m + 24}{m^2 + 5m + 6}$

33. $\dfrac{z^2 - z - 6}{z^2 - 2z - 8} \cdot \dfrac{z^2 + 7z + 12}{z^2 - 9}$

34. $\dfrac{y^2 + y - 2}{y^2 + 3y - 4} \div \dfrac{y + 2}{y + 3}$

35. $\dfrac{2m^2 - 5m - 12}{m^2 - 10m + 24} \div \dfrac{4m^2 - 9}{m^2 - 9m + 18}$

36. $\dfrac{2m^2 + 7m + 3}{m^2 - 9} \cdot \dfrac{m^2 - 3m}{2m^2 + 11m + 5}$

37. $\dfrac{m^2 + 2mp - 3p^2}{m^2 - 3mp + 2p^2} \div \dfrac{m^2 + 4mp + 3p^2}{m^2 + 2mp - 8p^2}$

38. $\dfrac{r^2 + rs - 12s^2}{r^2 - rs - 20s^2} \div \dfrac{r^2 - 2rs - 3s^2}{r^2 + rs - 30s^2}$

39. $\dfrac{(x + 1)^3(x + 4)}{x^2 + 5x + 4} \div \dfrac{x^2 + 2x + 1}{x^2 + 3x + 2}$

40. $\dfrac{(q - 3)^4(q + 2)}{q^2 + 3q + 2} \div \dfrac{q^2 - 6q + 9}{q^2 + 4q + 4}$

41. $\left(\dfrac{x^2 + 10x + 25}{x^2 + 10x} \cdot \dfrac{10x}{x^2 + 15x + 50} \right) \div \dfrac{x + 5}{x + 10}$

42. $\left(\dfrac{m^2 - 12m + 32}{8m} \cdot \dfrac{m^2 - 8m}{m^2 - 8m + 16} \right) \div \dfrac{m - 8}{m - 4}$

43. $\dfrac{3a - 3b - a^2 + b^2}{4a^2 - 4ab + b^2} \cdot \dfrac{4a^2 - b^2}{2a^2 - ab - b^2}$

44. $\dfrac{4r^2 - t^2 + 10r - 5t}{2r^2 + rt + 5r} \cdot \dfrac{4r^3 + 4r^2t + rt^2}{2r + t}$

45. $\dfrac{-x^3 - y^3}{x^2 - 2xy + y^2} \div \dfrac{3y^2 - 3xy}{x^2 - y^2}$

46. $\dfrac{b^3 - 8a^3}{4a^3 + 4a^2b + ab^2} \div \dfrac{4a^2 + 2ab + b^2}{-a^3 - ab^3}$

Review Exercises   *Write the prime factored form of each number. See Section 4.1.*

47. 12   **48.** 50   **49.** 72   **50.** 105

Factor each expression. See Sections 4.1 and 4.3.

51. $x^2 - 4x$ **52.** $p^2 + 6p$ **53.** $2y^2 + 5y - 3$ **54.** $8z^2 + 2z - 15$

5.3   The Least Common Denominator

Objectives

1 Find least common denominators.

2 Rewrite rational expressions with given denominators.

1   Just as with rational numbers, adding or subtracting rational expressions (to be discussed in the next section), often requires a **least common denominator;** the least expression that all denominators divide into without a remainder. For example, the least common denominator for 2/9 and 5/12 is 36, since 36 is the smallest number that both 9 and 12 divide into. Least common denominators are found by a procedure similar to that used in Chapter 4 for finding the greatest common factor, except that

> least common denominators are found by **multiplying together each different factor the *greatest* number of times it appears in any denominator.**

In Example 1, the least common denominator is found both for numerical denominators and algebraic denominators.

Example 1   Find the least common denominator for each pair of fractions.

   (a) $\dfrac{1}{24}, \dfrac{7}{15}$ **(b)** $\dfrac{1}{8x}, \dfrac{3}{10x}$

   Write each denominator in factored form, with numerical coefficients in prime factored form.

$$24 = 2 \cdot 2 \cdot 2 \cdot 3 \qquad\qquad 8x = 2 \cdot 2 \cdot 2 \cdot x$$
$$15 = 3 \cdot 5 \qquad\qquad 10x = 2 \cdot 5 \cdot x$$

The least common denominator is found by taking each different factor the greatest number of times it appears as a factor in any of the denominators.

The factor 2 appears three times in one product and not at all in the other, so the greatest number of times 2 appears is three. The greatest number of times both 3 and 5 appear is one.

$$\text{least common denominator}$$
$$= 2 \cdot 2 \cdot 2 \cdot 3 \cdot 5 = 120$$

Here 2 appears three times in one product and once in the other, so the greatest number of times the 2 appears is three. The greatest number of times the 5 appears is one, and the greatest number of times x appears in either product is one.

$$\text{least common denominator}$$
$$= 2 \cdot 2 \cdot 2 \cdot 5 \cdot x = 40x \quad \clubsuit$$

Example 2   Find the least common denominator for $\dfrac{5}{6r^2}$ and $\dfrac{3}{4r^3}$.

Factor each denominator.

$$6r^2 = 2 \cdot 3 \cdot r^2 \qquad 4r^3 = 2 \cdot 2 \cdot r^3$$

The greatest number of times 2 appears is two, the greatest number of times 3 appears is one, and the greatest number of times r appears is three; therefore,

$$\text{least common denominator} = 2 \cdot 2 \cdot 3 \cdot r^3 = 12r^3. \quad \clubsuit$$

Example 3   Find the least common denominator.

(a) $\dfrac{6}{5m}, \dfrac{4}{m^2 - 3m}$

Factor each denominator.

$$5m = 5 \cdot m \qquad m^2 - 3m = m(m - 3)$$

Take each different factor the greatest number of times it appears as a factor.

$$\text{least common denominator} = 5 \cdot m \cdot (m - 3) = 5m(m - 3)$$

Since m is *not* a factor of $m - 3$, both factors, m and $m - 3$, must appear in the least common denominator.

(b) $\dfrac{1}{r^2 - 4r - 5}, \dfrac{1}{r^2 - r - 20}$

Factor each denominator.

$$r^2 - 4r - 5 = (r - 5)(r + 1)$$
$$r^2 - r - 20 = (r - 5)(r + 4)$$

The least common denominator is

$$(r - 5)(r + 1)(r + 4).$$

(c) $\dfrac{1}{q - 5}, \dfrac{3}{5 - q}$

The expression $5 - q$ can be written as $-1(q - 5)$, since

$$-1(q - 5) = -q + 5 = 5 - q.$$

Because of this, either $q - 5$ or $5 - q$ can be used as the least common denominator. ✛

The steps in finding a least common denominator are listed below.

Finding the Least Common Denominator

Step 1 Completely factor all denominators.
Step 2 Take each different factor the *greatest* number of times that it appears as a factor in any of the denominators.
Step 3 The least common denominator is the product of all factors to the greatest power found in Step 2.

2 Once the least common denominator has been found, use the fundamental property to rewrite rational expressions with the least common denominator. The next example shows how to do this with both numerical and algebraic fractions.

Example 4 Rewrite each expression with the indicated denominator.

(a) $\dfrac{3}{8} = \dfrac{}{40}$ 　　　　　　　　　　(b) $\dfrac{9k}{25} = \dfrac{}{50k}$

For each example, decide what quantity the denominator on the left must be multiplied by to get the denominator on the right. (Find this quantity by dividing.)

Multiply by $\dfrac{5}{5}$ to get a denominator of 40. 　｜ 　Get a denominator of $50k$ by multiplying by $\dfrac{2k}{2k}$.

$$\dfrac{3}{8} = \dfrac{3}{8} \cdot \dfrac{5}{5} = \dfrac{15}{40}$$ 　｜ 　$$\dfrac{9k}{25} = \dfrac{9k}{25} \cdot \dfrac{2k}{2k} = \dfrac{18k^2}{50k}$$ ✛

Example 5 Rewrite the following rational expression with the indicated denominator.

$$\dfrac{12p}{p^2 + 8p} = \dfrac{}{p(p + 8)(p - 4)}$$

Factor $p^2 + 8p$ as $p(p + 8)$. This shows that the first denominator must be multiplied by $p - 4$.

$$\dfrac{12p}{p^2 + 8p} = \dfrac{12p}{p(p + 8)} \cdot \dfrac{p - 4}{p - 4} = \dfrac{12p(p - 4)}{p(p + 8)(p - 4)}$$ ✛

5.3   Exercises

Find the least common denominator for the following fractions. See Examples 1–3.

1. $\dfrac{5}{12}, \dfrac{7}{10}$

2. $\dfrac{1}{4}, \dfrac{5}{6}$

3. $\dfrac{7}{15}, \dfrac{11}{20}, \dfrac{5}{24}$

4. $\dfrac{9}{10}, \dfrac{12}{25}, \dfrac{11}{35}$

5. $\dfrac{17}{100}, \dfrac{13}{120}, \dfrac{29}{180}$

6. $\dfrac{17}{250}, \dfrac{1}{300}, \dfrac{127}{360}$

7. $\dfrac{9}{x^2}, \dfrac{8}{x^5}$

8. $\dfrac{2}{m^7}, \dfrac{3}{m^8}$

9. $\dfrac{2}{5p}, \dfrac{5}{6p}$

10. $\dfrac{4}{15k}, \dfrac{3}{4k}$

11. $\dfrac{7}{15y^2}, \dfrac{5}{36y^4}$

12. $\dfrac{4}{25m^3}, \dfrac{7}{10m^4}$

13. $\dfrac{1}{5a^2b^3}, \dfrac{2}{15a^5b}$

14. $\dfrac{7}{3r^4s^5}, \dfrac{2}{9r^6s^8}$

15. $\dfrac{7}{6p}, \dfrac{3}{4p - 8}$

16. $\dfrac{7}{8k}, \dfrac{13}{12k - 24}$

17. $\dfrac{5}{32r^2}, \dfrac{9}{16r - 32}$

18. $\dfrac{13}{18m^3}, \dfrac{8}{9m - 36}$

19. $\dfrac{7}{6r - 12}, \dfrac{4}{9r - 18}$

20. $\dfrac{4}{5p - 30}, \dfrac{5}{6p - 36}$

21. $\dfrac{5}{12p + 60}, \dfrac{1}{p^2 + 5p}$

22. $\dfrac{1}{r^2 + 7r}, \dfrac{3}{5r + 35}$

23. $\dfrac{3}{8y + 16}, \dfrac{2}{y^2 + 3y + 2}$

24. $\dfrac{2}{9m - 18}, \dfrac{9}{m^2 - 7m + 10}$

25. $\dfrac{2}{m - 3}, \dfrac{4}{3 - m}$

26. $\dfrac{7}{8 - a}, \dfrac{2}{a - 8}$

27. $\dfrac{9}{p - q}, \dfrac{3}{q - p}$

28. $\dfrac{4}{z - x}, \dfrac{8}{x - z}$

29. $\dfrac{6}{a^2 + 6a}, \dfrac{4}{a^2 + 3a - 18}$

30. $\dfrac{3}{y^2 - 5y}, \dfrac{2}{y^2 - 2y - 15}$

31. $\dfrac{5}{k^2 + 2k - 35}, \dfrac{3}{k^2 + 3k - 40}$

32. $\dfrac{9}{z^2 + 4z - 12}, \dfrac{1}{z^2 + z - 30}$

33. $\dfrac{3}{2y^2 + 7y - 4}, \dfrac{7}{2y^2 - 7y + 3}$

34. $\dfrac{2}{5a^2 + 13a - 6}, \dfrac{6}{5a^2 - 22a + 8}$

35. $\dfrac{1}{6r^2 - r - 15}, \dfrac{4}{3r^2 - 8r + 5}$

36. $\dfrac{8}{2m^2 - 11m + 14}, \dfrac{2}{2m^2 - m - 21}$

Rewrite each expression with the given denominator. See Examples 4 and 5.

37. $\dfrac{7}{11} = \dfrac{}{66}$

38. $\dfrac{5}{8} = \dfrac{}{56}$

39. $\dfrac{-11}{m} = \dfrac{}{8m}$

40. $\dfrac{-5}{z} = \dfrac{}{6z}$

41. $\dfrac{12}{35y} = \dfrac{}{70y^3}$

42. $\dfrac{17}{9r} = \dfrac{}{36r^2}$

43. $\dfrac{15m^2}{8k} = \dfrac{}{32k^4}$

44. $\dfrac{5t^2}{3y} = \dfrac{—}{9y^2}$

45. $\dfrac{19z}{2z - 6} = \dfrac{}{6z - 18}$

46. $\dfrac{2r}{5r - 5} = \dfrac{}{15r - 15}$

47. $\dfrac{-2a}{9a - 18} = \dfrac{}{18a - 36}$

48. $\dfrac{-5y}{6y + 18} = \dfrac{}{24y + 72}$

49. $\dfrac{6}{k^2 - 4k} = \dfrac{}{k(k - 4)(k + 1)}$

50. $\dfrac{15}{m^2 - 9m} = \dfrac{}{m(m - 9)(m + 8)}$

51. $\dfrac{36r}{r^2 - r - 6} = \dfrac{}{(r - 3)(r + 2)(r + 1)}$

52. $\dfrac{4m}{m^2 - 8m + 15} = \dfrac{}{(m - 5)(m - 3)(m + 2)}$

53. $\dfrac{a + 2b}{2a^2 + ab - b^2} = \dfrac{}{2a^3b + a^2b^2 - ab^3}$

54. $\dfrac{m - 4}{6m^2 + 7m - 3} = \dfrac{}{12m^3 + 14m^2 - 6m}$

55. $\dfrac{4r - t}{r^2 + rt + t^2} = \dfrac{}{t^3 - r^3}$

56. $\dfrac{3x - 1}{x^2 + 2x + 4} = \dfrac{}{x^3 - 8}$

57. $\dfrac{2(z - y)}{y^2 + yz + z^2} = \dfrac{}{y^4 - z^3y}$

58. $\dfrac{2p + 3q}{p^2 + 2pq + q^2} = \dfrac{}{(p + q)(p^3 + q^3)}$

Review Exercises *Add or subtract as indicated. See Section 1.1.*

59. $\dfrac{2}{3} + \dfrac{5}{3}$

60. $\dfrac{9}{5} + \dfrac{2}{5}$

61. $\dfrac{1}{2} + \dfrac{2}{5}$

62. $\dfrac{5}{4} + \dfrac{1}{3}$

63. $\dfrac{8}{10} - \dfrac{2}{15}$

64. $\dfrac{7}{12} - \dfrac{6}{15}$

65. $\dfrac{5}{24} + \dfrac{1}{18}$

66. $\dfrac{11}{36} + \dfrac{7}{45}$

5.4 Addition and Subtraction of Rational Expressions

Objectives

1 Add rational expressions having the same denominator.

2 Add rational expressions having different denominators.

3 Subtract rational expressions.

1 The sum of two rational expressions is found with a procedure similar to that for adding two fractions.

Adding Rational Expressions

If P/Q and R/Q are rational expressions, then

$$\frac{P}{Q} + \frac{R}{Q} = \frac{P + R}{Q}.$$

Again, the first example shows how the addition of rational expressions compares with that of rational numbers.

Example 1 Add.

(a) $\dfrac{4}{7} + \dfrac{2}{7}$

(b) $\dfrac{3x}{x+1} + \dfrac{2x}{x+1}$

Since the denominators are the same, the sum is found by adding the two numerators and keeping the same (common) denominator.

$$\frac{4}{7} + \frac{2}{7} = \frac{4+2}{7}$$

$$= \frac{6}{7}$$

$$\frac{3x}{x+1} + \frac{2x}{x+1} = \frac{3x+2x}{x+1}$$

$$= \frac{5x}{x+1} \quad \clubsuit$$

2 Use the steps given below to add two rational expressions with different denominators. These are the same steps that are used to add fractions with different denominators.

Adding with Different Denominators	*Step 1* Find the least common denominator. *Step 2* Rewrite each rational expression as a fraction with the least common denominator as the denominator. *Step 3* Add the numerators to get the numerator of the sum. The least common denominator is the denominator of the sum. *Step 4* Write the answer in lowest terms.

Example 2 Add.

(a) $\dfrac{1}{12} + \dfrac{7}{15}$

(b) $\dfrac{2}{3y} + \dfrac{1}{4y}$

First find the least common denominator, using the methods of the last section.

least common denominator
$= 2^2 \cdot 3 \cdot 5 = 60$

least common denominator
$= 2^2 \cdot 3 \cdot y = 12y$

Now rewrite each rational expression as a fraction with the least common denominator, either 60 or $12y$, as the denominator.

$$\frac{1}{12} + \frac{7}{15} = \frac{1(5)}{12(5)} + \frac{7(4)}{15(4)}$$

$$= \frac{5}{60} + \frac{28}{60}$$

$$\frac{2}{3y} + \frac{1}{4y} = \frac{2(4)}{3y(4)} + \frac{1(3)}{4y(3)}$$

$$= \frac{8}{12y} + \frac{3}{12y}$$

Since the fractions now have common denominators, add the numerators. (Write in lowest terms if necessary.)

$$\frac{5}{60} + \frac{28}{60} = \frac{5 + 28}{60}$$

$$= \frac{33}{60} = \frac{11}{20}$$

$$\frac{8}{12y} + \frac{3}{12y} = \frac{8 + 3}{12y}$$

$$= \frac{11}{12y} \quad \clubsuit$$

Example 3   Add $\dfrac{2x}{x^2 - 1}$ and $\dfrac{-1}{x + 1}$.

Find the least common denominator by factoring both denominators.

$$x^2 - 1 = (x + 1)(x - 1); \quad x + 1 \text{ cannot be factored}$$

Write the sum with denominators in factored form as

$$\frac{2x}{(x + 1)(x - 1)} + \frac{-1}{x + 1}.$$

The least common denominator is $(x + 1)(x - 1)$. Here only the second fraction must be changed. Multiply the numerator and denominator of the second fraction by $x - 1$.

$$\frac{2x}{(x + 1)(x - 1)} + \frac{-1(x - 1)}{(x + 1)(x - 1)} \qquad \text{Multiply by } \frac{x - 1}{x - 1}$$

With both denominators now the same, add the numerators.

$$\frac{2x - (x - 1)}{(x + 1)(x - 1)} = \frac{2x - x + 1}{(x + 1)(x - 1)} \qquad \text{Distributive property}$$

$$= \frac{x + 1}{(x + 1)(x - 1)}$$

$$= \frac{1}{x - 1} \qquad \text{Write in lowest terms} \quad \clubsuit$$

Example 4   Add $\dfrac{2x}{x^2 + 5x + 6}$ and $\dfrac{x + 1}{x^2 + 2x - 3}$.

Begin by factoring the denominators completely.

$$\frac{2x}{(x + 2)(x + 3)} + \frac{x + 1}{(x + 3)(x - 1)}$$

The least common denominator is $(x + 2)(x + 3)(x - 1)$. By the fundamental property of rational expressions,

$$\frac{2x}{(x + 2)(x + 3)} + \frac{x + 1}{(x + 3)(x - 1)}$$

$$= \frac{2x(x - 1)}{(x + 2)(x + 3)(x - 1)} + \frac{(x + 1)(x + 2)}{(x + 3)(x - 1)(x + 2)}.$$

Since the two rational expressions above have the same denominator, add their numerators.

$$\frac{2x(x-1)}{(x+2)(x+3)(x-1)} + \frac{(x+1)(x+2)}{(x+3)(x-1)(x+2)}$$

$$= \frac{2x(x-1)+(x+1)(x+2)}{(x+2)(x+3)(x-1)} \qquad \text{Add numerators}$$

$$= \frac{2x^2-2x+x^2+3x+2}{(x+2)(x+3)(x-1)} \qquad \text{Distributive property}$$

$$= \frac{3x^2+x+2}{(x+2)(x+3)(x-1)} \qquad \text{Combine terms}$$

Since $3x^2 + x + 2$ cannot be factored, the rational expression cannot be reduced. It is usually best to leave the denominator in factored form since it is then easier to identify common factors in the numerator and denominator. ✦

3   To *subtract* rational expressions, use the following rule.

Subtracting Rational Expressions	If P/Q and R/Q are rational expressions, then $$\frac{P}{Q} - \frac{R}{Q} = \frac{P-R}{Q}.$$

Example 5   Subtract: $\dfrac{2m}{m-1} - \dfrac{2}{m-1}$.

By the definition of subtraction,

$$\frac{2m}{m-1} - \frac{2}{m-1} = \frac{2m-2}{m-1}$$

$$= \frac{2(m-1)}{m-1} \qquad \text{Factor the numerator}$$

$$= 2. \qquad \text{Write in lowest terms} ✦$$

Example 6   Subtract: $\dfrac{9}{x-2} - \dfrac{3}{x}$.

The least common denominator is $x(x-2)$.

$$\frac{9}{x-2} - \frac{3}{x} = \frac{9x}{x(x-2)} - \frac{3(x-2)}{x(x-2)} \qquad \text{Get a least common denominator}$$

$$= \frac{9x-3(x-2)}{x(x-2)} \qquad \text{Subtract numerators}$$

$$= \frac{9x-3x+6}{x(x-2)} \qquad \text{Distributive property}$$

$$= \frac{6x+6}{x(x-2)} ✦$$

Example 7   Find $\dfrac{6x}{x^2 - 2x + 1} - \dfrac{1}{x^2 - 1}$.

Begin by factoring.

$$\frac{6x}{x^2 - 2x + 1} - \frac{1}{x^2 - 1} = \frac{6x}{(x - 1)(x - 1)} - \frac{1}{(x - 1)(x + 1)}$$

From the factored denominators, identify the common denominator, $(x - 1)(x - 1)(x + 1)$. Use the factor $x - 1$ twice, since it appears twice in the first denominator.

$$\frac{6x}{(x - 1)(x - 1)} - \frac{1}{(x - 1)(x + 1)}$$

$$= \frac{6x(x + 1)}{(x - 1)(x - 1)(x + 1)} - \frac{1(x - 1)}{(x - 1)(x - 1)(x + 1)} \qquad \text{Fundamental property}$$

$$= \frac{6x(x + 1) - 1(x - 1)}{(x - 1)(x - 1)(x + 1)} \qquad \text{Subtract}$$

$$= \frac{6x^2 + 6x - x + 1}{(x - 1)(x - 1)(x + 1)} \qquad \begin{array}{l}\text{Distributive}\\\text{property}\end{array}$$

$$= \frac{6x^2 + 5x + 1}{(x - 1)(x - 1)(x + 1)}$$

The result may be written as

$$\frac{6x^2 + 5x + 1}{(x - 1)^2(x + 1)}. \qquad \text{❖}$$

Example 8   Subtract: $\dfrac{q}{q^2 - 4q - 5} - \dfrac{3}{2q^2 - 13q + 15}$.

Start by factoring each denominator.

$$\frac{q}{q^2 - 4q - 5} - \frac{3}{2q^2 - 13q + 15} = \frac{q}{(q + 1)(q - 5)} - \frac{3}{(q - 5)(2q - 3)}$$

Now rewrite each of the two rational expressions with the least common denominator, $(q + 1)(q - 5)(2q - 3)$. Then subtract numerators.

$$\frac{q}{(q + 1)(q - 5)} - \frac{3}{(q - 5)(2q - 3)}$$

$$= \frac{q(2q - 3)}{(q + 1)(q - 5)(2q - 3)} - \frac{3(q + 1)}{(q + 1)(q - 5)(2q - 3)}$$

$$= \frac{q(2q - 3) - 3(q + 1)}{(q + 1)(q - 5)(2q - 3)} \qquad \text{Subtract}$$

$$= \frac{2q^2 - 3q - 3q - 3}{(q + 1)(q - 5)(2q - 3)} \qquad \text{Distributive property}$$

$$= \frac{2q^2 - 6q - 3}{(q + 1)(q - 5)(2q - 3)} \qquad \text{❖}$$

5.4   Exercises

Find the sums or differences. Write the answers in lowest terms. See Examples 1 and 5.

1. $\dfrac{2}{p} + \dfrac{5}{p}$

2. $\dfrac{3}{r} + \dfrac{6}{r}$

3. $\dfrac{9}{k} - \dfrac{12}{k}$

4. $\dfrac{15}{z} - \dfrac{25}{z}$

5. $\dfrac{y}{y+1} + \dfrac{1}{y+1}$

6. $\dfrac{3m}{m-4} + \dfrac{-12}{m-4}$

7. $\dfrac{p-q}{3} - \dfrac{p+q}{3}$

8. $\dfrac{a+b}{2} - \dfrac{a-b}{2}$

9. $\dfrac{m^2}{m+6} + \dfrac{6m}{m+6}$

10. $\dfrac{y^2}{y-1} + \dfrac{-y}{y-1}$

11. $\dfrac{q^2-4q}{q-2} + \dfrac{4}{q-2}$

12. $\dfrac{z^2-10z}{z-5} + \dfrac{25}{z-5}$

Find the sums or differences. See Examples 2 and 6.

13. $\dfrac{m}{3} + \dfrac{1}{2}$

14. $\dfrac{p}{6} - \dfrac{2}{3}$

15. $\dfrac{4}{3} - \dfrac{1}{y}$

16. $\dfrac{8}{5} - \dfrac{2}{a}$

17. $\dfrac{5m}{6} - \dfrac{2m}{3}$

18. $\dfrac{3}{x} + \dfrac{5}{2x}$

19. $\dfrac{4+2k}{5} + \dfrac{2+k}{10}$

20. $\dfrac{5-4r}{8} - \dfrac{2-3r}{6}$

21. $\dfrac{m+2}{m} + \dfrac{m+3}{4m}$

22. $\dfrac{2x-5}{x} + \dfrac{x-1}{2x}$

23. $\dfrac{6}{y^2} - \dfrac{2}{y}$

24. $\dfrac{3}{p} + \dfrac{5}{p^2}$

25. $\dfrac{-1}{x^2} + \dfrac{3}{xy}$

26. $\dfrac{9}{p^2} + \dfrac{4}{px}$

27. $\dfrac{8}{x-2} - \dfrac{4}{x+2}$

Find the sums or differences. See Examples 3, 4, 7, and 8.

28. $\dfrac{6}{m-5} - \dfrac{2}{m+5}$

29. $\dfrac{2x}{x+y} - \dfrac{3x}{2x+2y}$

30. $\dfrac{1}{a-b} - \dfrac{a}{4a-4b}$

31. $\dfrac{1}{m^2-9} + \dfrac{1}{3m+9}$

32. $\dfrac{-6}{y^2-4} - \dfrac{3}{2y+4}$

33. $\dfrac{1}{m^2-1} - \dfrac{1}{m^2+3m+2}$

34. $\dfrac{1}{y^2-4} + \dfrac{3}{y^2+5y+6}$

35. $\dfrac{4}{2-q} + \dfrac{7}{q-2}$

36. $\dfrac{9}{8-y} + \dfrac{6}{y-8}$

37. $\dfrac{3}{4p-5} + \dfrac{9}{5-4p}$

38. $\dfrac{8}{3-7y} - \dfrac{2}{7y-3}$

39. $\dfrac{8}{m-2} + \dfrac{3}{5m} + \dfrac{7}{5m(m-2)}$

40. $\dfrac{-1}{7z} + \dfrac{3}{z+2} + \dfrac{4}{7z(z+2)}$

41. $\dfrac{4}{r^2-r} + \dfrac{6}{r^2+2r} - \dfrac{1}{r^2+r-2}$

42. $\dfrac{6}{k^2+3k} - \dfrac{1}{k^2-k} + \dfrac{2}{k^2+2k-3}$

43. $\dfrac{4y - 1}{2y^2 + 5y - 3} - \dfrac{y + 3}{6y^2 + y - 2}$

44. $\dfrac{2q + 1}{3q^2 + 10q - 8} - \dfrac{3q + 5}{2q^2 + 5q - 12}$

45. $\dfrac{x + 3y}{x^2 + 2xy + y^2} + \dfrac{x - y}{x^2 + 4xy + 3y^2}$

46. $\dfrac{m}{m^2 - 1} + \dfrac{m - 1}{m^2 + 2m + 1}$

47. $\dfrac{r + y}{18r^2 + 12ry - 3ry - 2y^2} + \dfrac{3r - y}{36r^2 - y^2}$

48. $\dfrac{2x - z}{2x^2 - 4xz + 5xz - 10z^2} - \dfrac{x + z}{x^2 - 4z^2}$

49. $\dfrac{3p - 2}{1 - 4q^4} - \dfrac{p^2 + 2}{2pq^2 - 8q^2 + p - 4}$

50. $\dfrac{t + 5}{64t^3 + 1} + \dfrac{1 - 4t}{16t^2 - 4t + 1}$

Perform the indicated operations in Exercises 51–54.

51. $\left(\dfrac{-k}{2k^2 - 5k - 3} + \dfrac{3k - 2}{2k^2 - k - 1} \right) \dfrac{2k + 1}{k - 1}$

52. $\left(\dfrac{3p + 1}{2p^2 + p - 6} - \dfrac{5p}{3p^2 - p} \right) \dfrac{2p - 3}{p + 2}$

53. $\dfrac{k^2 + 4k + 16}{k + 4} \left(\dfrac{-5}{16 - k^2} + \dfrac{2k + 3}{k^3 - 64} \right)$

54. $\dfrac{m - 5}{2m + 5} \left(\dfrac{-3m}{m^2 - 25} - \dfrac{m + 4}{125 - m^3} \right)$

55. A rectangle has a width of $\dfrac{y + 1}{3}$ and a length of $\dfrac{y - 2}{4}$. (a) Find the perimeter. (b) Find the area.

56. The dimensions (length and width) of a rectangle are $\dfrac{9}{2p}$ and $\dfrac{4}{p^2}$. (a) Find the perimeter. (b) Find the area.

Review Exercises *Simplify each expression, using the order of operations as necessary. See Section 1.1.*

57. $\dfrac{\dfrac{2}{3} + \dfrac{5}{3}}{\dfrac{3}{4} - \dfrac{1}{4}}$

58. $\dfrac{\dfrac{5}{8} - \dfrac{3}{8}}{\dfrac{2}{5} + \dfrac{1}{5}}$

59. $\dfrac{\dfrac{5}{6} - \dfrac{2}{3}}{\dfrac{3}{4} + \dfrac{1}{12}}$

60. $\dfrac{\dfrac{9}{7} - \dfrac{3}{14}}{\dfrac{1}{3} + \dfrac{5}{12}}$

5.5 Complex Fractions

Objectives

Simplify complex fractions by:

1️⃣ simplifying numerator and denominator;

2️⃣ multiplying by the least common denominator.

Some rational expressions have fractions in the numerator or denominator.

> A rational expression with fractions in the numerator, denominator, or both, is called a **complex fraction.**

Examples of complex fractions include

$$\frac{3 + \dfrac{4}{x}}{5}, \quad \frac{\dfrac{3x^2 - 5x}{6x^2}}{2x - \dfrac{1}{x}}, \quad \text{and} \quad \frac{3 + x}{5 - \dfrac{2}{x}}.$$

The parts of a complex fraction are named as follows.

$$\begin{array}{l} \dfrac{2}{p} - \dfrac{1}{q} \\ \hline \dfrac{3}{p} + \dfrac{5}{q} \end{array} \quad \begin{array}{l} \leftarrow \text{Numerator of complex fraction} \\ \leftarrow \text{Main fraction bar} \\ \leftarrow \text{Denominator of complex fraction} \end{array}$$

1   Since a fraction represents a quotient, one method of simplifying complex fractions is to rewrite both the numerator and denominator as single fractions, and then to perform the indicated division.

Example 1   Simplify each complex fraction.

(a) $\dfrac{\dfrac{2}{3} + \dfrac{5}{9}}{\dfrac{1}{4} + \dfrac{1}{12}}$ **(b)** $\dfrac{6 + \dfrac{3}{x}}{\dfrac{x}{4} + \dfrac{1}{8}}$

First, write each numerator as a single fraction.

$$\frac{2}{3} + \frac{5}{9} = \frac{2(3)}{3(3)} + \frac{5}{9} \qquad\qquad 6 + \frac{3}{x} = \frac{6}{1} + \frac{3}{x}$$

$$= \frac{6}{9} + \frac{5}{9} = \frac{11}{9} \qquad\qquad = \frac{6x}{x} + \frac{3}{x} = \frac{6x + 3}{x}$$

Do the same thing with each denominator.

$$\frac{1}{4} + \frac{1}{12} = \frac{1(3)}{4(3)} + \frac{1}{12} \qquad\qquad \frac{x}{4} + \frac{1}{8} = \frac{x(2)}{4(2)} + \frac{1}{8}$$

$$= \frac{3}{12} + \frac{1}{12} = \frac{4}{12} \qquad\qquad = \frac{2x}{8} + \frac{1}{8} = \frac{2x + 1}{8}$$

The original complex fraction can now be written as

$$\dfrac{\dfrac{11}{9}}{\dfrac{4}{12}}, \qquad\qquad \dfrac{\dfrac{6x+3}{x}}{\dfrac{2x+1}{8}}.$$

Now use the rule for division and the fundamental property.

$$\dfrac{11}{9} \div \dfrac{4}{12} = \dfrac{11}{9} \cdot \dfrac{12}{4} \qquad\qquad \dfrac{6x+3}{x} \div \dfrac{2x+1}{8} = \dfrac{6x+3}{x} \cdot \dfrac{8}{2x+1}$$

$$= \dfrac{11 \cdot 3 \cdot 4}{3 \cdot 3 \cdot 4} \qquad\qquad\qquad\quad = \dfrac{3(2x+1)}{x} \cdot \dfrac{8}{2x+1}$$

$$= \dfrac{11}{3} \qquad\qquad\qquad\qquad\qquad = \dfrac{24}{x} \quad \blacklozenge$$

Example 2 Simplify the complex fraction $\dfrac{\dfrac{xp}{q^3}}{\dfrac{p^2}{qx^2}}$.

Here, the numerator and denominator are already single fractions, so use the division rule and then the fundamental property.

$$\dfrac{xp}{q^3} \div \dfrac{p^2}{qx^2} = \dfrac{xp}{q^3} \cdot \dfrac{qx^2}{p^2} = \dfrac{x^3}{q^2p} \quad \blacklozenge$$

Example 3 Simplify the complex fraction $\dfrac{\dfrac{x}{x+y}}{\dfrac{1}{x}+\dfrac{1}{y}}$.

Since the numerator is already a single fraction, it is only necessary to simplify the denominator. Do this by adding $1/x$ and $1/y$.

$$\dfrac{1}{x} + \dfrac{1}{y} = \dfrac{y}{xy} + \dfrac{x}{xy} = \dfrac{x+y}{xy}$$

Now divide.

$$\dfrac{x}{x+y} \div \dfrac{x+y}{xy} = \dfrac{x}{x+y} \cdot \dfrac{xy}{x+y} = \dfrac{x^2y}{(x+y)^2} \quad \blacklozenge$$

2 As an alternative method, complex fractions may be simplified by multiplying both numerator and denominator by the least common denominator of all the denominators appearing in the complex fraction.

In the next example, this second method is used to simplify the same complex fractions as in Example 1 above.

Example 4   Simplify each complex fraction.

(a) $\dfrac{\dfrac{2}{3} + \dfrac{5}{9}}{\dfrac{1}{4} + \dfrac{1}{12}}$

(b) $\dfrac{6 + \dfrac{3}{x}}{\dfrac{x}{4} + \dfrac{1}{8}}$

Find the least common denominator for all the denominators in the complex fraction.

The least common denominator for 3, 9, 4, and 12 is 36.

The least common denominator for x, 4, and 8 is $8x$.

Multiply numerator and denominator of the complex fraction by the least common denominator.

$$\frac{\dfrac{2}{3} + \dfrac{5}{9}}{\dfrac{1}{4} + \dfrac{1}{12}} = \frac{36\left(\dfrac{2}{3} + \dfrac{5}{9}\right)}{36\left(\dfrac{1}{4} + \dfrac{1}{12}\right)}$$

$$= \frac{36\left(\dfrac{2}{3}\right) + 36\left(\dfrac{5}{9}\right)}{36\left(\dfrac{1}{4}\right) + 36\left(\dfrac{1}{12}\right)}$$

$$= \frac{24 + 20}{9 + 3}$$

$$= \frac{44}{12} = \frac{11}{3}$$

$$\frac{6 + \dfrac{3}{x}}{\dfrac{x}{4} + \dfrac{1}{8}} = \frac{8x\left(6 + \dfrac{3}{x}\right)}{8x\left(\dfrac{x}{4} + \dfrac{1}{8}\right)}$$

$$= \frac{8x(6) + 8x\left(\dfrac{3}{x}\right)}{8x\left(\dfrac{x}{4}\right) + 8x\left(\dfrac{1}{8}\right)}$$

$$= \frac{48x + 24}{2x^2 + x}$$

$$= \frac{24(2x + 1)}{x(2x + 1)} = \frac{24}{x} \quad \blacklozenge$$

Example 5   Simplify the complex fraction $\dfrac{\dfrac{3}{5m} - \dfrac{2}{m^2}}{\dfrac{9}{2m} + \dfrac{3}{4m^2}}$.

The least common denominator for $5m$, m^2, $2m$, and $4m^2$ is $20m^2$. Multiply numerator and denominator by $20m^2$.

$$\frac{\dfrac{3}{5m} - \dfrac{2}{m^2}}{\dfrac{9}{2m} + \dfrac{3}{4m^2}} = \frac{20m^2\left(\dfrac{3}{5m} - \dfrac{2}{m^2}\right)}{20m^2\left(\dfrac{9}{2m} + \dfrac{3}{4m^2}\right)}$$

$$= \frac{20m^2\left(\dfrac{3}{5m}\right) - 20m^2\left(\dfrac{2}{m^2}\right)}{20m^2\left(\dfrac{9}{2m}\right) + 20m^2\left(\dfrac{3}{4m^2}\right)} = \frac{12m - 40}{90m + 15} \quad \blacklozenge$$

The two methods for simplifying a complex fraction are summarized below.

Simplifying Complex Fractions	*Method 1* Simplify the numerator and denominator of the complex fraction separately. Then divide the simplified numerator by the simplified denominator.
	Method 2 Multiply numerator and denominator of the complex fraction by the least common denominator of all the denominators appearing in the complex fraction.

5.5 Exercises

Simplify each complex fraction. Use either method. See Examples 1–5.

1. $\dfrac{\dfrac{5}{8} + \dfrac{2}{3}}{\dfrac{7}{3} - \dfrac{1}{4}}$

2. $\dfrac{\dfrac{6}{5} - \dfrac{1}{9}}{\dfrac{2}{5} + \dfrac{5}{3}}$

3. $\dfrac{1 - \dfrac{3}{8}}{2 + \dfrac{1}{4}}$

4. $\dfrac{3 + \dfrac{5}{4}}{1 - \dfrac{7}{8}}$

5. $\dfrac{\dfrac{p}{q^2}}{\dfrac{p^2}{q}}$

6. $\dfrac{\dfrac{a}{x}}{\dfrac{a^2}{2x}}$

7. $\dfrac{\dfrac{x}{y}}{\dfrac{x^2}{y}}$

8. $\dfrac{\dfrac{p^4}{r}}{\dfrac{p^2}{r^2}}$

9. $\dfrac{\dfrac{m^3p^4}{5m}}{\dfrac{8mp^5}{p^2}}$

10. $\dfrac{\dfrac{9z^5x^3}{2x}}{\dfrac{8z^2x^5}{3z}}$

11. $\dfrac{\dfrac{x+1}{4}}{\dfrac{x-3}{x}}$

12. $\dfrac{\dfrac{m+6}{m}}{\dfrac{m-6}{2}}$

13. $\dfrac{\dfrac{3}{q}}{\dfrac{1-q}{6q^2}}$

14. $\dfrac{\dfrac{6}{x}}{\dfrac{1+x}{8x^5}}$

15. $\dfrac{\dfrac{2x+3}{12}}{\dfrac{x-1}{15}}$

16. $\dfrac{\dfrac{15}{k+1}}{\dfrac{10}{3(k+1)}}$

17. $\dfrac{\dfrac{3}{y} + 1}{\dfrac{3+y}{2}}$

18. $\dfrac{6 + \dfrac{2}{r}}{\dfrac{r+2}{3}}$

19. $\dfrac{\dfrac{1}{x} + x}{\dfrac{x^2+1}{8}}$

20. $\dfrac{\dfrac{3}{m} - m}{\dfrac{3-m^2}{4}}$

21. $\dfrac{x + \dfrac{1}{x}}{\dfrac{4}{x} + y}$

22. $\dfrac{y - \dfrac{6}{y}}{y + \dfrac{2}{y}}$

23. $\dfrac{\dfrac{p+3}{p}}{\dfrac{1}{p} + \dfrac{1}{5}}$

24. $\dfrac{r + \dfrac{1}{r}}{\dfrac{1}{r} - r}$

25. $\dfrac{\dfrac{2}{p^2} - \dfrac{3}{5p}}{\dfrac{4}{p} + \dfrac{1}{4p}}$

26. $\dfrac{\dfrac{2}{m^2} - \dfrac{3}{m}}{\dfrac{2}{5m^2} + \dfrac{1}{3m}}$

27. $\dfrac{\dfrac{5}{x^2y} - \dfrac{2}{xy^2}}{\dfrac{3}{x^2y^2} + \dfrac{4}{xy}}$

28. $\dfrac{\dfrac{1}{m^3p} + \dfrac{2}{mp^2}}{\dfrac{4}{mp} + \dfrac{1}{m^2p}}$

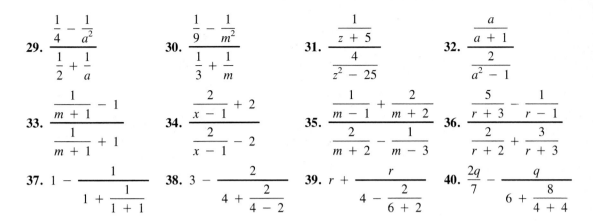

29. $\dfrac{\dfrac{1}{4} - \dfrac{1}{a^2}}{\dfrac{1}{2} + \dfrac{1}{a}}$

30. $\dfrac{\dfrac{1}{9} - \dfrac{1}{m^2}}{\dfrac{1}{3} + \dfrac{1}{m}}$

31. $\dfrac{\dfrac{1}{z+5}}{\dfrac{4}{z^2-25}}$

32. $\dfrac{\dfrac{a}{a+1}}{\dfrac{2}{a^2-1}}$

33. $\dfrac{\dfrac{1}{m+1} - 1}{\dfrac{1}{m+1} + 1}$

34. $\dfrac{\dfrac{2}{x-1} + 2}{\dfrac{2}{x-1} - 2}$

35. $\dfrac{\dfrac{1}{m-1} + \dfrac{2}{m+2}}{\dfrac{2}{m+2} - \dfrac{1}{m-3}}$

36. $\dfrac{\dfrac{5}{r+3} - \dfrac{1}{r-1}}{\dfrac{2}{r+2} + \dfrac{3}{r+3}}$

37. $1 - \dfrac{1}{1 + \dfrac{1}{1+1}}$

38. $3 - \dfrac{2}{4 + \dfrac{2}{4-2}}$

39. $r + \dfrac{r}{4 - \dfrac{2}{6+2}}$

40. $\dfrac{2q}{7} - \dfrac{q}{6 + \dfrac{8}{4+4}}$

Review Exercises   *Use the distributive property to simplify. See Section 1.10.*

41. $9\left(\dfrac{4x}{3} + \dfrac{2}{9}\right)$

42. $8\left(\dfrac{3r}{4} - \dfrac{9}{8}\right)$

43. $12\left(\dfrac{11p^2}{3} - \dfrac{9p}{4}\right)$

44. $6\left(\dfrac{5z^2}{2} + \dfrac{8z}{3}\right)$

Solve each equation. See Section 2.4.

45. $3x - 5 = 7x + 3$

46. $9z + 2 = 7z - 6$

47. $8(2q + 5) - 1 = 7q$

48. $6(z - 3) + 5 = 8z$

49. $9 - (5 - 3y) = 6$

50. $-(8 - a) + 5a = -7$

5.6   Equations Involving Rational Expressions

Objectives

1 Solve equations involving rational expressions.

2 Solve for a specified variable.

1   An equation with fractions is solved by rewriting the equation using the multiplication property of equality. The goal is to get a new equation that does not have fractions. Choose as the multiplier the least common denominator of all denominators in the fractions of the equation.

Example 1   Solve the equation $\dfrac{x}{3} + \dfrac{x}{4} = 10 + x$.

Since the least common denominator of the two fractions is 12, begin by multiplying both sides of the equation by 12.

$$12\left(\frac{x}{3} + \frac{x}{4}\right) = 12(10 + x)$$

$$12\left(\frac{x}{3}\right) + 12\left(\frac{x}{4}\right) = 12(10) + 12x \qquad \text{Distributive property}$$

$$\frac{12x}{3} + \frac{12x}{4} = 120 + 12x$$

$$4x + 3x = 120 + 12x$$

This equation has no fractions. Solve it by the methods used to solve linear equations given earlier.

$$7x = 120 + 12x$$

$$-5x = 120$$

$$x = -24$$

Check this solution by substituting -24 for x in the original equation.

$$\frac{x}{3} + \frac{x}{4} = 10 + x$$

$$\frac{-24}{3} + \frac{-24}{4} = 10 + (-24)$$

$$-8 + (-6) = -14 \qquad \text{True}$$

This true statement shows that -24 is the solution. ✤

Example 2 Solve $\dfrac{p}{2} - \dfrac{p-1}{3} = 1$.

Multiply both sides by the least common denominator, 6.

$$6\left(\frac{p}{2} - \frac{p-1}{3}\right) = 6 \cdot 1$$

$$6\left(\frac{p}{2}\right) - 6\left(\frac{p-1}{3}\right) = 6 \qquad \text{Distributive property}$$

$$3p - 2(p - 1) = 6$$

Be careful to put parentheses around $p - 1$; otherwise an incorrect solution may be found. Continue simplifying and solve the equation.

$$3p - 2p + 2 = 6 \qquad \text{Distributive property}$$

$$p + 2 = 6$$

$$p = 4$$

Check to see that 4 is correct by replacing p with 4 in the original equation. ✤

When solving equations that have a variable in the denominator, remember that the number 0 cannot be used as a denominator. Therefore, the solution cannot be a number that will make the denominator equal 0.

Example 3   Solve $\dfrac{x}{x-2} = \dfrac{2}{x-2} + 2$.

Multiply both sides by the least common denominator, $x - 2$.

$$(x-2)\left(\frac{x}{x-2}\right) = (x-2)\left(\frac{2}{x-2}\right) + (x-2)(2)$$

$$x = 2 + 2x - 4 \qquad \text{Distributive property}$$

$$x = -2 + 2x$$

$$-x = -2$$

$$x = 2$$

The proposed solution is 2. However, 2 cannot be a solution because 2 makes both denominators equal 0. This equation has no solution. Equations with no solutions are one of the main reasons that it is important to always check proposed solutions.   ✛

Solving Equations with Rational Expressions	*Step 1* Multiply both sides of the equation by the least common denominator. (This clears the equation of fractions.) *Step 2* Solve the resulting equation. *Step 3* Check each solution by substituting it in the original equation.

Example 4   Solve $\dfrac{2m}{m^2 - 4} + \dfrac{1}{m-2} = \dfrac{2}{m+2}$.

Since $m^2 - 4 = (m+2)(m-2)$, use $(m+2)(m-2)$ as the least common denominator. Multiply both sides by $(m+2)(m-2)$.

$$(m+2)(m-2)\left(\frac{2m}{m^2-4} + \frac{1}{m-2}\right) = (m+2)(m-2)\frac{2}{m+2}$$

$$(m+2)(m-2)\frac{2m}{(m+2)(m-2)} + (m+2)(m-2)\frac{1}{m-2}$$

$$= (m+2)(m-2)\frac{2}{m+2}$$

$$2m + m + 2 = 2(m-2)$$

$$3m + 2 = 2m - 4$$

$$m = -6$$

Verify by substitution in the original equation that -6 is a solution for the given equation.   ✛

Example 5 Solve $\dfrac{2}{x^2 - x} = \dfrac{1}{x^2 - 1}$.

Begin by finding a least common denominator. Since $x^2 - x$ can be factored as $x(x - 1)$, and $x^2 - 1$ can be factored as $(x + 1)(x - 1)$, the least common denominator is $x(x + 1)(x - 1)$. Multiply both sides of the equation by $x(x + 1)(x - 1)$.

$$x(x + 1)(x - 1)\frac{2}{x(x - 1)} = x(x + 1)(x - 1)\frac{1}{(x + 1)(x - 1)}$$
$$2(x + 1) = x$$
$$2x + 2 = x$$
$$2 = -x$$
$$x = -2$$

To be sure that -2 is a solution, substitute -2 for x in the original equation. Since -2 satisfies the equation, -2 is the solution. ✤

Example 6 Solve $\dfrac{1}{x - 1} + \dfrac{1}{2} = \dfrac{2}{x^2 - 1}$.

The least common denominator is $2(x + 1)(x - 1)$. Multiply both sides of the equation by this common denominator.

$$2(x + 1)(x - 1)\left(\frac{1}{x - 1} + \frac{1}{2}\right) = 2(x + 1)(x - 1)\frac{2}{(x + 1)(x - 1)}$$
$$2(x + 1)(x - 1)\frac{1}{x - 1} + 2(x + 1)(x - 1)\frac{1}{2}$$
$$= 2(x + 1)(x - 1)\frac{2}{(x + 1)(x - 1)}$$
$$2(x + 1) + (x + 1)(x - 1) = 4$$
$$2x + 2 + x^2 - 1 = 4$$
$$x^2 + 2x + 1 = 4$$
$$x^2 + 2x - 3 = 0$$

Factoring gives

$$(x + 3)(x - 1) = 0.$$

Solving this equation suggests that $x = -3$ or $x = 1$. But 1 makes a denominator of the original equation equal 0, so 1 is not a solution. However, -3 is a solution, as can be shown by substituting -3 for x in the original equation. ✤

Example 7 Solve $\dfrac{1}{k^2 + 4k + 3} + \dfrac{1}{2k + 2} = \dfrac{3}{4k + 12}$.

Factor the three denominators to get the common denominator, which is $4(k + 1)(k + 3)$. Multiply both sides by this product.

$$4(k + 1)(k + 3)\left(\frac{1}{(k + 1)(k + 3)} + \frac{1}{2(k + 1)}\right) = 4(k + 1)(k + 3)\frac{3}{4(k + 3)}$$

$$4(k + 1)(k + 3)\frac{1}{(k + 1)(k + 3)} + 4(k + 1)(k + 3)\frac{1}{2(k + 1)}$$

$$= 4(k + 1)(k + 3)\frac{3}{4(k + 3)}$$

$$4 + 2(k + 3) = 3(k + 1)$$

$$4 + 2k + 6 = 3k + 3$$

$$2k + 10 = 3k + 3$$

$$7 = k$$

Check to see that 7 actually is a solution for the given equation. ✦

2 Solving a formula for a specified variable was discussed in Chapter 2. In the next example this process is applied to formulas with fractions.

Example 8 Solve the formula $\dfrac{1}{R} = \dfrac{1}{r_1} + \dfrac{1}{r_2}$ for r_2.

Here r_1 (read "r-sub-one") and r_2 are different variables. The least common denominator for R, r_1, and r_2 is Rr_1r_2, so multiply both sides of the formula by Rr_1r_2.

$$Rr_1r_2\left(\frac{1}{R}\right) = Rr_1r_2\left(\frac{1}{r_1} + \frac{1}{r_2}\right)$$

$$Rr_1r_2\left(\frac{1}{R}\right) = Rr_1r_2\left(\frac{1}{r_1}\right) + Rr_1r_2\left(\frac{1}{r_2}\right)$$

$$r_1r_2 = Rr_2 + Rr_1$$

Get the terms containing r_2 (the desired variable) alone on one side of the equals sign. Do this here by subtracting Rr_2 from both sides.

$$r_1r_2 - Rr_2 = Rr_1$$

Factor out the common factor of r_2 on the left.

$$r_2(r_1 - R) = Rr_1$$

Finally, divide both sides of the equation by $r_1 - R$.

$$r_2 = \frac{Rr_1}{r_1 - R}$$ ✦

5.6   Exercises

Solve each equation and check your answers. See Examples 1 and 2.

1. $\dfrac{1}{4} = \dfrac{x}{2}$

2. $\dfrac{2}{m} = \dfrac{5}{12}$

3. $\dfrac{9}{k} = \dfrac{3}{4}$

4. $\dfrac{15}{f} = \dfrac{30}{8}$

5. $\dfrac{7}{y} = \dfrac{8}{3}$

6. $\dfrac{2}{z} = \dfrac{11}{5}$

7. $\dfrac{6}{x} - \dfrac{4}{x} = 5$

8. $\dfrac{3}{x} + \dfrac{2}{x} = 5$

9. $\dfrac{x}{2} - \dfrac{x}{4} = 6$

10. $\dfrac{4}{y} + \dfrac{2}{3} = 1$

11. $\dfrac{9}{m} = 5 - \dfrac{1}{m}$

12. $\dfrac{3x}{5} + 2 = \dfrac{1}{4}$

13. $\dfrac{2t}{7} - 5 = t$

14. $\dfrac{1}{2} + \dfrac{2}{m} = 1$

15. $\dfrac{x+1}{2} = \dfrac{x+2}{3}$

16. $\dfrac{t-4}{3} = t + 2$

17. $\dfrac{3m}{2} + m = 5$

18. $\dfrac{2k+3}{k} = \dfrac{3}{2}$

19. $\dfrac{5-y}{y} + \dfrac{3}{4} = \dfrac{7}{y}$

20. $\dfrac{x}{x-4} = \dfrac{2}{x-4} + 5$

21. $\dfrac{m-2}{5} = \dfrac{m+8}{10}$

22. $\dfrac{2p+8}{9} = \dfrac{10p+4}{27}$

23. $\dfrac{5r-3}{7} = \dfrac{15r-2}{28}$

24. $\dfrac{2y-1}{y} + 2 = \dfrac{1}{2}$

25. $\dfrac{a}{2} - \dfrac{17+a}{5} = 2a$

26. $\dfrac{m-2}{4} + \dfrac{m+1}{3} = \dfrac{10}{3}$

27. $\dfrac{y+2}{5} + \dfrac{y-5}{3} = \dfrac{7}{5}$

28. $\dfrac{a+7}{8} - \dfrac{a-2}{3} = \dfrac{4}{3}$

29. $\dfrac{m+2}{5} - \dfrac{m-6}{7} = 2$

30. $\dfrac{p}{2} - \dfrac{p-1}{4} = \dfrac{5}{4}$

31. $\dfrac{r}{6} - \dfrac{r-2}{3} = \dfrac{-4}{3}$

32. $\dfrac{5y}{3} - \dfrac{2y-1}{4} = \dfrac{1}{4}$

33. $\dfrac{8k}{5} - \dfrac{3k-4}{2} = \dfrac{5}{2}$

Solve each equation and check your answers. See Examples 3–5.

34. $\dfrac{8}{2k-4} + \dfrac{3}{5k-10} = \dfrac{23}{5}$

35. $\dfrac{1}{3p+15} + \dfrac{5}{4p+20} = \dfrac{19}{24}$

36. $\dfrac{m}{2m+2} = \dfrac{-2m}{4m+4} + \dfrac{2m-3}{m+1}$

37. $\dfrac{5p+1}{3p+3} = \dfrac{5p-5}{5p+5} + \dfrac{3p-1}{p+1}$

38. $\dfrac{x+1}{x-3} = \dfrac{4}{x-3} + 6$

39. $\dfrac{p}{p-2} + 4 = \dfrac{2}{p-2}$

40. $\dfrac{2}{k-3} - \dfrac{3}{k+3} = \dfrac{12}{k^2-9}$

41. $\dfrac{1}{r+5} - \dfrac{3}{r-5} = \dfrac{-10}{r^2-25}$

42. $\dfrac{4}{p} - \dfrac{2}{p+1} = 3$

43. $\dfrac{6}{r} + \dfrac{1}{r-2} = 3$

44. $\dfrac{2}{m-1} + \dfrac{1}{m+1} = \dfrac{5}{4}$

45. $\dfrac{5}{z-2} + \dfrac{10}{z+2} = 7$

Solve for the specified variable. See Example 8.

46. $P = \dfrac{kT}{V}$; for T

47. $I = \dfrac{kE}{R}$; for R

48. $N = \dfrac{kF}{d}$; for d

49. $F = \dfrac{k}{r}$; for r

50. $F = \dfrac{k}{d - D}$; for D

51. $I = \dfrac{E}{R + r}$; for R

52. $I = \dfrac{E}{R + r}$; for r

53. $S = \dfrac{a}{1 - r}$; for r

54. $h = \dfrac{2A}{B + b}$; for b

55. $\dfrac{1}{x} = \dfrac{1}{y} - \dfrac{1}{z}$; for y

56. $\dfrac{3}{k} = \dfrac{1}{p} + \dfrac{1}{q}$; for q

57. $9x + \dfrac{3}{z} = \dfrac{5}{y}$; for z

58. $2a - \dfrac{5}{b} + \dfrac{1}{c} = 0$; for b

59. $\dfrac{1}{R} = \dfrac{1}{r_1} + \dfrac{1}{r_2}$; for r_1

Solve each equation and check the answers. See Examples 6 and 7.

60. $\dfrac{2}{y} = \dfrac{y}{5y - 12}$

61. $\dfrac{8x + 3}{x} = 3x$

62. $\dfrac{x + 4}{x^2 - 3x + 2} - \dfrac{5}{x^2 - 4x + 3} = \dfrac{x - 4}{x^2 - 5x + 6}$

63. $\dfrac{3y}{y^2 + 5y + 6} = \dfrac{5y}{y^2 + 2y - 3} - \dfrac{2}{y^2 + y - 2}$

64. $\dfrac{3}{r^2 + r - 2} - \dfrac{1}{r^2 - 1} = \dfrac{7}{2(r^2 + 3r + 2)}$

65. $\dfrac{m}{m^2 + m - 2} + \dfrac{m}{m^2 - 1} = \dfrac{m}{m^2 + 3m + 2}$

Review Exercises Write a mathematical expression for each phrase. See Section 2.5.

66. Sharon goes 800 kilometers in p hours; find her rate.

67. Joann drives for 10 hours, going d kilometers. Find her rate.

68. Sam goes 780 kilometers at z kilometers per hour. Find his time.

69. Walt can do a job in x hours. What portion of the job is done in 1 hour?

70. Kathy needs a hours to tune up her car. How much of the job does she do in 1 hour?

Supplementary Exercises on Rational Expressions

A common error when working with rational expressions is to confuse *operations* on rational expressions with the *solution of equations* with rational expressions. For example, the four possible operations on the rational expressions

$$\frac{1}{x} \quad \text{and} \quad \frac{1}{x - 2}$$

can be performed as follows.

Add: $\quad \dfrac{1}{x} + \dfrac{1}{x - 2} = \dfrac{x - 2}{x(x - 2)} + \dfrac{x}{x(x - 2)} = \dfrac{x - 2 + x}{x(x - 2)} = \dfrac{2x - 2}{x(x - 2)}$

Subtract: $\quad \dfrac{1}{x} - \dfrac{1}{x - 2} = \dfrac{x - 2}{x(x - 2)} - \dfrac{x}{x(x - 2)} = \dfrac{x - 2 - x}{x(x - 2)} = \dfrac{-2}{x(x - 2)}$

Multiply: $\quad \dfrac{1}{x} \cdot \dfrac{1}{x - 2} = \dfrac{1}{x(x - 2)}$

Divide: $\quad \dfrac{1}{x} \div \dfrac{1}{x - 2} = \dfrac{1}{x} \cdot \dfrac{x - 2}{1} = \dfrac{x - 2}{x}$

On the other hand, the *equation*

$$\frac{1}{x} + \frac{1}{x - 2} = \frac{3}{4}$$

is solved by multiplying both sides by the least common denominator, $4x(x - 2)$, giving an equation with no denominators.

$$4x(x - 2)\frac{1}{x} + 4x(x - 2)\frac{1}{x - 2} = 4x(x - 2)\frac{3}{4}$$

$$4x - 8 + 4x = 3x^2 - 6x$$

$$0 = 3x^2 - 14x + 8$$

$$0 = (3x - 2)(x - 4)$$

$$x = \frac{2}{3} \quad \text{or} \quad x = 4$$

In each of the following exercises, decide whether the given rational expressions should be added, subtracted, multiplied, or divided, and perform the operation, or else solve the given equation.

1. $\dfrac{6}{m} + \dfrac{2}{m}$

2. $\dfrac{b^2c^3}{b^5c^4} \cdot \dfrac{c^5}{b^7}$

3. $\dfrac{2}{x^2 + 2x - 3} \div \dfrac{8x^2}{3x - 3}$

4. $\dfrac{4}{m - 2} = 1$

5. $\dfrac{2r^2 - 3r - 9}{2r^2 - r - 6} \cdot \dfrac{r^2 + 2r - 8}{r^2 - 2r - 3}$

6. $\dfrac{1}{m^2 - 3m} + \dfrac{4}{m^2 - 9}$

7. $\dfrac{p + 3}{8} = \dfrac{p - 2}{9}$

8. $\dfrac{4t^2 - t}{6t^2 + 10t} \div \dfrac{8t^2 + 2t - 1}{3t^2 + 11t + 10}$

9. $\dfrac{5}{y-1} + \dfrac{2}{3y-3}$

10. $\dfrac{1}{z} + \dfrac{1}{z+2} = \dfrac{8}{15}$

11. $\dfrac{2}{r-1} + \dfrac{1}{r} = \dfrac{5}{2}$

12. $\dfrac{2}{y} - \dfrac{7}{5y}$

13. $\dfrac{4}{9z} - \dfrac{3}{2z}$

14. $\dfrac{r-3}{2} = \dfrac{2r-5}{5}$

15. $\dfrac{1}{m^2 + 5m + 6} + \dfrac{2}{m^2 + 4m + 3}$

16. $\dfrac{2k^2 - 3k}{20k^2 - 5k} \div \dfrac{2k^2 - 5k + 3}{4k^2 + 11k - 3}$

5.7   Applications of Rational Expressions

Objectives

Solve word problems with rational expressions involving

1 numbers;

2 distance;

3 work.

4 Solve problems about variation.

1   Rational expressions are often useful in solving word problems. Example 1, involving numbers, is used to show how problems with fractions are set up and solved.

Example 1   If the same number is added to both the numerator and denominator of the fraction 3/4, the result is 5/6. Find the number.

If x represents the number added to numerator and denominator, then

$$\frac{3+x}{4+x}$$

represents the result of adding the same number to both the numerator and denominator. Since this result is 5/6,

$$\frac{3+x}{4+x} = \frac{5}{6}.$$

Solve this equation by multiplying both sides by the least common denominator, $6(4 + x)$.

$$6(4+x)\frac{3+x}{4+x} = 6(4+x)\frac{5}{6}$$

$$6(3+x) = 5(4+x)$$

$$18 + 6x = 20 + 5x$$

$$x = 2$$

Check the solution in the words of the original problem: if 2 is added to both the numerator and denominator of 3/4, the result is 5/6, as required.   ✚

2   The next example shows how to solve word problems involving distance.

Example 2   An excursion boat travels on a river with a current of 3 miles per hour. The boat takes as long to go 12 miles downstream as to go 8 miles upstream. What is the speed of the boat in still water?

This problem requires the distance formula,

$$d = rt \quad \text{(distance} = \text{rate} \cdot \text{time)}.$$

Let x represent the speed of the boat in still water. Since the current pushes the boat when the boat is going downstream, the speed of the boat downstream will be the sum of the speed of the boat and the speed of the current, or $x + 3$ miles per hour. Also, the boat's speed going upstream is $x - 3$ miles per hour. The information given in the problem is summarized in this chart.

	d	r	t
Downstream	12	$x + 3$	
Upstream	8	$x - 3$	

Fill in the last column, representing time by solving the formula $d = rt$ for t.

$$d = rt$$
$$\frac{d}{r} = t \quad \text{Divide by } r$$

Then the time upstream is the distance divided by the rate, or

$$\frac{8}{x - 3},$$

and the time downstream is also the distance divided by the rate, or

$$\frac{12}{x + 3}.$$

Now complete the chart.

	d	r	t	
Downstream	12	$x + 3$	$\dfrac{12}{x + 3}$	⟵ Times
Upstream	8	$x - 3$	$\dfrac{8}{x - 3}$	⟵ are equal

According to the original problem, the time upstream equals the time downstream. The two times from the chart must therefore be equal, giving the equation

$$\frac{12}{x + 3} = \frac{8}{x - 3}.$$

Solve this equation by multiplying both sides by $(x + 3)(x - 3)$.

$$(x + 3)(x - 3)\frac{12}{x + 3} = (x + 3)(x - 3)\frac{8}{x - 3}$$

$$12(x - 3) = 8(x + 3)$$

$$12x - 36 = 8x + 24$$

$$4x = 60$$

$$x = 15$$

The speed of the boat in still water is 15 miles per hour. Check this solution by first finding the speed of the boat downstream, which is $15 + 3 = 18$ miles per hour. Traveling 12 miles would take

$$d = rt$$

$$12 = 18t$$

$$t = \frac{2}{3} \text{ hour.}$$

On the other hand, the speed of the boat upstream is $15 - 3 = 12$ miles per hour, and traveling 8 miles would take

$$d = rt$$

$$8 = 12t$$

$$t = \frac{2}{3} \text{ hour.}$$

The time upstream equals the time downstream, as required.   ✦

3   The third example shows a word problem about the length of time needed to do a job. This type of problem is often called a work problem.

Example 3   With a riding lawn mower, John, the grounds keeper in a large park, can cut the lawn in 8 hours. With a small mower, his assistant Walt needs 14 hours to cut the same lawn. If both John and Walt work on the lawn, how long will it take to cut it?

Let x be the number of hours that it takes John and Walt to cut the lawn, working together. Certainly x will be less than 8, since John alone can cut the lawn in 8 hours. In one hour, John can do 1/8 of the lawn, and in one hour Walt can do 1/14 of the lawn. Since it takes them x hours to cut the lawn when working together, in one hour together they can do $1/x$ of the lawn. The amount of the lawn cut by John in one hour plus the amount cut by Walt in one hour must equal the amount they can do together in one hour, or

Amount by Walt

Amount by John → $\dfrac{1}{8} + \dfrac{1}{14} = \dfrac{1}{x}$. ← Amount together

Since $56x$ is the least common denominator for 8, 14, and x, multiply both sides of the equation by $56x$.

$$56x\left(\frac{1}{8} + \frac{1}{14}\right) = 56x \cdot \frac{1}{x}$$

$$56x \cdot \frac{1}{8} + 56x \cdot \frac{1}{14} = 56x \cdot \frac{1}{x}$$

$$7x + 4x = 56$$

$$11x = 56$$

$$x = \frac{56}{11}$$

Working together, John and Walt can cut the lawn in 56/11 hours, or 5 1/11 hours, about 5 hours and 5 minutes.   ✤

4   Equations with fractions often result when discussing **variation.** Two variables **vary directly** if one is a constant multiple of the other.

Direct Variation

> y **varies directly** as x if there exists a constant k such that
>
> $$y = kx.$$

Example 4   Suppose y varies directly as x, and $y = 20$ when $x = 4$. Find y when $x = 9$.
   Since y varies directly as x, there is a constant k such that $y = kx$. Also, $y = 20$ when $x = 4$. Substituting these values into $y = kx$ gives

$$y = kx$$

$$20 = k \cdot 4,$$

from which $k = 5.$

Since $y = kx$ and $k = 5$,

$$y = 5x.$$

When $x = 9$,

$$y = 5x = 5 \cdot 9 = 45.$$

Thus, $y = 45$ when $x = 9$. ✜

Other common types of variation are defined below, where k represents a constant.

Variation		
	y varies directly as the square of x	$y = kx^2$
	m varies inversely as p	$m = \dfrac{k}{p}$
	r varies inversely as the square of s	$r = \dfrac{k}{s^2}$

Example 5   Suppose z varies inversely as the square of y, and $z = 8$ when $y = 1/2$. Find z when $y = 3/4$.

Since z varies inversely as the square of y, there is a constant k such that

$$z = \frac{k}{y^2}.$$

Find k by replacing z with 8 and y with 1/2.

$$8 = \frac{k}{\left(\dfrac{1}{2}\right)^2}$$

$$8 = \frac{k}{\dfrac{1}{4}}$$

Multiply both sides by 1/4 to get

$$k = 2,$$

so that

$$z = \frac{2}{y^2}.$$

When $y = 3/4$,

$$z = \frac{2}{\left(\dfrac{3}{4}\right)^2} = \frac{2}{\dfrac{9}{16}} = \frac{2 \cdot 16}{9} = \frac{32}{9}. \quad ✜$$

5.7 Exercises

Solve each problem. See Example 1.

1. One half of a number is 3 more than one sixth of the same number. What is the number?

2. The numerator of the fraction 4/7 is increased by an amount so that the value of the resulting fraction is 27/21. By what amount was the numerator increased?

3. In a certain fraction, the denominator is 5 larger than the numerator. If 3 is added to both the numerator and the denominator, the result is 3/4. Find the original fraction.

4. The denominator of a certain fraction is three times the numerator. If 1 is added to the numerator and subtracted from the denominator, the result equals 1/2. Find the original fraction.

5. One number is 3 more than another. If the smaller is added to two thirds the larger, the result is four fifths the sum of the original numbers. Find the numbers.

6. The sum of a number and its reciprocal is 5/2. Find the number.

7. If twice the reciprocal of a number is subtracted from the number, the result is $-7/3$. Find the number.

8. The sum of the reciprocals of two consecutive integers is 5/6. Find the integers.

9. A man and his son worked four days at a job. The son's daily wage was 2/5 that of the father. If together they earned $672, what were their daily wages?

10. The profits from a student show are to be given to two scholarships so that one scholarship receives 3/2 as much money as the other. If the total amount given to the two scholarships is $390, find the amount that goes to the scholarship that receives the lesser amount.

11. A new instructor is paid 3/4 the salary of an experienced professor. In a certain college, the total salary paid an instructor and a professor was $56,000. Find the salary paid the professor.

12. A child takes 5/8 the number of pills that an adult takes for the same illness. Together the child and the adult use 26 pills. Find the number used by the adult.

Solve each problem. See Example 2.

13. Sam can row 4 miles per hour in still water. It takes as long to row 8 miles upstream as 24 miles downstream. How fast is the current?

	d	r	t
Upstream	8	$4 - x$	
Downstream	24	$4 + x$	

14. Mary flew from Philadelphia to Des Moines at 180 miles per hour and from Des Moines to Philadelphia at 150 miles per hour. The trip at the slower speed took 1 hour longer than the trip at the faster speed. Find the distance between the two cities. (Assume there was no wind in either direction.)

	d	r	t
P to D	x	180	
D to P	x	150	

15. On a business trip, Arlene traveled to her destination at an average speed of 60 miles per hour. Coming home, her average speed was 50 miles per hour and the trip took 1/2 hour longer. How far did she travel each way?

16. A boat goes 210 miles downriver in the same time it can go 140 miles upriver. The speed of the current is 5 miles per hour. Find the speed of the boat in still water.

Solve each problem. See Example 3.

17. Paul can tune up his Toyota in 2 hours. His friend Marco can do the job in 3 hours. How long would it take them if they worked together?

18. George can paint a room, working alone, in 8 hours. Jenny can paint the same room, working alone, in 6 hours. How long will it take them if they work together?

19. Machine A can make all the bolts for an order in 7 hours, but machine B takes 12 hours. How long will it take the two machines working together?

20. One pipe can fill a swimming pool in 6 hours, and another pipe can do it in 9 hours. How long will it take the two pipes working together to fill the pool 3/4 full?

21. Dennis can do a job in 4 days. When Dennis and Sue work together, the job takes 2 1/3 days. How long would the job take Sue working alone?

22. An inlet pipe can fill a swimming pool in 9 hours, and an outlet pipe can empty the pool in 12 hours. Through an error, both pipes are left open. How long will it take to fill the pool?

23. A cold water faucet can fill a sink in 12 minutes, and a hot water faucet can fill it in 15. The drain can empty the sink in 25 minutes. If both faucets are on and the drain is open, how long will it take to fill the sink?

24. Refer to Exercise 22. Assume the error was discovered after both pipes had been running for 3 hours, and the outlet pipe was then closed. How much more time would then be required to fill the pool? (*Hint:* How much of the job had been done when the error was discovered?)

Solve the following problems about variation. See Examples 4 and 5.

25. If x varies directly as y, and $x = 9$ when $y = 2$, find x when y is 6.

26. If z varies directly as x, and $z = 15$ when $x = 4$, find z when x is 8.

27. If m varies directly as p^2, and $m = 20$ when $p = 2$, find m when p is 5.

28. If a varies directly as b^2, and $a = 48$ when $b = 4$, find a when $b = 7$.

29. If p varies inversely as q^2, and $p = 4$ when $q = 1/2$, find p when $q = 3/2$.

30. If z varies inversely as x^2, and $z = 9$ when $x = 2/3$, find z when $x = 5/4$.

31. The circumference of a circle varies directly as the radius. A circle with a radius of 7 centimeters has a circumference of 43.96 centimeters. Find the circumference if the radius changes to 11 centimeters.

32. The pressure exerted by a certain liquid at a given point varies directly as the depth of the point beneath the surface of the liquid. The pressure at 10 feet is 50 pounds per square inch. What is the pressure at 20 feet?

33. The force required to compress a spring varies directly as the change in the length of the spring. If a force of 12 pounds is required to compress a certain spring 3 inches, how much force is required to compress the spring 5 inches?

34. The illumination produced by a light source varies inversely as the square of the distance from the source. If the illumination produced 4 feet from a light source is 75 footcandles, find the illumination produced 9 feet from the same source.

Solve each problem.

35. An experienced employee can enter tax data from standard returns into a computer twice as fast as a new employee. Working together, it takes the employees 2 hours to enter a fixed amount of data. How long would it take the experienced employee working alone to enter the same amount of data?

36. If r varies inversely as s, and $r = 7$ when $s = 8$, find r when $s = 12$.

37. If three times a number is added to twice its reciprocal, the answer is 5. Find the number.

38. Rae flew her airplane 500 miles against the wind in the same time it took her to fly it 600 miles with the wind. If the speed of the wind was 10 miles per hour, what is the speed of her plane in still air?

39. If y varies inversely as x, and $y = 10$ when $x = 3$, find y when $x = 12$.

40. The current in a simple electrical circuit varies inversely as the resistance. If the current is 50 amps (an *amp* is a unit for measuring current) when the resistance is 10 ohms (an *ohm* is a unit for measuring resistance), find the current if the resistance is 5 ohms.

41. The distance from Seattle, Washington, to Victoria, British Columbia, is about 148 miles by ferry. It takes about 4 hours less to travel by ferry from Victoria to Vancouver, British Columbia, a distance of about 74 miles. What is the average speed of the ferry?

42. If twice a number is subtracted from three times its reciprocal, the result is 1. Find the number.

43. For a body falling freely from rest (disregarding air resistance), the distance the body falls varies directly as the square of the time. If an object is dropped from the top of a tower 400 feet high and hits the ground in 5 seconds, how far did it fall in the first 3 seconds?

44. One painter can paint a house three times faster than another. Working together, they can paint a house in 4 days. How long would it take the faster painter working alone?

Review Exercises   *Find the value of y when (a) x* = 2 *and (b) x* = −4. *See Sections 1.4 and 2.4.*

45. $y = 4x - 7$ **46.** $y = 3 - 2x$ **47.** $4x - y = 1$ **48.** $y + 3x = 2$

49. $3x + 7y = 10$ **50.** $2x - 3y = 5$ **51.** $x = -3y$ **52.** $y = 2x$

Chapter 5   *Summary*

Multiplying Rational Expressions	The product of the rational expressions P/Q and R/S is $$\frac{P}{Q} \cdot \frac{R}{S} = \frac{PR}{QS}.$$
Dividing Rational Expressions	If P/Q and R/S are any two rational expressions, with $R/S \neq 0$, then their quotient is $$\frac{P}{Q} \div \frac{R}{S} = \frac{P}{Q} \cdot \frac{S}{R} = \frac{PS}{QR}.$$
Adding and Subtracting Rational Expressions	If P/Q and R/Q are rational expressions, then $$\frac{P}{Q} + \frac{R}{Q} = \frac{P + R}{Q} \quad \text{and} \quad \frac{P}{Q} - \frac{R}{Q} = \frac{P - R}{Q}.$$
Variation	y varies directly as x $y = kx$
	y varies directly as the square of x $y = kx^2$
	m varies inversely as p $m = \dfrac{k}{p}$
	r varies inversely as the square of s $r = \dfrac{k}{s^2}$

Chapter 5 *Review Exercises*

❶ [5.1] *Find any values for which the following expressions are undefined.*

1. $\dfrac{2}{7x}$

2. $\dfrac{3}{m-5}$

3. $\dfrac{r-3}{r^2-2r-8}$

4. $\dfrac{3z+5}{2z^2+5z-3}$

Find the numerical value of each expression when (a) $x = 3$ and (b) $x = -1$.

5. $\dfrac{x^2}{x+2}$

6. $\dfrac{5x+3}{2x-1}$

7. $\dfrac{8x}{x^2-2}$

8. $\dfrac{x-5}{x-3}$

Write each expression in lowest terms.

9. $\dfrac{15p^2}{5p}$

10. $\dfrac{6y^2z^3}{9y^4z^2}$

11. $\dfrac{9x^2-16}{6x+8}$

12. $\dfrac{m-5}{5-m}$

[5.2] *Find each product or quotient. Write each answer in lowest terms.*

13. $\dfrac{10p^5}{5} \div \dfrac{3p^7}{20}$

14. $\dfrac{8z^2}{(4z)^3} \div \dfrac{4z^5}{32z}$

15. $\dfrac{7y+14}{8y-5} \div \dfrac{4y+8}{16y-10}$

16. $\dfrac{3k+5}{k+2} \cdot \dfrac{k^2-4}{18k^2-50}$

17. $\dfrac{2p^2+3p-2}{p^2+5p+6} \cdot \dfrac{p^2-2p-15}{2p^2-7p-15}$

18. $\dfrac{8r^2+23r-3}{64r^2-1} \div \dfrac{r^2-4r-21}{64r^2+16r+1}$

[5.3] *Find the least common denominator for the following fractions.*

19. $\dfrac{1}{15}, \dfrac{7}{30}, \dfrac{4}{45}$

20. $\dfrac{3}{8y}, \dfrac{7}{12y^2}, \dfrac{1}{16y^3}$

21. $\dfrac{1}{y^2+2y}, \dfrac{4}{y^2+6y+8}$

22. $\dfrac{3}{z^2+z-6}, \dfrac{2}{z^2+4z+3}$

Rewrite each rational expression with the given denominator.

23. $\dfrac{4}{9}, \dfrac{}{45}$

24. $\dfrac{12}{m}, \dfrac{}{5m}$

25. $\dfrac{3}{8m^2}, \dfrac{}{24m^3}$

26. $\dfrac{12}{y-4}, \dfrac{}{8y-32}$

27. $\dfrac{-2k}{3k+15}, \dfrac{}{15k+75}$

28. $\dfrac{12y}{y^2-y-2}, \dfrac{}{(y-2)(y+1)(y-4)}$

[5.4] *Add or subtract as indicated. Write all answers in lowest terms.*

29. $\dfrac{11}{3r} - \dfrac{8}{3r}$

30. $\dfrac{b}{b+5} + \dfrac{5}{b+5}$

31. $\dfrac{7}{k} - \dfrac{2}{5}$

32. $\dfrac{3+5m}{2} - \dfrac{m}{4}$

33. $\dfrac{2}{y+1} - \dfrac{3}{y-1}$

34. $\dfrac{2}{r-5} + \dfrac{3}{5-r}$

35. $\dfrac{10}{p^2-2p} - \dfrac{2}{p^2-5p+6}$

[5.5] *Simplify each complex fraction.*

36. $\dfrac{\dfrac{5k-1}{k}}{\dfrac{4k+3}{8k}}$

37. $\dfrac{\dfrac{6}{r}-1}{\dfrac{6-r}{4r}}$

38. $\dfrac{\dfrac{1}{a+b}-1}{\dfrac{1}{a+b}+1}$

[5.6] *Solve each equation. Check your answers.*

39. $\dfrac{2}{p}=\dfrac{5}{8}$

40. $\dfrac{3}{k}-\dfrac{2}{k}=7$

41. $\dfrac{y}{2}-\dfrac{y}{5}=6$

42. $\dfrac{3}{4}r-1=-4$

43. $\dfrac{z+3}{8}=\dfrac{z-2}{3}$

44. $\dfrac{2}{z}=\dfrac{z+1}{z+3}$

45. $\dfrac{3y-1}{y-2}=\dfrac{5}{y-2}+1$

46. $\dfrac{3}{m-2}+\dfrac{1}{m-1}=\dfrac{7}{m^2-3m+2}$

47. $\dfrac{p+2}{p^2-1}+\dfrac{2p}{p^2+2p+1}=\dfrac{2}{p-1}$

Solve for the specified variable.

48. $m=\dfrac{Rv}{t};$ for t

49. $x=\dfrac{3y}{2y+z};$ for y

[5.7] *Solve each word problem.*

50. When half a number is subtracted from two thirds of a number, the result is 2. Find the number.

51. A certain fraction has a numerator that is 4 more than the denominator. If 6 is subtracted from the denominator, the result is 3. Find the original fraction.

52. The commission received by a salesperson for selling a small car is 2/3 that received for selling a large car. On a recent day, Linda sold one of each, earning a commission of $300. Find the commission for each type of car.

53. Kerrie flew her plane 400 kilometers with the wind in the same time it took her to go 200 kilometers against the wind. The speed of the wind is 50 kilometers per hour. Find the speed of the plane in still air.

54. A man can plant his garden in 5 hours, working alone. His daughter can do the same job in 8 hours. How long would it take them if they worked together?

55. One painter can paint a house twice as fast as another. Working together, they can paint the house in 1 1/3 days. How long would it take the faster painter working alone?

56. If m varies directly as q^2, and $m=8$ when $q=4$, find m when $q=6$.

57. If r varies inversely as s, and $r=9$ when $s=1/2$, find r when $s=2$.

58. If z varies inversely as y^2, and $z=5$ when $y=2$, find z when $y=3/4$.

❷ *Perform the indicated operations.*

59. $\dfrac{3p + q}{5} - \dfrac{p - q}{3}$

60. $\dfrac{6 - y}{6 + y} \div \dfrac{y - 6}{y + 6}$

61. $\dfrac{z + \dfrac{1}{x}}{z - \dfrac{1}{x}}$

62. $\dfrac{z^2 + z - 2}{z^2 + 7z + 10} \div \dfrac{z - 3}{z + 5}$

63. $\dfrac{8}{r^2} - \dfrac{3}{2r}$

Solve the following.

64. $\dfrac{1}{k} + \dfrac{3}{r} = \dfrac{5}{z};$ for r

65. $\dfrac{5 + m}{m} + \dfrac{3}{4} = \dfrac{-2}{m}$

66. When Mary and Sue work together on a job, they can do it in 3 3/7 days. Mary can do the job working alone in 8 days. How long would it take Sue working alone?

67. $\dfrac{y}{3} - \dfrac{y - 2}{8} = -1$

68. If x varies directly as y, and $x = 12$ when $y = 5$, find x when $y = 3$.

69. Five times a number is added to three times the reciprocal of the number, giving 17/2. Find the number.

70. A boat goes 7 miles per hour in still water. It takes as long to go 20 miles upstream as 50 miles downstream. Find the speed of the current.

Chapter 5 *Test*

1. Find any values for which $\dfrac{8k + 1}{k^2 - 4k + 3}$ is undefined.

2. Find the numerical value of $\dfrac{6r + 1}{2r^2 - 3r - 20}$ when (a) $r = -1$ and (b) $r = 4$.

Write each rational expression in lowest terms.

3. $\dfrac{8m^2p^2}{6m^3p^5}$

4. $\dfrac{5y^3 - 5y}{2y + 2}$

Multiply or divide, as indicated. Write all answers in lowest terms.

5. $\dfrac{a^6b}{a^3} \cdot \dfrac{b^2}{a^2b^3}$

6. $\dfrac{8y - 16}{9} \div \dfrac{6 - 3y}{5}$

7. $\dfrac{6m^2 - m - 2}{8m^2 + 10m + 3} \cdot \dfrac{4m^2 + 7m + 3}{3m^2 + 5m + 2}$

8. $\dfrac{5a^2 + 7a - 6}{2a^2 + 3a - 2} \div \dfrac{5a^2 + 17a - 12}{2a^2 + 5a - 3}$

Find the least common denominator for the following fractions.

9. $\dfrac{3}{10p^2}, \dfrac{1}{25p^3}, \dfrac{-7}{30p^5}$

10. $\dfrac{r-1}{2r^2+7r+6}, \dfrac{2r+1}{2r^2-7r-15}$

Rewrite each rational expression with the given denominator.

11. $\dfrac{11}{7r}, \dfrac{}{49r^2}$

12. $\dfrac{5}{8m-16}, \dfrac{}{24m^2-48m}$

Add or subtract as indicated. Write all answers in lowest terms.

13. $\dfrac{5}{x} - \dfrac{6}{x}$

14. $\dfrac{-3}{a+1} + \dfrac{5}{6a+6}$

15. $\dfrac{m^2}{m-3} + \dfrac{m+1}{3-m}$

16. $\dfrac{3}{2k^2+3k-2} - \dfrac{k}{k^2+3k+2}$

Simplify each complex fraction.

17. $\dfrac{\dfrac{2p}{k^2}}{\dfrac{3p^2}{k^3}}$

18. $\dfrac{\dfrac{1}{p+4} - 2}{\dfrac{1}{p+4} + 2}$

Solve each equation.

19. $\dfrac{3}{2p} + \dfrac{12}{5p} = \dfrac{13}{20}$

20. $\dfrac{p}{p-2} = \dfrac{2}{p-2} + 3$

21. $\dfrac{2}{z^2-2z-3} = \dfrac{3}{z-3} + \dfrac{2}{z+1}$

For each problem, write an equation and solve it.

22. If four times a number is added to the reciprocal of twice the number, the result is 3. Find the number.

23. The current in a river is 5 miles per hour. A boat can go 125 miles downstream in the same time as 75 miles upstream. Find the speed of the boat in still water.

24. A man can paint a room in his house, working alone, in 5 hours. His wife can do the job in 4 hours. How long will it take them to paint the room if they work together?

25. If x varies directly as y, and $x = 8$ when $y = 12$, find x when $y = 28$.

Graphing Linear Equations

Objectives

1   Write a solution as an ordered pair.
2   Decide whether a given ordered pair is a solution of a given equation.
3   Complete ordered pairs for a given equation.
4   Understand coordinate systems.
5   Plot ordered pairs.

The equations discussed so far, such as

$$3x + 5 = 12 \quad \text{or} \quad 2x^2 + x + 5 = 0,$$

have contained only one variable. Equations in two variables, like

$$y = 4x + 5 \quad \text{or} \quad 2x + 3y = 6,$$

are discussed in this chapter. Both of these equations are examples of *linear equations* in two variables.

Linear Equation	A **linear equation** in two variables is an equation that can be put in the form

$$Ax + By = C,$$

where A, B, and C are real numbers and A and B are not both 0.

1 A solution of a linear equation requires two numbers, one for each variable. For example, the equation $y = 4x + 5$ is satisfied if x is replaced with 2 and y is replaced with 13, since

$$13 = 4(2) + 5.$$

Thus, $x = 2$ and $y = 13$ is a solution of the equation $y = 4x + 5$. The phrase "$x = 2$ and $y = 13$" is abbreviated

$$(2, 13).$$

This abbreviation gives the x-value, 2, and the y-value, 13, as a pair of numbers, written inside parentheses. The x-value is always given first. A pair of numbers written in this order is called an **ordered pair.**

Of course, letters other than x and y may be used in an equation. For example, one solution to the equation $3p - q = -11$ is $p = -2$ and $q = 5$. If the ordered pairs are (p, q), then this solution is written as $(-2, 5)$.

2 The next example shows how to decide whether an ordered pair is a solution of an equation.

Example 1 Decide whether the given ordered pair is a solution of the given equation.

(a) $(3, 2)$; $2x + 3y = 12$

Decide whether $(3, 2)$ is a solution of the equation $2x + 3y = 12$ by substituting 3 for x and 2 for y in the given equation.

$$2x + 3y = 12$$
$$2(3) + 3(2) = 12 \qquad \text{Let } x = 3; \text{ let } y = 2$$
$$6 + 6 = 12$$
$$12 = 12 \qquad \text{True}$$

This result is true, so $(3, 2)$ is a solution of $2x + 3y = 12$.

(b) $(-2, -7)$; $2m + 3n = 12$

$$2(-2) + 3(-7) = 12 \qquad \text{Let } m = -2; \text{ let } n = -7$$
$$-4 + (-21) = 12$$
$$-25 = 12 \qquad \text{False}$$

This result is false, so $(-2, -7)$ is *not* a solution of $2m + 3n = 12$.   ✦

3 By choosing a number for one variable in a linear equation, the value of the other variable can be found, as shown in the next example.

Example 2 Complete the given ordered pairs for the equation $y = 4x + 5$.

(a) $(7, \quad)$

In this ordered pair, $x = 7$. (Remember that x always come first.) Find the corresponding value of y by replacing x with 7 in the equation $y = 4x + 5$.

$$y = 4(7) + 5 = 28 + 5 = 33$$

This gives the ordered pair $(7, 33)$.

(b) $(\quad, 13)$

In this ordered pair, $y = 13$. Find the value of x by replacing y with 13 in the equation and then solving for x.

$$
\begin{array}{ll}
y = 4x + 5 & \\
13 = 4x + 5 & \text{Let } y = 13 \\
8 = 4x & \text{Subtract 5 from both sides} \\
2 = x & \text{Divide both sides by 4}
\end{array}
$$

The ordered pair is $(2, 13)$. ✛

Example 3 Complete the given ordered pairs for the equation $5x - y = 24$.

Equation	*Ordered Pairs*
$5x - y = 24$	$(5, \quad)$ $(-3, \quad)$ $(0, \quad)$

Find the y-value for the ordered pair $(5, \quad)$ by replacing x with 5 in the given equation and solving for y.

$$
\begin{array}{ll}
5x - y = 24 & \\
5(5) - y = 24 & \text{Let } x = 5 \\
25 - y = 24 & \\
-y = -1 & \text{Subtract 25 from both sides} \\
y = 1 &
\end{array}
$$

This gives the ordered pair $(5, 1)$. ⌐

Complete the ordered pair $(-3, \quad)$ by letting $x = -3$ in the given equation. Also, complete $(0, \quad)$ by letting $x = 0$.

If	$x = -3$,		If	$x = 0$,
then	$5x - y = 24$		then	$5x - y = 24$
becomes	$5(-3) - y = 24$		becomes	$5(0) - y = 24$
	$-15 - y = 24$			$0 - y = 24$
	$-y = 39$			$-y = 24$
	$y = -39.$			$y = -24.$

The completed ordered pairs are as follows.

Equation	*Ordered Pairs*
$5x - y = 24$	$(5, 1)$ $(-3, -39)$ $(0, -24)$ ✛

Example 4 Complete the given ordered pairs for the equation $x = 5$.

$$\begin{array}{cc} \textit{Equation} & \textit{Ordered Pairs} \\ x = 5 & (\ , -2)\ (\ , 6)\ (\ , 3) \end{array}$$

The given equation is $x = 5$. No matter which value of y is chosen, the value of x is always the same, 5. Each ordered pair can be completed by placing 5 in the first position.

$$\begin{array}{cc} \textit{Equation} & \textit{Ordered Pairs} \\ x = 5 & (5, -2)\ (5, 6)\ (5, 3) \end{array}$$ ✚

When an equation such as $x = 5$ is discussed along with equations of two variables, think of $x = 5$ as an equation in two variables by rewriting $x = 5$ as $x + 0y = 5$. This form of the equation shows that for any value of y, the value of x is 5.

Each of the equations discussed in this section has many ordered pairs as solutions. Each choice of a real number for one variable will lead to a particular real number for the other variable. This is true of linear equations in general: linear equations in two variables have an infinite number of solutions.

4 In Chapter 2 a number line was used to graph solutions of equations in one variable. Techniques for graphing the solutions of an equation in two variables are shown in this section. Since the solutions of such an equation are ordered pairs of numbers in the form (x, y), two number lines are needed, one for x and one for y. These two number lines are drawn as shown in Figure 6.1. The horizontal number line is called the **x-axis.** The vertical line is called the **y-axis.** Together, the x-axis and y-axis form a **coordinate system.**

The coordinate system is divided into four regions, called **quadrants.** These quadrants are numbered counterclockwise, as shown in Figure 6.1. Points on the axes themselves are not in any quadrant. The point at which the x-axis and y-axis meet is called the **origin.** The origin is labeled 0 in Figure 6.1.

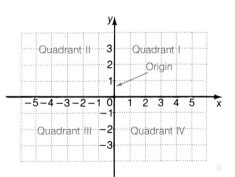

Figure 6.1

5   By referring to the two axes, every point on the plane can be associated with an ordered pair. The numbers in the ordered pair are called the **coordinates** of the point. For example, locate the point associated with the ordered pair (2, 3), by starting at the origin. Since the *x*-coordinate is 2, go 2 units to the right along the *x*-axis. Then since the *y*-coordinate is 3, turn and go up 3 units on a line parallel to the *y*-axis. This is called **plotting** the point (2, 3). (See Figure 6.2.) From now on the point (2, 3) will be used to refer to the point with *x*-coordinate 2 and *y*-coordinate 3.

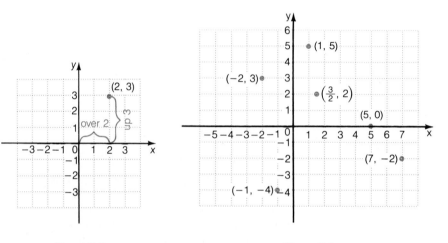

Figure 6.2   Figure 6.3

Example 5   Plot the given points on a coordinate system.

    (a) (1, 5)   **(b)** (−2, 3)   **(c)** (−1, −4)

    (d) (7, −2)   **(e)** $\left(\dfrac{3}{2}, 2\right)$ **(f)** (5, 0)

    Locate the point (−1, −4), for example, by first going 1 unit to the left along the *x*-axis. Then turn and go 4 units down, parallel to the *y*-axis. Plot the point (3/2, 2), by going 3/2 (or 1 1/2) units to the right along the *x*-axis. Then turn and go 2 units up, parallel to the *y*-axis. Figure 6.3 shows the graphs of the points in this example.   ✚

Example 6   A company has found that the cost to produce *x* small calculators is

$$y = 25x + 250,$$

where *y* represents the cost in cents. Complete the given ordered pairs for this equation.

             (1,) (2,) (3,)

To complete the ordered pair (1,   ), let $x = 1$.

$$y = 25x + 250$$
$$y = 25(1) + 250 \qquad \text{Let } x = 1$$
$$y = 25 + 250$$
$$y = 275$$

This gives the ordered pair (1, 275), which says that the cost to produce 1 calculator is 275 cents or $2.75. Complete the ordered pairs (2,   ) and (3,   ) as follows.

$$y = 25x + 250 \qquad\qquad y = 25x + 250$$
$$y = 25(2) + 250 \qquad\qquad y = 25(3) + 250$$
$$y = 50 + 250 \qquad\qquad y = 75 + 250$$
$$y = 300 \qquad\qquad\qquad y = 325$$

This gives the ordered pairs (2, 300) and (3, 325).

Check that other ordered pairs that satisfy the equation $y = 25x + 250$ are (4, 350), (5, 375), (6, 400), (7, 425), (8, 450), and (9, 475). All of the ordered pairs from this example are plotted in Figure 6.4. A different scale was used on the y-axis, since the y-values in the ordered pairs are much larger than the x-values. Here, each square represents 100 in the vertical direction.   ✦

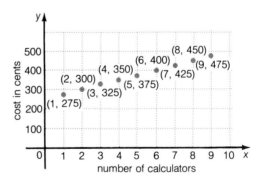

Figure 6.4

6.1   Exercises

Decide whether the given ordered pair is a solution of the given equation. See Example 1.

1. $x + y = 9$;   (2, 7)

2. $3x + y = 8$;   (0, 8)

3. $2x - y = 6$;   (2, -2)

4. $2x + y = 5$;   (2, 1)

5. $4x - 3y = 6$;   (1, 2)

6. $5x - 3y = 1$;   (0, 1)

7. $y = 3x$;   (1, 3)

8. $x = -4y$;   (8, -2)

9. $x = -6$;   (-6, 8)

10. $y = 2$;   (9, 2)

11. $x + 4 = 0$;   (-5, 1)

12. $x - 6 = 0$;   (5, -1)

Complete the given ordered pairs for the equation $y = 3x + 5$. See Example 2.

13. $(2, \quad)$ **14.** $(5, \quad)$ **15.** $(8, \quad)$ **16.** $(0, \quad)$

17. $(-3, \quad)$ **18.** $(-4, \quad)$ **19.** $(\quad, 14)$ **20.** $(\quad, -10)$

Complete the ordered pairs, using the given equations. See Example 3.

Equation	Ordered Pairs		
21. $y = 2x + 1$	$(3, \quad)$	$(0, \quad)$	$(-1, \quad)$
22. $y = 3x - 5$	$(2, \quad)$	$(0, \quad)$	$(-3, \quad)$
23. $y = 8 - 3x$	$(2, \quad)$	$(0, \quad)$	$(-3, \quad)$
24. $y = -2 - 5x$	$(4, \quad)$	$(0, \quad)$	$(-4, \quad)$
25. $2a + b = 9$	$(0, \quad)$	$(3, \quad)$	$(12, \quad)$
26. $-3m + n = 4$	$(1, \quad)$	$(0, \quad)$	$(-2, \quad)$

Complete the ordered pairs using the given equations. See Example 4.

Equation	Ordered Pairs		
27. $x = -4$	$(\quad, 6)$	$(\quad, 2)$	$(\quad, -3)$
28. $y = -8$	$(4, \quad)$	$(0, \quad)$	$(-4, \quad)$
29. $x + 9 = 0$	$(\quad, 8)$	$(\quad, 3)$	$(\quad, 0)$
30. $y - 4 = 0$	$(9, \quad)$	$(-5, \quad)$	$(0, \quad)$
31. $y = -3x$	$(-2, \quad)$	$(0, \quad)$	$(\quad, -6)$
32. $x = 4y$	$(0, \quad)$	$(\quad, 3)$	$(-8, \quad)$

Give the x- and y-coordinates of the following points labeled in the figure.

33. A **34.** B **35.** C **36.** D **37.** E **38.** F

Plot the following ordered pairs on a coordinate system. See Example 5.

39. $(6, 1)$ **40.** $(4, -2)$ **41.** $(-4, -5)$ **42.** $(-2, 4)$

43. $(-5, -1)$ **44.** $(-3, 5)$ **45.** $(3, -5)$ **46.** $(4, 0)$

47. $(-2, 0)$ **48.** $(0, 6)$ **49.** $(0, -5)$ **50.** $(0, 0)$

Without plotting the given point, state the quadrant in which each point lies.

51. (2, 3)   **52.** (2, −3)   **53.** (−2, 3)   **54.** (−2, −3)
55. (−1, −1)   **56.** (4, 7)   **57.** (−3, 6)   **58.** (1, −5)
59. (5, −4)   **60.** (9, −1)   **61.** (0, 0)   **62.** (−2, 0)

Complete the ordered pairs using the given equation. Then plot the ordered pairs. See Examples 2–5.

Equation	*Ordered Pairs*
63. $y = 2x + 6$	(0,) (2,) (, 0) (, 2)
64. $y = 8 − 4x$	(0,) (3,) (, 0) (, 16)
65. $3x + 5y = 15$	(0,) (10,) (, 0) (, 6)
66. $2x − 5y = 10$	(0,) (10,) (, 0) (, −6)
67. $y = 3x$	(0,) (−2,) (4,) (, −3)
68. $x + 2y = 0$	(0,) (, 3) (4,) (, −1)
69. $y + 2 = 0$	(5,) (0,) (−3,) (−2,)
70. $x − 4 = 0$	(, 7) (, 0) (, −4) (, 4)

In statistics, ordered pairs are used to decide whether two quantities (such as the height and weight of an individual) are related in such a way that one can be predicted when given the other. Ordered pairs that give these quantities for a number of individuals are plotted on a graph (called a scatter diagram*). If the points lie approximately on a line, the variables have a linear relationship.*

71. Make a scatter diagram by plotting the following ordered pairs of heights and weights for six women on the given axes: (62, 105),   (65, 130),   (67, 142),   (63, 115),   (66, 120),   (60, 98).

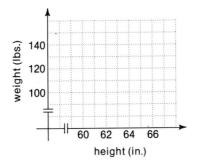

72. In Exercise 71, is there a linear relationship between height and weight? (Do the points lie approximately on a straight line?)

73. Make a scatter diagram by plotting the following points, which give the annual cost (in thousands of dollars) to attend a private college for selected years: (1984, 9.2),   (1985, 9.8),   (1986, 10.5),   (1987, 11.0). Put years on the horizontal axis. Is this a linear relationship?

74. Repeat Exercise 73 using the following points which give the annual cost (in thousands of dollars) to attend a public college for selected years: (1984, 4.6), (1985, 5.0), (1986, 5.3), (1987, 5.8). Is this a linear relationship?

Review Exercises   *Solve each equation. See Section 2.4.*

75. $3m + 5 = 13$

76. $2k - 7 = 9$

77. $3 - y = 12$

78. $-4 - x = 6$

79. $7 + 5p = 12$

80. $9 - 3r = 15$

81. $8 - q = -7$

82. $-3 + 8z = -19$

6.2   Graphing Linear Equations

Objectives

1 Graph linear equations by completing and plotting ordered pairs.

2 Find intercepts.

3 Graph linear equations with just one intercept.

4 Graph linear equations of the form y = a number or x = a number.

1   There are an infinite number of ordered pairs that satisfy an equation in two variables. For example, to find ordered pairs that are solutions of the equation $x + 2y = 7$, choose as many values of x (or y) as desired, and then complete each ordered pair.

For instance, if $x = 1$ is chosen, then

$$x + 2y = 7$$
$$1 + 2y = 7 \qquad \text{Let } x = 1$$
$$2y = 6$$
$$y = 3,$$

so the ordered pair (1, 3) is a solution of the equation. In the same way, $(-3, 5)$, (3, 2), (0, 7/2), $(-2, 9/2)$, $(-1, 4)$, (6, 1/2), and (7, 0) are all solutions of the equation $x + 2y = 7$. These ordered pairs have been plotted in Figure 6.5.

Notice that the points plotted in this figure all lie on a straight line. The line that goes through these points is shown in Figure 6.6. In fact, all ordered pairs satisfying the equation $x + 2y = 7$ produce points that lie on this same straight line. This graph gives a "picture" of all the solutions of the equation $x + 2y = 7$. Only a portion of the line is shown here, but it extends indefinitely in both directions as suggested by the arrowhead on each end of the line.

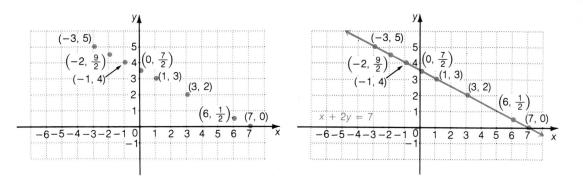

Figure 6.5   **Figure 6.6**

It can be shown that

the graph of any linear equation in two variables is a straight line.

(Remember that the word *line* appears in the name "*linear* equation.")

Since two distinct points determine a line, a straight line can be graphed by finding any two different points on the line. However, it is a good idea to plot a third point as a check.

Example 1   Graph the linear equation $3x + 2y = 6$.

For most linear equations, two different points on the graph can be found by first letting $x = 0$, and then letting $y = 0$.

If $x = 0$:	If $y = 0$:
$3x + 2y = 6$	$3x + 2y = 6$
$3(0) + 2y = 6$	$3x + 2(0) = 6$
$0 + 2y = 6$	$3x + 0 = 6$
$2y = 6$	$3x = 6$
$y = 3.$	$x = 2.$

This gives the ordered pairs $(0, 3)$ and $(2, 0)$. Get a third point (as a check), by letting x or y equal some other number. For example, let $x = -2$. (Any other number could have been used instead.) Replace x with -2 in the given equation.

$$3x + 2y = 6$$
$$3(-2) + 2y = 6$$
$$-6 + 2y = 6$$
$$2y = 12$$
$$y = 6$$

Now plot these three ordered pairs, (0, 3), (2, 0), and (−2, 6), which should lie on a line, and draw a line through them. This line, shown in Figure 6.7, is the graph of $3x + 2y = 6$. A **table of values** showing the three ordered pairs is given next to Figure 6.7. Making a table of values is a useful way to organize the ordered pairs used to graph a linear equation. ✦

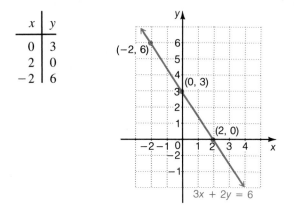

x	y
0	3
2	0
−2	6

Figure 6.7

Example 2 Graph the linear equation $4x - 5y = 20$.

At least two different points are needed to draw the graph. First let $x = 0$ and then let $y = 0$ to complete two ordered pairs.

$$4x - 5y = 20 \qquad\qquad 4x - 5y = 20$$
$$4(0) - 5y = 20 \qquad\qquad 4x - 5(0) = 20$$
$$-5y = 20 \qquad\qquad\qquad 4x = 20$$
$$y = -4 \qquad\qquad\qquad\quad x = 5$$

The ordered pairs are (0, −4) and (5, 0). Get a third ordered pair (as a check), by choosing some number other than 0 for x or y; for example, choose $y = 2$. Replacing y with 2 in the equation $4x - 5y = 20$ leads to the ordered pair (15/2, 2). These ordered pairs are shown in the table next to Figure 6.8. ✦

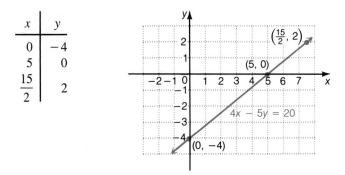

x	y
0	−4
5	0
$\dfrac{15}{2}$	2

Figure 6.8

2   In Figure 6.8 the graph crosses the y-axis at $(0, -4)$ and the x-axis at $(5, 0)$. For this reason, -4 is called the **y-intercept** and 5 is called the **x-intercept** of the graph. The intercepts are particularly useful for graphing linear equations, as shown in Examples 1 and 2. The intercepts are found by replacing each variable, in turn, with 0 in the equation and solving for the value of the other variable.

Finding Intercepts	Find the x-intercept by letting $y = 0$ in the given equation and solving for x.
	Find the y-intercept by letting $x = 0$ in the given equation and solving for y.

Example 3   Find the intercepts for the graph of $2x + y = 4$. Draw the graph.
Find the y-intercept by letting $x = 0$; find the x-intercept by letting $y = 0$.

$$2x + y = 4 \qquad\qquad 2x + y = 4$$
$$2(0) + y = 4 \qquad\qquad 2x + 0 = 4$$
$$0 + y = 4 \qquad\qquad 2x = 4$$
$$y = 4 \qquad\qquad x = 2$$

The y-intercept is 4. The x-intercept is 2. The graph with the two intercepts shown in color is given in Figure 6.9. Get a third point as a check. For example, choosing $x = 1$ gives $y = 2$. These three ordered pairs are shown in the table with Figure 6.9. Plot $(0, 4)$, $(2, 0)$, and $(1, 2)$ and draw a line through them. This line, shown in Figure 6.9, is the graph of $2x + y = 4$.   ✦

x	y
0	4
2	0
1	2

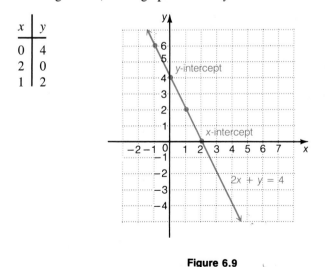

Figure 6.9

3   In the earlier examples, the x- and y-intercepts were used to help draw the graphs. This is not always possible, as the following examples show. Example 4 shows what to do when the x- and y-intercepts are the same point.

Example 4   Graph the linear equation $x = 3y$.

Letting $x = 0$ gives $y = 0$ and the ordered pair $(0, 0)$. If $y = 0$ then $x = 0$, giving $(0, 0)$, the same pair. Find another ordered pair satisfying $x = 3y$ by choosing, for example, $x = 6$, which gives $y = 2$ and the ordered pair $(6, 2)$. For a check point, choose -6 for x to get -2 for y. This gives the ordered pair $(-6, -2)$. These three ordered pairs were used to get the graph shown in Figure 6.10. ✛

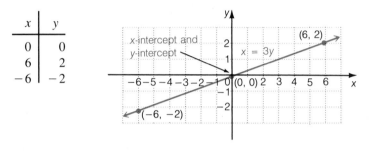

x	y
0	0
6	2
-6	-2

Figure 6.10

Generalizing from Example 4 gives the following result.

**Line Through
the Origin**

For any real numbers A and B, the graph of a linear equation of the form

$$x = By \quad \text{or} \quad y = Ax$$

goes through the origin, $(0, 0)$.

4   The equation $y = -4$ is a linear equation in which the coefficient of x is 0. (Write $y = -4$ as $y = 0x - 4$ to see this.) Also, $x = 3$ is a linear equation in which the coefficient of y is 0. These equations lead to horizontal or vertical straight lines, as the next examples show.

Example 5   Graph the linear equation $y = -4$.

As the equation states, for any value of x, y is always equal to -4. To get ordered pairs that are solutions of this equation, choose any numbers for x but always let y equal -4. Three ordered pairs that can be used are $(-2, -4)$, $(0, -4)$, and $(3, -4)$. Drawing a line through these points gives the horizontal line shown in Figure 6.11. ✛

Horizontal Line

The graph of the linear equation $y = k$, where k is a real number, is the horizontal line going through the point $(0, k)$.

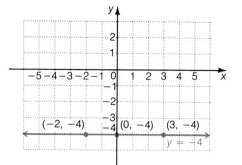

Figure 6.11

Example 6   Graph the linear equation $x = 3$.

   All the ordered pairs that are solutions of this equation have an x-value of 3. Any number can be used for y. Three ordered pairs that satisfy the equation are $(3, 3)$, $(3, 0)$, and $(3, -2)$. Drawing a line through these points gives the vertical line shown in Figure 6.12.   ✛

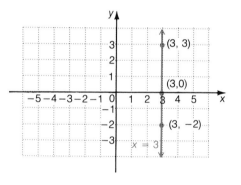

Figure 6.12

Vertical Line	The graph of the linear equation $x = k$, where k is a real number, is the vertical line going through the point $(k, 0)$.

   In particular, it is useful to notice that the horizontal line $y = 0$ is the x-axis and the vertical line $x = 0$ is the y-axis.

   The different forms of straight-line equations and the methods of graphing them are summarized on the next page.

Graphing Straight Lines

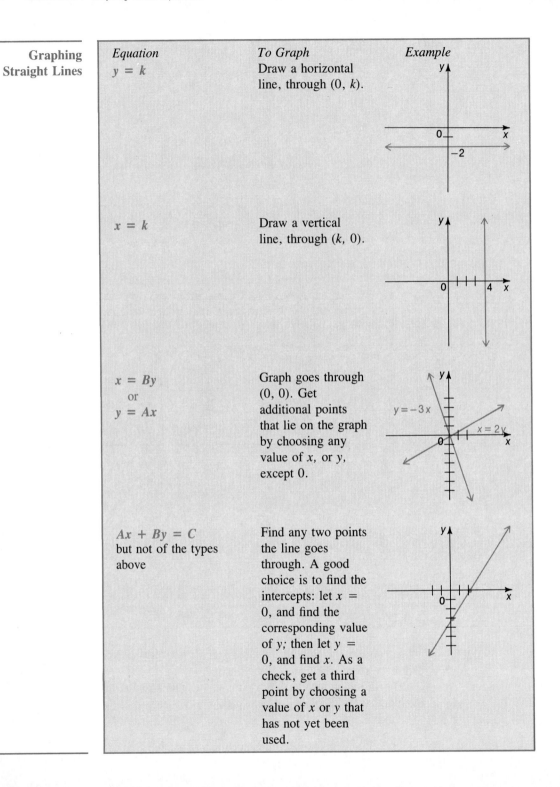

Equation	*To Graph*	*Example*
$y = k$	Draw a horizontal line, through $(0, k)$.	
$x = k$	Draw a vertical line, through $(k, 0)$.	
$x = By$ or $y = Ax$	Graph goes through $(0, 0)$. Get additional points that lie on the graph by choosing any value of x, or y, except 0.	
$Ax + By = C$ but not of the types above	Find any two points the line goes through. A good choice is to find the intercepts: let $x = 0$, and find the corresponding value of y; then let $y = 0$, and find x. As a check, get a third point by choosing a value of x or y that has not yet been used.	

6.2 Exercises

Complete the ordered pairs for each equation. Then graph the equation by plotting the points and drawing a line through them. See Examples 1 and 2.

1. $x + y = 5$ (0,   )   (, 0)   (2,   )

2. $y = x - 3$ (0,   )   (, 0)   (5,   )

3. $y = x + 4$ (0,   )   (, 0)   (-2,   )

4. $y + 5 = x$ (0,   )   (, 0)   (6,   )

5. $y = 3x - 6$ (0,   )   (, 0)   (3,   )

6. $x = 2y + 1$ (0,   )   (, 0)   (3,   )

7. $2x + 5y = 20$ (0,   )   (, 0)   (5,   )

8. $3x - 4y = 12$ (0,   )   (, 0)   (8,   )

9. $x + 5 = 0$ (, 2)   (, 0)   (, -3)

10. $y - 4 = 0$ (3,   )   (0,   )   (-2,   )

Find the intercepts for each equation. See Example 3.

11. $2x + 3y = 6$ **12.** $7x + 2y = 14$ **13.** $3x - 5y = 9$ **14.** $6x - 5y = 12$

15. $2y = 5x$ **16.** $x = -3y$ **17.** $-2x = 8$ **18.** $6y = 12$

Graph each linear equation. See Examples 1, 2, and 4–6.

19. $x - y = 2$ **20.** $x + y = 6$ **21.** $y = x + 4$ **22.** $y = x - 5$

23. $x + 2y = 6$ **24.** $3x - y = 6$ **25.** $4x = 3y - 12$ **26.** $5x = 2y - 10$

27. $3x = 6 - 2y$ **28.** $2x + 3y = 12$ **29.** $2x - 7y = 14$ **30.** $3x + 5y = 15$

31. $3x + 7y = 14$ **32.** $6x - 5y = 30$ **33.** $y = 2x$ **34.** $y = -3x$

35. $y + 6x = 0$ **36.** $x - 4y = 0$ **37.** $x + 2 = 0$ **38.** $y - 3 = 0$

39. $y = -1$ **40.** $x = 2$ **41.** $x = 0$ **42.** $y = 0$

Translate each of the statements of Exercises 43–48 into an equation. Then graph the equation.

43. The x-value is 2 more than the y-value.

44. The y-value is 3 less than the x-value.

45. The y-value is 3 less than twice the x-value.

46. The x-value is 4 more than three times the y-value.

47. If 3 is added to the y-value, the result is 4 less than twice the x-value.

48. If 6 is subtracted from 4 times the y-value, the result is three times the x-value.

49. The height h of a woman (in centimeters) can be estimated from the length of her radius bone r (from the wrist to the elbow) with the following formula:

$$h = 73.5 + 3.9r.$$

Estimate the heights of women with radius bones of the following lengths.

(a) 23 centimeters   (b) 25 centimeters   (c) 20 centimeters

(d) Graph $h = 73.5 + 3.9r$.

50. As a rough estimate, the weight of a man taller than about 60 inches is approximated by $y = 5.5x - 220$, where x is the height of the person in inches, and y is the weight in pounds. Estimate the weights of men whose heights are as follows.
    (a) 62 inches   (b) 64 inches   (c) 68 inches
    (d) 72 inches   (e) Graph $y = 5.5x - 220$.

51. The graph shows the average cost (in thousands of dollars) of attending college for the school years ending as shown. Estimated data is used for 1988 and 1989. Use the graph to estimate the following information.
    (a) The cost of attending a private school in 1986.
    (b) The cost of attending a public school in 1987.
    (c) In 1989, how much more will a private school cost than a public school?
    (d) During what year did the cost of a public school increase from \$5300 to \$5800?

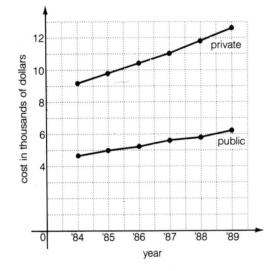

52. The demand for an item is closely related to its price. As price goes up, demand goes down. On the other hand, when price goes down, demand goes up. Suppose the demand for a certain small calculator is 500 when its price is \$30 and 4000 when its price is \$15.
    (a) Let x be the price and y be the demand (in thousands) for the calculator. Graph the two given pairs of prices and demand.
    (b) Assume the relationship is linear. Draw a line through the two points from part (a). From your graph estimate the demand if the price drops to \$10.

Review Exercises   *Find each quotient. See Section 1.9.*

53. $\dfrac{4 - 2}{8 - 5}$

54. $\dfrac{-3 - 5}{2 - 7}$

55. $\dfrac{-2 - (-4)}{3 - (-1)}$

56. $\dfrac{5 - (-7)}{-4 - (-1)}$

57. $\dfrac{-2 - (-5)}{-9 - 12}$

58. $\dfrac{-6 - 3}{4 - (-5)}$

59. $\dfrac{-9 - 4}{-2 - 3}$

60. $\dfrac{12 - (-4)}{-3 - 5}$

61. $\dfrac{-2 - (-7)}{3 - 3}$

6.3   The Slope of a Line

Objectives

1 Find the slope of a line given two points.

2 Find the slope from the equation of a line.

3 Use the slope to determine whether two lines are parallel, perpendicular, or neither.

A straight line can be graphed if at least two different points on the line are known. A line also can be graphed by using one point that the line goes through, along with the "steepness" of the line. The steepness of a line is measured by comparing the vertical change in the line (the rise) to the horizontal change (the run) going along the line from one fixed point to another.

1   Figure 6.13 shows a line with the points (x_1, y_1) and (x_2, y_2). (Read x_1 as x-sub-one and x_2 as x-sub-two.) Moving along the line from the point (x_1, y_1) to the point (x_2, y_2) causes y to change by $y_2 - y_1$ units while x changes by $x_2 - x_1$ units. The ratio of the change in y to the change in x gives the steepness of the line, called the *slope* of the line. The slope of a line is defined as follows.

Slope Formula

The **slope** of the line through the points (x_1, y_1) and (x_2, y_2) is

$$m = \frac{\text{change in } y}{\text{change in } x} = \frac{y_2 - y_1}{x_2 - x_1}.$$

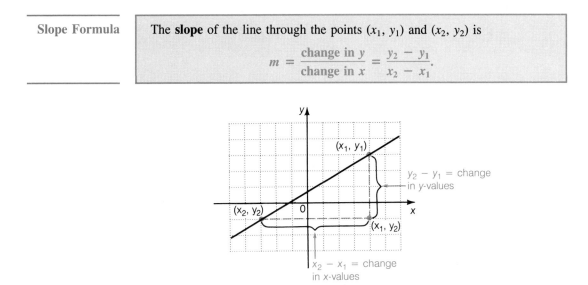

Figure 6.13

The slope of a line tells how fast y changes for each unit of change in x; that is, the slope gives the rate of change in y for each unit of change in x.

The idea of slope is useful in many everyday situations. For example, a highway with a 10% grade (or slope) rises 1 meter for every 10 meters horizontally. Architects specify the pitch of a roof by indicating the slope: a 5/12 roof means that the roof rises 5 feet for every 12 feet in the horizontal direction. The slope of a stairwell also indicates the ratio of the vertical rise to the horizontal run. See Figure 6.14.

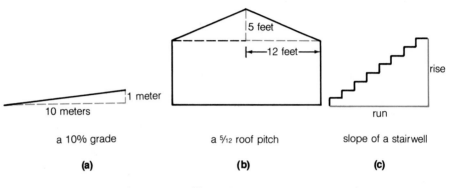

a 10% grade

(a)

a ⁵⁄₁₂ roof pitch

(b)

slope of a stairwell

(c)

Figure 6.14

Example 1   Find the slope of each of the following lines.

(a) The line through $(-4, 7)$ and $(1, -2)$

Use the definition of slope. Let $(-4, 7) = (x_2, y_2)$ and $(1, -2) = (x_1, y_1)$. Then

$$\text{slope} = \frac{\text{change in } y}{\text{change in } x}$$

$$= \frac{y_2 - y_1}{x_2 - x_1}$$

$$= \frac{7 - (-2)}{-4 - 1} = \frac{9}{-5} = -\frac{9}{5}.$$

See Figure 6.15.

(b) The line through $(12, -5)$ and $(-9, -2)$

$$\text{slope} = \frac{-5 - (-2)}{12 - (-9)} = \frac{-3}{21} = -\frac{1}{7}$$

The same slope is found by subtracting in reverse order.

$$\frac{-2 - (-5)}{-9 - 12} = \frac{3}{-21} = -\frac{1}{7} \quad \clubsuit$$

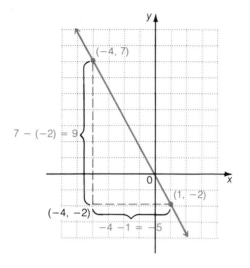

Figure 6.15

As shown in Example 1(b), it makes no difference which point is (x_1, y_1) or (x_2, y_2); however,

it is important to always start with the *x*- and *y*-values of *one* point (either one) and subtract the corresponding values of the *other* point.

In Example 1, part (a) showed the graph of a line with negative slope; the line goes down as x increases. As Figure 6.16(a) shows, this is generally true of lines with negative slopes. Lines with positive slopes go up as x increases, as shown in Figure 6.16(b).

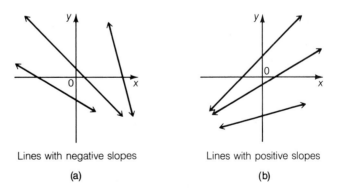

Lines with negative slopes   Lines with positive slopes

(a)   (b)

Figure 6.16

The next examples show how to determine slopes of horizontal and vertical lines.

Example 2   Find the slope of the line through $(-8, 4)$ and $(2, 4)$.
Use the definition of slope.

$$\text{slope} = \frac{4 - 4}{-8 - 2} = \frac{0}{-10} = 0$$

The graph in Figure 6.17 shows that the line through these two points is horizontal. Since all points on a horizontal line have the same y-value, *the slope of any horizontal line is 0.*   ✚

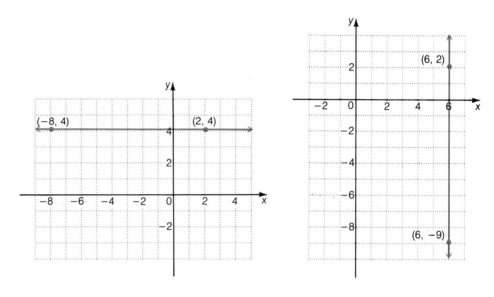

Figure 6.17   **Figure 6.18**

Example 3   Find the slope of the line through $(6, 2)$ and $(6, -9)$.

$$\text{slope} = \frac{2 - (-9)}{6 - 6} = \frac{11}{0} \qquad \text{Undefined}$$

Division by 0 is not possible, so the slope is undefined. The graph in Figure 6.18 shows that the line through these two points is vertical. Since all points on a vertical line have the same x-value, *the slope of any vertical line is undefined.*   ✚

Slopes of Horizontal and Vertical Lines	**Horizontal lines,** with equations of the form $y = k$, have a slope of 0. **Vertical lines,** with equations of the form $x = k$, have undefined slope.

2   The slope of a line can be found directly from its equation. For example, the slope of the line

$$y = -3x + 5$$

can be found from two points on the line. Get these two points by choosing two different values of x, say -2 and 4, and finding the corresponding y-values.

If $x = -2$:	If $x = 4$:
$y = -3(-2) + 5$	$y = -3(4) + 5$
$y = 6 + 5$	$y = -12 + 5$
$y = 11.$	$y = -7.$

The ordered pairs are $(-2, 11)$ and $(4, -7)$. Now find the slope, m, using the definition of slope.

$$m = \frac{11 - (-7)}{-2 - 4} = \frac{18}{-6} = -3$$

The slope, -3, is the same number as the coefficient of x in the equation $y = -3x + 5$. It can be shown that this always happens, as long as the equation is solved for y. This fact is used to find the slope of a line from its equation.

Slope of a Line from Its Equation	*Step 1* Solve the equation for y. *Step 2* The slope is given by the coefficient of x.

Since the slope of a line is always given by the coefficient of x (when the equation is solved for y), *any* two ordered pairs that satisfy the equation can be used to find the slope of the line. The result is always the same slope.

Example 4   Find the slope of each of the following lines.

(a) $2x - 5y = 4$

Solve the equation for y.

$$2x - 5y = 4$$
$$-5y = -2x + 4$$
$$y = \frac{2}{5}x - \frac{4}{5}$$

The slope is given by the coefficient of x, so the slope is

$$m = \frac{2}{5}.$$

(b) $8x + 4y = 1$

Solve the equation for y.

$$8x + 4y = 1$$
$$4y = -8x + 1$$
$$y = -2x + \frac{1}{4}$$

The slope of this line is given by the coefficient of x, -2. ✚

3   Slopes can be used to tell whether two lines are parallel. For example, Figure 6.19 shows the graph of $x + 2y = 4$ and the graph of $x + 2y = -6$. These lines appear to be parallel. Solve for y to find that both $x + 2y = 4$ and $x + 2y = -6$ have a slope of $-1/2$. It turns out that parallel lines always have equal slopes.

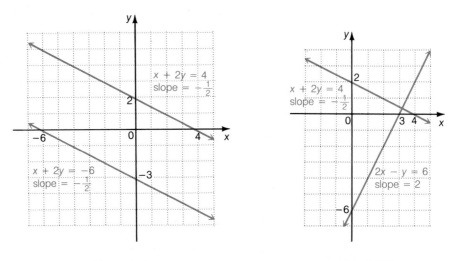

Figure 6.19   Figure 6.20

Also, Figure 6.20 shows the graph of $x + 2y = 4$ and the graph of $2x - y = 6$. These lines appear to be perpendicular (meet at a 90° angle). Solving for y shows that the slope of $x + 2y = 4$ is $-1/2$, while the slope of $2x - y = 6$ is 2. The product of $-1/2$ and 2 is

$$\left(-\frac{1}{2}\right)(2) = -1.$$

It turns out that the product of the slopes of two perpendicular lines is always -1.

Parallel and Perpendicular Lines	Two lines with the same slope are parallel; two lines that have slopes with a product of -1 are perpendicular.

Example 5   Decide whether each pair of lines is *parallel, perpendicular,* or *neither.*

(a) $x + 2y = 7$
$-2x + y = 3$

Find the slope of each line by first solving each equation for y.

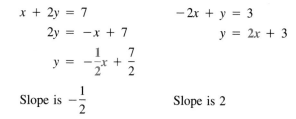

$x + 2y = 7$	$-2x + y = 3$
$2y = -x + 7$	$y = 2x + 3$
$y = -\dfrac{1}{2}x + \dfrac{7}{2}$	
Slope is $-\dfrac{1}{2}$	Slope is 2

Since the slopes are not equal, the lines are not parallel. Check the product of the slopes: $(-1/2)(2) = -1$. The two lines are perpendicular because the product of their slopes is -1.

(b) $3x - y = 4$
$6x - 2y = 9$

Find the slopes. Both lines have a slope of 3, so the lines are parallel.

(c) $4x + 3y = 6$
$2x - y = 5$

Here the slopes are $-4/3$ and 2. These lines are neither parallel nor perpendicular.   ✚

6.3 Exercises

Find the slope of each of the following lines. See Example 1.

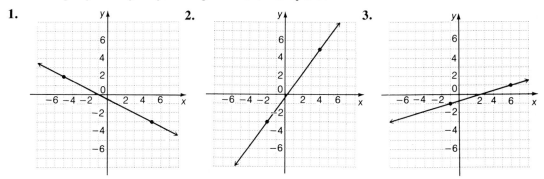

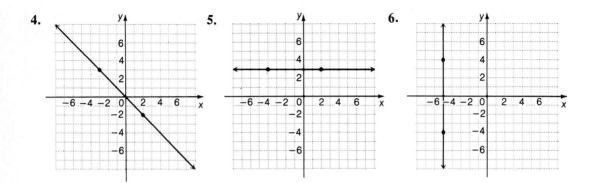

Find the slope of the line going through each pair of points. Round to the
nearest thousandth in Exercises 16 and 17. See Examples 1–3.

7. $(-4, 1)$, $(2, 8)$ **8.** $(3, 7)$, $(5, 2)$ **9.** $(-1, 2)$, $(-3, -7)$

10. $(5, -4)$, $(-5, -9)$ **11.** $(8, 0)$, $(0, 5)$ **12.** $(0, -3)$, $(2, 0)$

13. $(-1, 6)$, $(4, 6)$ **14.** $(5, 3)$, $(5, -2)$ **15.** $(-9, 1)$, $(-9, 0)$

16. $(1.23, 4.80)$, $(2.56, -3.75)$ **17.** $(0.03, 1.57)$, $(3.54, -2.01)$

Find the slope of each of the following lines. See Example 4.

18. $y = 5x + 2$ **19.** $y = -x + 4$ **20.** $y = x + 1$

21. $y = 6 - 5x$ **22.** $y = 3 + 9x$ **23.** $2x + y = 5$

24. $4x - y = 8$ **25.** $-6x + 4y = 1$ **26.** $3x - 2y = 5$

27. $2x + 5y = 4$ **28.** $9x + 7y = 5$ **29.** $y + 4 = 0$

Decide whether each pair of lines is parallel, perpendicular, *or* neither. *See*
Example 5.

30. $x + y = 5$ **31.** $y - x = 3$ **32.** $y - x = 4$
  $x - y = 1$ $y - x = 5$ $y + x = 3$

33. $2x - 5y = 4$ **34.** $3x - 2y = 4$ **35.** $3x - 5y = 2$
  $4x - 10y = 1$ $2x + 3y = 1$ $5x + 3y = -1$

36. $4x - 3y = 4$ **37.** $x - 4y = 2$ **38.** $8x - 9y = 2$
  $8x - 6y = 0$ $2x + 4y = 1$ $3x + 6y = 1$

39. $5x - 3y = 8$ **40.** $6x + y = 12$ **41.** $2x - 5y = 11$
  $3x - 5y = 10$ $x - 6y = 12$ $4x + 5y = 2$

Find the slope of the following lines.

42. $\dfrac{2}{3}y = \dfrac{5}{4}x - 3$ **43.** $\dfrac{3}{4}y - \dfrac{2}{5}x = 6$

44. $\dfrac{y}{2} + \dfrac{x}{4} = 12$ **45.** $\dfrac{5}{4}x + \dfrac{1}{4}y = -3$

Find the slope (or pitch) of the roofs shown in the figures. Measurements are given in feet.

46.   **47.**

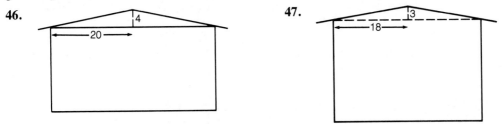

What is the slope (or grade) of the hills shown in the figures? Measurements are given in meters.

48.   **49.**

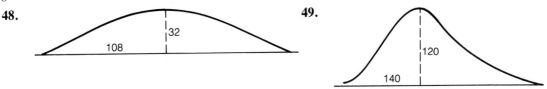

Review Exercises   *Solve for y. Simplify your answers. See Sections 2.2 and 2.6.*

50. $y - 2 = 3(x - 4)$ **51.** $y + 1 = 2(x - 5)$

52. $y - (-3) = -2(x - 1)$ **53.** $y - (-4) = -(x + 1)$

54. $y - (-5) = 4[x - (-1)]$ **55.** $y - 2 = -3[x - (-1)]$

56. $y - \dfrac{2}{3} = -\dfrac{3}{4}(x + 2)$ **57.** $y - \left(-\dfrac{3}{5}\right) = -\dfrac{1}{2}[x - (-3)]$

58. $y - \left(-\dfrac{5}{8}\right) = \dfrac{3}{8}(x - 5)$

6.4   The Equation of a Line

Objectives

1 Write an equation of a line given its slope and *y*-intercept.

2 Graph a line given its slope and a point on the line.

3 Write an equation of a line given its slope and any point on the line.

4 Write an equation of a line given two points on the line.

The last section showed how to find the slope of a line from the equation of the line. For example, the slope of the line having the equation $y = 2x + 3$ is 2, the coefficient of *x*. What does the number 3 represent? If $x = 0$, the equation becomes

$$y = 2(0) + 3 = 0 + 3 = 3.$$

Since $y = 3$ corresponds to $x = 0$, the number 3 is the y-intercept of the graph of $y = 2x + 3$. An equation like $y = 2x + 3$ that is solved for y is said to be in **slope-intercept form** because both the slope and the y-intercept of the line can be found from the equation.

Slope-Intercept Form	When the equation of a line is in the form $$y = mx + b,$$ for any real numbers m and b, the slope of the line is m and the y-intercept is b.

1 Given the slope and y-intercept of a line, the slope-intercept form can be used to find an equation of the line.

Example 1 Find an equation of each of the following lines.

(a) With slope 5 and y-intercept 3

Use the slope-intercept form. Let $m = 5$ and $b = 3$.

$$y = mx + b$$
$$y = 5x + 3$$

(b) With slope 2/3 and y-intercept -1

Here $m = 2/3$ and $b = -1$.

$$y = mx + b$$
$$y = \frac{2}{3}x - 1 \quad \clubsuit$$

2 The slope and y-intercept can be used to graph a line. For example, to graph $y = \frac{2}{3}x - 1$, first locate the y-intercept, -1, on the y-axis. From the definition of slope and the fact that the slope of this line is 2/3,

$$\frac{\text{change in } y\text{-values}}{\text{change in } x\text{-values}} = \frac{2}{3}.$$

Get another point P on the graph of the line by counting from the y-intercept 2 units up and then going 3 units to the right. The line is then drawn through point P and the y-intercept, as shown in Figure 6.21. This method can be extended to graph a line given its slope and any point on the line.

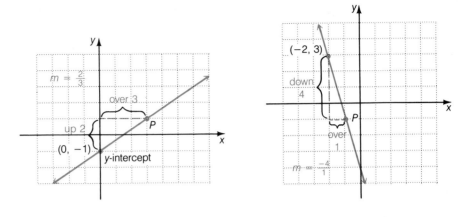

Figure 6.21   **Figure 6.22**

Example 2   Graph the line through $(-2, 3)$ with slope -4.
First, locate the point $(-2, 3)$. Write the slope as

$$\frac{\text{change in } y\text{-values}}{\text{change in } x\text{-values}} = -4 = \frac{-4}{1}.$$

$(4/-1$ could be used instead.) Another point on the line is located by counting 4 units down (because of the negative sign) and then 1 unit to the right. Finally, draw the line through this new point P and the given point $(-2, 3)$. See Figure 6.22.   ✦

3   An equation of a line also can be found from any point on the line and the slope of the line. Let (x_1, y_1) represent the given point on the line. Let (x, y) represent any other point on the line. Then by the definition of slope,

$$\frac{y - y_1}{x - x_1} = m$$

or

$$y - y_1 = m(x - x_1).$$

This result is the *point-slope form* of the equation of a line.

Point-Slope Form

> The **point-slope form** of the equation of a line with slope m going through (x_1, y_1) is
>
> $$y - y_1 = m(x - x_1).$$

Example 3   Find an equation of each of the following lines. Write the equation in the form $Ax + By = C$.

(a) Through $(-2, 4)$, with slope -3

The given point is $(-2, 4)$ so $x_1 = -2$ and $y_1 = 4$. Also, $m = -3$. Substitute these values into the point-slope form.

$$y - y_1 = m(x - x_1)$$
$$y - 4 = -3[x - (-2)]$$
$$y - 4 = -3(x + 2)$$
$$y - 4 = -3x - 6 \qquad \text{Distributive property}$$
$$y = -3x - 2 \qquad \text{Add 4}$$
$$3x + y = -2 \qquad \text{Add } 3x$$

The last equation is in the form $Ax + By = C$.

(b) Through (4, 2), with slope 3/5

Use $x_1 = 4$, $y_1 = 2$, and $m = 3/5$ in the point-slope form.

$$y - y_1 = m(x - x_1)$$

$$y - 2 = \frac{3}{5}(x - 4)$$

Multiply both sides by 5 to clear of fractions.

$$5(y - 2) = 5 \cdot \frac{3}{5}(x - 4)$$

$$5(y - 2) = 3(x - 4)$$
$$5y - 10 = 3x - 12 \qquad \text{Distributive property}$$
$$5y = 3x - 2 \qquad \text{Add 10}$$
$$-3x + 5y = -2 \qquad \text{Subtract } 3x \quad \clubsuit$$

4 The point-slope form also can be used to find an equation of a line when two points on the line are known.

Example 4 Find an equation of the line through the points $(-2, 5)$ and $(3, 4)$. Write the equation in the form $Ax + By = C$.

First find the slope of the line, using the definition of slope.

$$\text{slope} = \frac{5 - 4}{-2 - 3} = \frac{1}{-5} = -\frac{1}{5}$$

Now use either $(-2, 5)$ or $(3, 4)$ and the point-slope form. Using $(3, 4)$ gives

$$y - y_1 = m(x - x_1)$$

$$y - 4 = -\frac{1}{5}(x - 3)$$

$$5(y - 4) = -1(x - 3) \qquad \text{Multiply by 5}$$
$$5y - 20 = -x + 3 \qquad \text{Distributive property}$$
$$5y = -x + 23$$
$$x + 5y = 23.$$

The same result would be found by using $(-2, 5)$ for (x_1, y_1). $\quad \clubsuit$

A summary of the types of linear equations is given here.

Linear Equations		
$Ax + By = C$	**General form** (neither A nor B is 0)	
	Slope is $-A/B$	
	x-intercept is C/A	
	y-intercept is C/B	
$x = k$	**Vertical line**	
	Slope is undefined	
	x-intercept is k	
$y = k$	**Horizontal line**	
	Slope is 0	
	y-intercept is k	
$y = mx + b$	**Slope-intercept form**	
	Slope is m	
	y-intercept is b	
$y - y_1 = m(x - x_1)$	**Point-slope form**	
	Slope is m	
	Line goes through (x_1, y_1)	

6.4 Exercises

Write an equation for each line given its slope and y-intercept. See Example 1.

1. $m = 3$, *y*-intercept 5

2. $m = -2$, *y*-intercept 4

3. $m = -1$, *y*-intercept -6

4. $m = \dfrac{5}{3}$, *y*-intercept $\dfrac{1}{2}$

5. $m = \dfrac{2}{5}$, *y*-intercept $-\dfrac{1}{4}$

6. $m = 8$, *y*-intercept 0

7. $m = 0$, *y*-intercept -5

8. $m = -2.15$, *y*-intercept .832

9. $m = 4.61$, *y*-intercept -2.38

Graph the line going through the given point and having the given slope. (In Exercises 21–24, recall the type of lines having 0 slope and undefined slope.) See Example 2.

10. $(2, 5)$, $m = \dfrac{1}{2}$

11. $(-4, -3)$, $m = -\dfrac{2}{5}$

12. $(-1, -1)$, $m = -\dfrac{3}{8}$

13. $(0, 2)$, $m = \dfrac{7}{4}$

14. $(-3, 0)$, $m = \dfrac{5}{9}$

15. $(6, 4)$, $m = 2$

16. $(1, 8)$, $m = 1$

17. $(-4, 7)$, $m = -3$

18. $(2, 9)$, $m = -4$

19. $(4, -1)$, $m = 2$

20. $(3, -5)$, $m = 1$

21. $(1, 2)$, $m = 0$

22. $(-4, -8)$, $m = 0$

23. $(3, 5)$, undefined slope

24. $(2, 3)$, undefined slope

Write an equation for the line passing through the given point and having the given slope. Write the equation in the form Ax + By = C. See Example 3.

25. $(5, 3)$, $m = 2$

26. $(1, 4)$, $m = 3$

27. $(2, -8)$, $m = -2$

28. $(-1, 7)$, $m = -4$

29. $(3, 5)$, $m = \dfrac{2}{3}$

30. $(2, -4)$, $m = \dfrac{4}{5}$

31. $(-3, -2)$, $m = -\dfrac{3}{4}$

32. *x*-intercept -8, $m = -\dfrac{5}{9}$

33. *x*-intercept 6, $m = -\dfrac{8}{11}$

Write equations of the lines passing through each pair of points. Write the equations in the form Ax + By = C. See Example 4.

34. $(7, 4)$, $(8, 5)$

35. $(-2, 1)$, $(3, 4)$

36. $(-8, -2)$, $(-1, -7)$

37. $(3, -4)$, $(-2, -1)$

38. $(-7, -5)$, $(-9, -2)$

39. $(0, 2)$, $(3, 0)$

40. $(4, 0)$, $(0, -2)$

41. $(2, -5)$, $(-4, 7)$

42. $(3, -7)$, $(-5, 0)$

43. $\left(\dfrac{1}{2}, \dfrac{3}{2}\right)$, $\left(-\dfrac{1}{4}, \dfrac{5}{4}\right)$

44. $\left(-\dfrac{2}{3}, \dfrac{8}{3}\right)$, $\left(\dfrac{1}{3}, \dfrac{7}{3}\right)$

45. $\left(-1, \dfrac{5}{8}\right)$, $\left(\dfrac{1}{8}, 2\right)$

The cost y of an item is often expressed as y = mx + b, where x is the number of items produced. The number b gives the fixed cost (the cost that is the same no matter how many items are produced), and the number m is the variable cost (the cost to produce an additional item). Write the cost equation for each of the following.

46. Fixed cost 50, variable cost 9

47. Fixed cost 100, variable cost 12

48. Fixed cost 70.5, variable cost 3.5

49. Refer to Exercise 47 and find the total cost to make (a) 50 items and (b) 125 items.

50. Refer to Exercise 48 and find the total cost to make (a) 25 items and (b) 110 items.

The sales of a company for a given year can be written as an ordered pair in which the first number gives the year (perhaps since the company started business) and the second number gives the sales for that year. If the sales increase at a steady rate, a linear equation for sales can be found. Sales for two years are given for each of two companies below.

Company A		Company B	
Year	Sales	Year	Sales
x	*y*	*x*	*y*
1	24	1	18
5	48	4	27

51. (a) Write two ordered pairs of the form (year, sales) for Company A.

   (b) Write an equation of the line through the two pairs in part (a). This is a sales equation for Company A.

52. (a) Write two ordered pairs of the form (year, sales) for Company B.

   (b) Write an equation of the line through the two pairs in part (a). This is a sales equation for Company B.

Review Exercises   *Solve each inequality. See Section 2.9.*

53. $m + 3 < 10$ **54.** $k - 5 \geq 10$ **55.** $2p + 7 \leq 8$ **56.** $3p - 9 \geq 4$

57. $6 - 2p < 12$ **58.** $5 - 3p < -8$ **59.** $-4 - p > 7$ **60.** $-8 - 3p < 4$

6.5   *Graphing Linear Inequalities*

Objectives

1   Graph $\leq$ or $\geq$ linear inequalities.

2   Graph $<$ or $>$ linear inequalities.

3   Graph inequalities with a boundary through the origin.

1   Section 6.2 discussed methods for graphing linear equations, such as the equation $2x + 3y = 6$. Now this discussion is extended to linear inequalities, such as

$$2x + 3y \leq 6.$$

(Recall that $\leq$ is read "is less than or equal to.")
The inequality $2x + 3y \leq 6$ means that

$$2x + 3y < 6 \qquad \text{or} \qquad 2x + 3y = 6.$$

As shown earlier in this chapter, the graph of $2x + 3y = 6$ is a line. This line will be the boundary of a region that includes the ordered pairs that are solutions to the inequality $2x + 3y < 6$. The line and the region make up the complete solution. To find the required region, solve the given inequality for y.

$$2x + 3y \leq 6$$
$$3y \leq -2x + 6$$
$$y \leq -\frac{2}{3}x + 2$$

By this last statement, ordered pairs in which y is *less than or equal to* $(-2/3)x + 2$ will be solutions to the inequality. The ordered pairs in which y is equal to $(-2/3)x + 2$ are on a line, so the pairs in which y is less than $(-2/3)x + 2$ will be *below* that line. To indicate the solution, shade the region below the line, as in Figure 6.23. The shaded region, along with the original line, is the desired graph.

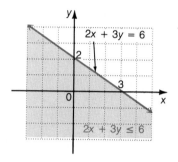

Figure 6.23

A quick way to find the correct region to shade is to use a test point. Choose any point *not on the line*. Because (0, 0) is easy to substitute into the inequality it is often a good choice, and will be used here. Substitute 0 for *x* and 0 for *y* in the given inequality to see whether the resulting statement is true or false. In the example above,

$$2x + 3y \le 6$$
$$2(0) + 3(0) \le 6 \qquad \text{Let } x = 0 \text{ and } y = 0$$
$$0 + 0 \le 6$$
$$0 \le 6. \qquad \text{True}$$

Since the last statement is true, shade the region that includes the test point (0, 0). This agrees with the result shown in Figure 6.23.

2 Inequalities that do not include the equals sign are graphed in a similar way.

Example 1 Graph the inequality $x - y > 5$.

This inequality does not include the equals sign. Therefore, the points on the line $x - y = 5$ do not belong to the graph, but the line still serves as a boundary for two regions, one of which satisfies the inequality. To graph the inequality, first graph the equation $x - y = 5$. Use a dashed line to show that the points on the line are not solutions of the inequality $x - y > 5$. Choose a test point to see which region satisfies the inequality. Choosing (0, 0) gives

$$x - y > 5 \qquad \text{Original inequality}$$
$$0 - 0 > 5 \qquad \text{Let } x = 0 \text{ and } y = 0$$
$$0 > 5. \qquad \text{False}$$

Since $0 > 5$ is false, the graph of the inequality includes the region that does *not* contain (0, 0). Shade this region, as shown in Figure 6.24. This shaded region is the required graph. Check that the correct region is shaded by selecting a test point in the shaded region and substituting for x and y in the inequality $x - y > 5$. For example, use $(4, -3)$ from the shaded region, as follows.

$$x - y > 5$$
$$4 - (-3) > 5$$
$$7 > 5 \qquad \text{True}$$

This verifies that the correct region was shaded in Figure 6.24.   ✛

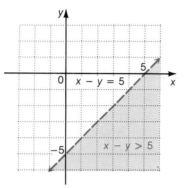

Figure 6.24

Example 2 Graph the inequality $2x - 5y \geq 10$.

First graph the equation $2x - 5y = 10$. Use a solid line to show that the points on the line are solutions of the inequality $2x - 5y \geq 10$. Choose any test point not on the line. Again, choose $(0, 0)$.

$$2x - 5y \geq 10 \qquad \text{Original inequality}$$
$$2(0) - 5(0) \geq 10 \qquad \text{Let } x = 0 \text{ and } y = 0$$
$$0 - 0 \geq 10$$
$$0 \geq 10 \qquad \text{False}$$

Since $0 \geq 10$ is false, shade the region *not* containing the test point $(0, 0)$. (See Figure 6.25.)   ✛

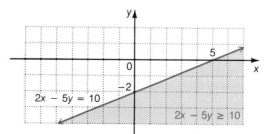

Figure 6.25

Example 3   Graph the inequality $x \leq 3$.

First, graph $x = 3$, a vertical line going through the point $(3, 0)$. Use a solid line. (Why?) Choose $(0, 0)$ as a test point.

$x \leq 3$	Original inequality
$0 \leq 3$	Let $x = 0$
$0 \leq 3$	True

Since $0 \leq 3$ is true, shade the region containing $(0, 0)$, as in Figure 6.26.   ✦

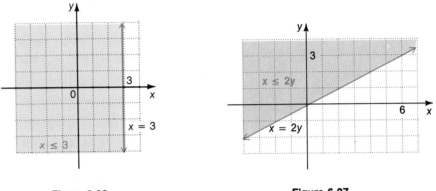

Figure 6.26   **Figure 6.27**

3   The next example shows how to graph an inequality having a boundary line going through the origin, an inequality in which $(0, 0)$ cannot be used as a test point.

Example 4   Graph the inequality $x \leq 2y$.

Begin by graphing $x = 2y$. Some ordered pairs that can be used to graph this line are $(0, 0)$, $(6, 3)$, and $(4, 2)$. Use a solid line. The point $(0, 0)$ cannot be used as a test point since $(0, 0)$ is on the line $x = 2y$. Instead, choose a test point off the line $x = 2y$. For example, choose $(1, 3)$, which is not on the line.

$x \leq 2y$	Original inequality
$1 \leq 2(3)$	Let $x = 1$ and $y = 3$
$1 \leq 6$	True

Since $1 \leq 6$ is true, shade the side of the graph containing the test point $(1, 3)$. (See Figure 6.27.)   ✦

A summary of the steps used to graph a linear inequality in two variables is given on the next page.

Graphing a Linear Inequality	*Step 1* Graph the line that is the boundary of the region. Use the methods of Section 6.2. Draw a solid line if the inequality has ≤ or ≥; draw a dashed line if the inequality has < or >.
	Step 2 Use any point off the line as a test point. Substitute for x and y in the inequality. If a true statement results, shade the region containing the test point. If a false statement results, shade the region not containing the test point.

6.5   Exercises

In Exercises 1–12, the straight line for each inequality has been drawn.
Complete each graph by shading the correct region. See Example 1.

1. $x + y \le 4$

2. $x + y \ge 2$

3. $x + 2y \le 7$

4. $2x + y \le 5$

5. $-3x + 4y < 12$

6. $4x - 5y > 20$

7. $5x + 3y > 15$

8. $6x - 5y < 30$

9. $x < 4$

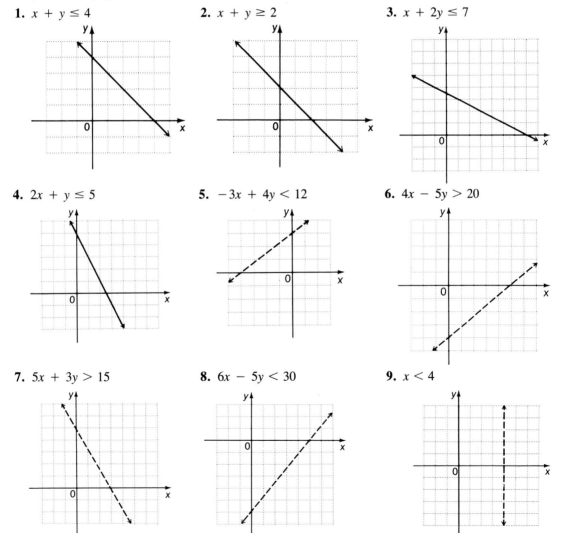

10. $y > -1$ **11.** $x \leq 4y$ **12.** $-2x > y$

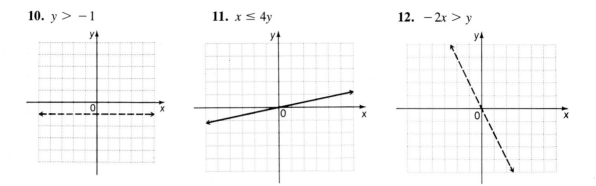

Graph each linear inequality. See Examples 1–4.

13. $x + y \leq 8$ **14.** $x + y \geq 4$ **15.** $x - y \leq -2$ **16.** $x - y \leq 3$

17. $x + 2y \geq 4$ **18.** $x + 3y \leq 6$ **19.** $2x + 3y > 6$ **20.** $3x + 4y > 12$

21. $3x - 4y < 12$ **22.** $2x - 3y < -6$ **23.** $3x + 7y \geq 21$ **24.** $2x + 5y \geq 10$

25. $x < 4$ **26.** $x < -2$ **27.** $y \leq 2$ **28.** $y \leq -3$

29. $x \geq -2$ **30.** $x \leq 3y$ **31.** $x \leq 5y$ **32.** $x \geq -2y$

33. $-4x \leq y$ **34.** $2x + 3y \geq 0$ **35.** $3x + 4y \leq 0$ **36.** $x + y < 0$

For each of the following problems: (a) Graph the inequality. Here $x \geq 0$ and $y \geq 0$, so graph only the part of the inequality in quadrant I. (b) Give some sample values of x and y that satisfy the inequality.

37. The Sweet Tooth Candy Company uses x pounds of chocolate for chocolate cookies and y pounds of chocolate for fudge. The company has 200 pounds of chocolate available, so that

$$x + y \leq 200.$$

38. A company will ship x units of merchandise to outlet I and y units of merchandise to outlet II. The company must ship a total of at least 500 units to these two outlets. This can be expressed by writing

$$x + y \geq 500.$$

39. A toy manufacturer makes stuffed bears and geese. It takes 20 minutes to sew a bear and 30 minutes to sew a goose. There is a total of 480 minutes of sewing time available to make x bears and y geese. These restrictions lead to the inequality

$$20x + 30y \leq 480.$$

40. Ms. Branson takes x vitamin C tablets each day at a cost of 10¢ each and y multivitamins each day at a cost of 15¢ each. She wants the total cost to be no more than 50¢ a day. This can be expressed by writing

$$10x + 15y \leq 50.$$

Review Exercises   *Find the value of $3x^2 - 2x + 5$ for each value of x. See Section 3.4.*

41. 2   **42.** -4 **43.** 3   **44.** 0

Find the value of $2x^4 - x^2$ for each value of x.

45. 3   **46.** -2 **47.** -1 **48.** -3

6.6   Functions

Objectives

1️⃣ Understand the definition of a relation.

2️⃣ Understand the definition of a function.

3️⃣ Decide whether an equation defines a function.

4️⃣ Find domains and ranges.

5️⃣ Use $f(x)$ notation.

In Section 6.1, the equation $y = 25x + 250$ was used to find the cost y to produce x calculators. Choosing values for x and using the equation to find the corresponding values of y led to a set of ordered pairs (x, y). In each ordered pair, y (the cost) was *related* to x (the number of calculators produced) by the equation $y = 25x + 250$.

1️⃣ Any set of ordered pairs is called a **relation.** The set of all first elements in the ordered pairs of a relation is the **domain** of the relation, and the set of all second elements in the ordered pairs is the **range** of the relation. Recall from Chapter 1 that sets are written with set braces, { }.

Example 1   **(a)** The relation $\{(0, 1), (2, 5), (3, 8), (4, 2)\}$ has domain $\{0, 2, 3, 4\}$ and range $\{1, 2, 5, 8\}$. The correspondence between the elements of the domain and the elements of the range is shown in Figure 6.28.

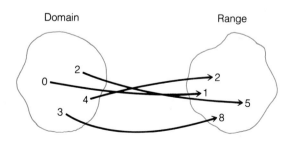

Figure 6.28

(b) Figure 6.29 shows a relation where the domain elements represent four mothers and the range elements represent eight children. The figure shows the correspondence between each mother and her children. The relation also could be written as the set of ordered pairs $\{(A, 8), (B, 1), (B, 2), (C, 3), (C, 4), (C, 7), (D, 5), (D, 6)\}$. ✤

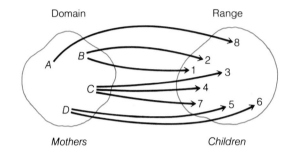

Domain   Range

Mothers   Children

Figure 6.29

2   A special type of relation, called a *function,* is particularly useful in applications.

Function

A **function** is a set of ordered pairs in which each first element corresponds to exactly one second element.

By this definition, the relation in Example 1(a) is a function. However, the relation in Example 1(b) is *not* a function, because at least one first element (mother) corresponds to more than one second element (child). Notice that if the ordered pairs in Example 1(b) are reversed, with the child as the first element and the mother as the second element, the result *is* a function.

The simple relations given here were defined by listing the ordered pairs or by showing the correspondence with a figure. Most useful functions have an infinite number of ordered pairs and must be defined with an equation that tells how to get the second element given the first element. It is customary to use an equation with x and y as the variables, where x represents the first element and y the second element in the ordered pairs.

Example 2   **(a)** In the United States, for letters weighing up to twelve ounces, the postage y has been a function of the weight x (in ounces), determined by the equation

$$y = .17(x - 1) + .22.$$

(b) The basic California sales tax is 6% of the price of an item. The tax y on a particular item is a function of the price x, so that

$$y = .06x.$$ ✢

3 Given a graph of an equation, the definition of a function can be used to decide whether the graph represents a function or not. By the definition of a function, each value of x must lead to exactly one value of y. In Figure 6.30, the indicated value of x leads to two values of y, so this graph is not the graph of a function. A vertical line can be drawn that cuts this graph in more than one point.

On the other hand, in Figure 6.31 any vertical line will cut the graph in no more than one point. Because of this, the graph in Figure 6.31 is the graph of a function.

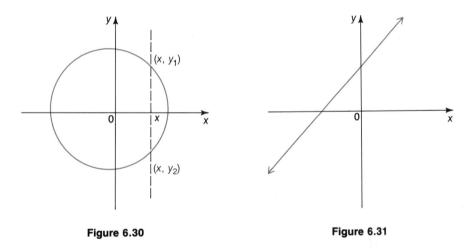

Figure 6.30   **Figure 6.31**

This method gives the **vertical line test** for a function.

Vertical Line Test	If a vertical line cuts a graph in more than one point, the graph is not the graph of a function.

As Figure 6.31 suggests, any nonvertical line is the graph of a function. For this reason, any linear equation of the form $Ax + By = C$, where $B \neq 0$, defines a function.

Example 3 Decide whether or not the following relations are functions.

(a) $y = 2x - 9$

This linear equation can be written as $1y - 2x = -9$. Since the graph of this equation is not a vertical line, the equation defines a function.

(b)

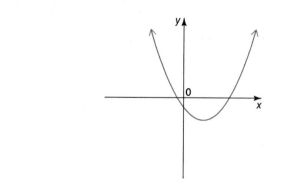

Figure 6.32

Use the vertical line test. Any vertical line will cross the graph in Figure 6.32 just once, so this is the graph of a function.

(c)

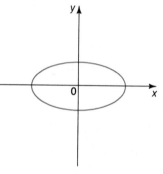

Figure 6.33

The vertical line test shows that the graph in Figure 6.33 is not the graph of a function; a vertical line can cross the graph twice.

(d) $x = 4$

The graph of $x = 4$ is a vertical line, so the equation does not define a function.

(e) $x = y^2$

If $x = 36$, then

$$x = y^2 \quad \text{becomes} \quad y^2 = 36.$$

The equation $y^2 = 36$ has *two* solutions: $y = 6$ and $y = -6$. Because the *one* x-value, 36, leads to *two* y-values, the equation $x = y^2$ does not define a function.

(f) $2x + y < 4$

Since this is an inequality, any choice of a value for x will lead to an *infinite* number of y-values. For example, if $x = 3$,

$$2x + y < 4$$
$$2(3) + y < 4 \qquad \text{Let } x = 3$$
$$6 + y < 4$$
$$y < -2.$$

If $x = 3$, y can take *any* value less than -2. Since $x = 3$ leads to more than one y-value, the inequality $2x + y < 4$ is not a function. ✦

4 By the definitions of domain and range given for relations, the set of all numbers that can be used as replacements for x in a function is the domain of the function, and the set of all possible values of y is the range of the function.

Example 4 Find the domain and range for the following functions.

(a) $y = 6x - 9$

Any number at all may be used for x, so the domain is the set of all real numbers. Also, any number may be used for y, so the range is also the set of all real numbers.

(b) $y = x^2$

Any number can be squared, so the domain is the set of all real numbers. However, since the square of a real number cannot be negative and since $y = x^2$, the values of y cannot be negative, making the range the set of all nonnegative numbers, written as $y \geq 0$. ✦

5 It is common to use the letters f, g, and h to name functions. For example, the function $y = 3x + 5$ is often written

$$f(x) = 3x + 5,$$

where $f(x)$ is read "f of x." The notation $f(x)$ is another way of writing y in a function. For the function $f(x) = 3x + 5$, if $x = 7$ then

$$f(7) = 3 \cdot 7 + 5 \qquad \text{Let } x = 7$$
$$= 21 + 5 = 26.$$

Read this result, $f(7) = 26$, as "f of 7 equals 26." The notation $f(7)$ means the value of y when x is 7. The statement $f(7) = 26$ says that the value of y is 26 when x is 7.

To find $f(-3)$, replace x with -3.

$$f(x) = 3x + 5$$
$$f(-3) = 3(-3) + 5 \qquad \text{Let } x = -3$$
$$f(-3) = -9 + 5 = -4$$

Note: The symbol $f(x)$ does *not* mean f times x.

$f(x)$ Notation	In the notation $f(x)$, f is the name of the function, x is the domain value, $f(x)$ is the range value for the domain value x.

Example 5   For the function $f(x) = x^2 - 3$, find the following.

(a) $f(2)$

     Replace x with 2.

$$f(x) = x^2 - 3$$
$$f(2) = 2^2 - 3 \qquad \text{Let } x = 2$$
$$f(2) = 4 - 3 = 1$$

(b) $f(0) = 0^2 - 3 = 0 - 3 = -3$

(c) $f(-3) = (-3)^2 - 3 = 9 - 3 = 6$ ✜

Example 6   Let $P(x) = 5x^2 - 4x + 3$. Find the following. (This function is named P instead of f or g to show that it is a polynomial.)

(a) $P(0)$

     Replace x with 0.

$$P(x) = 5x^2 - 4x + 3$$
$$P(0) = 5 \cdot 0^2 - 4 \cdot 0 + 3$$
$$P(0) = 3$$

(b) $P(-2) = 5 \cdot (-2)^2 - 4 \cdot (-2) + 3 = 20 + 8 + 3 = 31$

(c) $P(3) = 5 \cdot 3^2 - 4 \cdot 3 + 3 = 36$ ✜

6.6   *Exercises*

Decide which of the following relations are functions. See Examples 1 and 2.

1. $\{(-4, 3), (-2, 1), (0, 5), (-2, -4)\}$

2. $\{(3, 7), (1, 4), (0, -2), (-1, -1), (-2, 4)\}$

3. $\{(-2, 3), (-1, 2), (0, 0), (1, 2), (2, 3)\}$

4. $\{(1, 5), (5, 7), (5, 9), (7, 11)\}$

5.

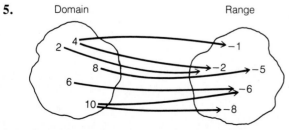

6.

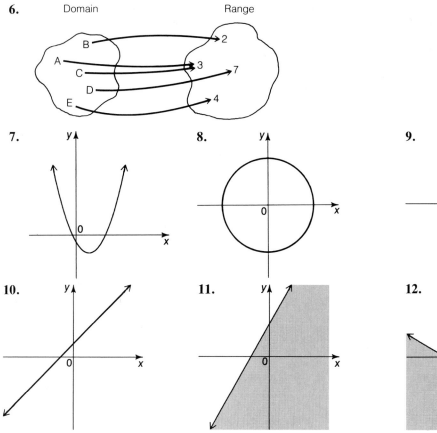

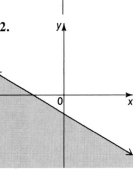

Decide which of the following are functions. See Example 3.

13. $y = 5x - 1$ **14.** $y = 4x + 5$ **15.** $2x + 3y = 6$ **16.** $4x - 3y = 12$

17. $y = x^2 + 3$ **18.** $y = 5 - x^2$ **19.** $x = y^2 - 4$ **20.** $y = x^2 + 6$

21. $2x + y < 6$ **22.** $3x - 4y > 2$ **23.** $y = \dfrac{1}{x}$ **24.** $y = \dfrac{1}{x + 2}$

Find the domain and range of the following functions. See Example 4.

25. $y = 2x + 5$ **26.** $y = 5x - 6$ **27.** $2x - y = 6$ **28.** $2x + 3y = 12$

29. $y = x^2 - 3$ **30.** $y = x^2 + 4$ **31.** $y = -x^2 + 6$ **32.** $y = -x^2 - 8$

For the following, find (a) f(2), (b) f(0), and (c) f(−3). See Example 5.

33. $f(x) = 3x + 2$ **34.** $f(x) = 4x - 1$ **35.** $f(x) = 4 - x$ **36.** $f(x) = 2 - 3x$

37. $f(x) = -4 - 4x$ **38.** $f(x) = -5 - 6x$ **39.** $f(x) = x^2 + 2$ **40.** $f(x) = x^2 - 5$

41. $f(x) = (x - 3)^2$ **42.** $f(x) = -(x - 4)^2$ **43.** $f(x) = -|x + 2|$ **44.** $f(x) = -|x - 3|$

For the following, find (a) P(0), (b) P(−3), and (c) P(2). See Example 6.

45. $P(x) = x^2 + 2x$ **46.** $P(x) = 2x^2 + 3x - 6$

47. $P(x) = -x^2 - 8x + 9$ **48.** $P(x) = -3x^2 + 4x - 2$

49. $P(x) = x^3 - 4x^2 + 1$ **50.** $P(x) = x^3 + 5x^2 - 11x + 2$

*Recall that $|x|$ represents the absolute value of x. For example, $|3| = 3, |-5| =$
5, and so on. Decide which of the following are functions.*

51. $y = |x|$ **52.** $x = |y|$ **53.** $x = |y| - 3$ **54.** $y = |x| + 8$

55. $y = |x - 1|$ **56.** $x = |y + 4|$ **57.** $|x + y| = 2$ **58.** $|x - y| = 4$

Review Exercises   Add the polynomials. See Section 3.3.

59. $x - 2y$ **60.** $5x - 7y$ **61.** $9a - 5b$ **62.** $-11p + 8q$

      $3x + 2y$ $12x + 7y$ $-9a + 7b$ $\underline{11p - 9q}$

Chapter 6   *Summary*

Slope Formula   The slope of the line through the points (x_1, y_1) and (x_2, y_2) is

$$m = \frac{\text{change in } y}{\text{change in } x} = \frac{y_2 - y_1}{x_2 - x_1}.$$

Linear Equations	$Ax + By = C$	**General form** (neither A nor B is 0) Slope is $-A/B$ x-intercept is C/A y-intercept is C/B
	$x = k$	**Vertical line** Slope is undefined x-intercept is k
	$y = k$	**Horizontal line** Slope is 0 y-intercept is k
	$y = mx + b$	**Slope-intercept form** Slope is m y-intercept is b
	$y - y_1 = m(x - x_1)$	**Point-slope form** Slope is m Line goes through (x_1, y_1)

| **Graphing a Linear Inequality** | *Step 1* | Graph the line that is the boundary of the region. Use the methods of Section 6.2. Draw a solid line if the inequality has $\leq$ or $\geq$; draw a dashed line if the inequality has $<$ or $>$. |
| | *Step 2* | Use any point off the line as a test point. Substitute for x and y in the inequality. If a true statement results, shade the region containing the test point. If a false statement results, shade the region not containing the test point. |

| **Function** | A **function** is a set of ordered pairs in which each first element corresponds to exactly one second element. |

Chapter 6   *Review Exercises*

[6.1] *Decide whether the given ordered pair is a solution of the given equation.*

1. $x + y = 7$; (3, 4) **2.** $2x + y = 5$; (1, 4) **3.** $x + 3y = 9$; (1, 3)

4. $2x + 5y = 7$; (1, 1) **5.** $3x - y = 4$; (1, -1) **6.** $5x - 3y = 16$; (2, -2)

Complete the given ordered pairs for each equation.

Equation	Ordered Pairs
7. $y = 3x - 2$	$(-1, \ )$ $(0, \ )$ $(\ , 5)$
8. $2y = 4x + 1$	$(0, \ )$ $(\ , 0)$ $(\ , 2)$
9. $x + 4 = 0$	$(\ , -3)$ $(\ , 0)$ $(\ , 5)$
10. $y - 5 = 0$	$(-2, \ )$ $(0, \ )$ $(8, \ )$

[6.2] *Graph each linear equation.*

11. $y = 2x + 3$ **12.** $x + y = 5$ **13.** $2x - y = 5$

14. $x + 2y = 0$ **15.** $y + 3 = 0$ **16.** $x - y = 0$

Find the intercepts for each equation.

17. $y = 2x - 5$ **18.** $2x + y = 7$ **19.** $5x - 2y = 0$

[6.3] *Find the slope of each line.*

20. Through (2, 3) and (-1, 1)

21. Through (0, 0) and (-1, -2)

22. Through (2, 5) and (2, -2)

23. $y = 3x - 1$ **24.** $y = 8$ **25.** $x = 2$ **26.** $2x - 5y = 3$

Decide whether each pair of lines is parallel, perpendicular, or neither.

27. $3x + 2y = 5$	**28.** $x - 3y = 8$	**29.** $4x + 3y = 10$	**30.** $x - 2y = 3$
$6x + 4y = 12$	$3x + y = 6$	$3x - 4y = 12$	$x + 2y = 3$

[6.4] *Write an equation for each line in the form* $Ax + By = C$.

31. $m = 3$, y-intercept -2

32. $m = -1$, y-intercept $\dfrac{3}{4}$

33. $m = \dfrac{2}{3}$, y-intercept 5

34. Through $(5, -2)$, $m = 1$

35. Through $(-1, 4)$, $m = \dfrac{2}{3}$

36. Through $(1, -1)$, $m = \dfrac{-3}{4}$

37. Through $(2, 1)$ and $(-2, 2)$

38. Through $(-2, 6)$ and $(3, 6)$

[6.5] *Complete the graph of each linear inequality by shading the correct region.*

39. $x - y \le 3$

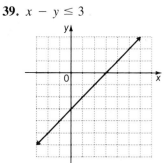

40. $3x - y \ge 5$

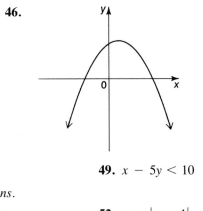

Graph each linear inequality.

41. $x + 2y \le 6$

42. $3x + 5y \ge 9$

[6.6] *Decide which of the following are functions.*

43. $\{(-2, 4), (0, 8), (2, 5), (2, 3)\}$

44. $\{(8, 3), (7, 4), (6, 5), (5, 6), (4, 7)\}$

45.

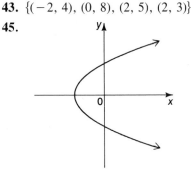

46.

47. $2x + 3y = 12$ **48.** $y = x^2$ **49.** $x - 5y < 10$

Find the domain and range of the following functions.

50. $4x - 3y = 12$ **51.** $y = x^2 + 1$ **52.** $y = |x - 1|$

Find (a) $f(2)$ and (b) $f(-1)$.

53. $f(x) = 3x + 2$ **54.** $f(x) = 2x^2 - 1$ **55.** $f(x) = |x + 3|$

56. Let $P(x) = -x^2 - 4x + 2$. Find $P(-1)$ and $P(-3)$.

② *Find the slope and the intercepts of each line.*

57. Through $(4, -1)$ and $(-2, -3)$

58. $y = \dfrac{2}{3}x + 5$

Which of the following are functions?

59.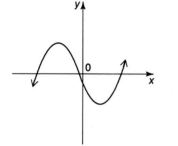

60. $y = |x|$

Complete the given ordered pairs for each equation.

61. $4x + 3y = 9$; $(0,\ \ )$, $(\ \ , 0)$, $(-2,\ \ )$

62. $x = 2y$; $(0,\ \ )$, $(8,\ \ )$, $(\ \ , -3)$

Write an equation for each line in the form $Ax + By = C$.

63. Through $(5, 0)$ and $(5, -1)$

64. $m = -\dfrac{1}{4}$, y-intercept $-\dfrac{5}{4}$

65. Through $(8, 6)$, $m = -3$

66. Through $(3, -5)$ and $(-4, -1)$

67. Find the domain and range of $y = 2x + 1$.

In Exercises 68–71, graph each relation.

68. $3y = x$ **69.** $y = 4x - 1$ **70.** $y \le -2$ **71.** $x - 2y > 0$

72. Find (a) $f(2)$ and (b) $f(-1)$ for $f(x) = -(x + 1)^2$.

Chapter 6 *Test*

Complete the ordered pairs using the given equations.

Equation		Ordered Pairs	
1. $y = 5x - 6$	$(0,\ \ )$	$(-2,\ \ )$	$(\ \ , 14)$
2. $2x + 7y = 21$	$(0,\ \ )$	$(\ \ , 0)$ $(3,\ \ )$	$(\ \ , 2)$
3. $x = 3y$	$(0,\ \ )$	$(\ \ , 2)$ $(8,\ \ )$	$(-12,\ \ )$
4. $y - 2 = 0$	$(5,\ \ )$	$(4,\ \ )$ $(0,\ \ )$	$(-3,\ \ )$

Graph each linear equation. Give the x- and y-intercepts.

5. $x + y = 4$ **6.** $2x + y = 6$ **7.** $3x - 4y = 18$

8. $2x + y = 0$ **9.** $x + 5 = 0$ **10.** $y = 2$

Find the slope of each line.

11. Through $(-2, 4)$ and $(5, 1)$ **12.** $y = -\dfrac{3}{4}x + 6$

13. $4x + 7y = 10$ **14.** $x - 5 = 0$

Write an equation for each line in the form $Ax + By = C$.

15. Through $(1, -3)$, $m = -4$ **16.** $m = 3$,   y-intercept -1

17. $m = -\dfrac{3}{4}$,   y-intercept 2 **18.** Through $(-2, -6)$ and $(-1, 3)$

Complete the graph of each linear inequality.

19. $2x + y \le 8$ **20.** $y < -3$

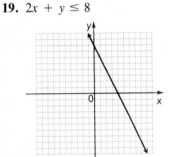 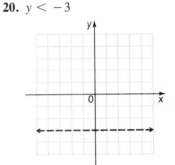

Decide which are functions.

21. 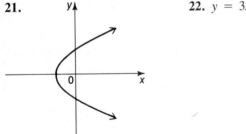   **22.** $y = 3x + 5$ **23.** $x = y^2 - 1$

24. Find the domain and range of $y = 2x^2 + 1$.

25. For $f(x) = 6x - 2$ find $f(-3)$ and $f(5)$.

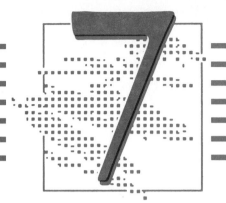

Linear Systems

Solving Systems of Linear Equations by Graphing

Objectives

1   Decide whether a given ordered pair is a solution of a system.

2   Solve linear systems by graphing.

3   Identify systems with no solutions or with an infinite number of solutions.

4   Identify inconsistent systems or systems with dependent equations without graphing.

A **system of linear equations** consists of two or more linear equations with the same variables. Examples of systems of linear equations include

$$2x + 3y = 4 \qquad x + 3y = 1 \qquad\qquad x - y = 1$$
$$3x - y = -5 \qquad\quad -y = 4 - 2x \quad \text{and} \qquad y = 3.$$

In the system on the right, think of $y = 3$ as an equation in two variables by writing it as $0x + y = 3$.

1   Applications often require solving a system of equations. The **solution of a system** of linear equations includes all the ordered pairs that make both equations true at the same time.

Example 1   Is $(4, -3)$ a solution of the following systems?

(a) $x + 4y = -8$
$3x + 2y = 6$

To decide whether $(4, -3)$ is a solution of the system, substitute 4 for x and -3 for y in each equation.

$$x + 4y = -8 \qquad\qquad 3x + 2y = 6$$
$$4 + 4(-3) = -8 \qquad\qquad 3(4) + 2(-3) = 6$$
$$4 + (-12) = -8 \qquad\qquad 12 + (-6) = 6$$
$$-8 = -8 \quad \text{True} \qquad\qquad 6 = 6 \quad \text{True}$$

Since $(4, -3)$ satisfies both equations, it is a solution.

(b) $2x + 5y = -7$
$3x + 4y = 2$

Again, substitute 4 for x and -3 for y in both equations.

$$2x + 5y = -7 \qquad\qquad 3x + 4y = 2$$
$$2(4) + 5(-3) = -7 \qquad\qquad 3(4) + 4(-3) = 2$$
$$8 + (-15) = -7 \qquad\qquad 12 + (-12) = 2$$
$$-7 = -7 \quad \text{True} \qquad\qquad 0 = 2 \quad \text{False}$$

Here $(4, -3)$ is not a solution since it does not satisfy the second equation. ✛

Several methods of solving a system of two linear equations in two variables are discussed in this chapter.

2   One way to find the solution of a system of two linear equations is to graph both equations on the same axes. The graph of each line shows the points whose coordinates satisfy the equation of that line. The coordinates of any point where the lines intersect give a solution of the system. Since two different straight lines usually intersect in exactly one point, there is usually exactly one solution for such a system.

Example 2   Solve each system of equations by graphing both equations on the same axes.

(a) $2x + 3y = 4$
$3x - y = -5$

Use methods from the previous chapter to graph both $2x + 3y = 4$ and $3x - y = -5$ on the same axes. (It is important to draw the graphs carefully, since the solution will be read from the graph.) The lines in Figure 7.1 suggest that the graphs intersect at the point $(-1, 2)$. Check this by substituting -1 for x and 2 for y in both equations. Since the ordered pair $(-1, 2)$ satisfies both equations, the solution of this system is $(-1, 2)$.

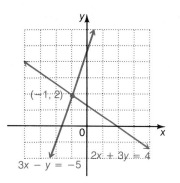

Figure 7.1

(b) $2x + y = 0$
$4x - 3y = 10$

Find the solution of the system by graphing the two lines on the same axes. As suggested by Figure 7.2, the solution is $(1, -2)$, the point at which the graphs of the two lines intersect. Check by substituting 1 for x and -2 for y in both equations of the system. ✤

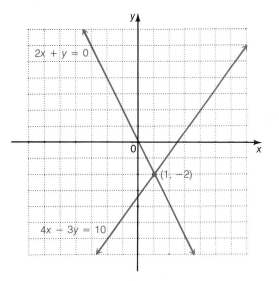

Figure 7.2

A difficulty with the graphing method of solution is that it may not be possible to determine from the graph the exact coordinates of the point that represents the solution.

3   Sometimes the graphs of the two equations in a system either do not intersect at all or are the same line, as in the systems of Example 3.

Example 3   Solve each system by graphing.

(a) $2x + y = 2$
$2x + y = 8$

The graphs of these lines are shown in Figure 7.3. The two lines are parallel and have no points in common. A system of equations like this, with no solution, is called an **inconsistent system.** A system of equations that has a solution is a **consistent system.**

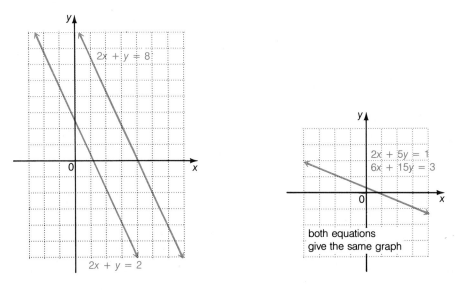

Figure 7.3   **Figure 7.4**

(b) $2x + 5y = 1$
$6x + 15y = 3$

The graphs of these two equations are the same line. See Figure 7.4. The second equation can be obtained by multiplying both sides of the first equation by 3. In this case, every point on the line is a solution of the system, and the solution is an infinite number of ordered pairs. We will write "infinite number of solutions" to indicate this case. Equations like these, that have the same graph, are called **dependent equations. Independent equations** have different graphs.   ✤

4   Example 3 showed that the graphs of an inconsistent system are parallel lines and the graphs of a system of dependent equations are the same line. These special kinds of systems can be recognized without graphing by using slopes.

Example 4   Describe each system without graphing.

(a) $3x + 2y = 6$
$-2y = 3x - 5$

Write each equation in slope-intercept form by solving for y.

$$3x + 2y = 6 \qquad\qquad -2y = 3x - 5$$
$$2y = -3x + 6 \qquad\qquad y = -\frac{3}{2}x + \frac{5}{2}$$
$$y = -\frac{3}{2}x + 3$$

Both equations have a slope of $-3/2$ but they have different y-intercepts. The previous section showed that lines with the same slope are parallel, so these equations have graphs that are parallel lines. The system has no solution.

(b) $2x - y = 4$
$x = \dfrac{y}{2} + 2$

Again, write the equations in slope-intercept form.

$$2x - y = 4 \qquad\qquad x = \frac{y}{2} + 2$$
$$-y = -2x + 4 \qquad\qquad \frac{y}{2} + 2 = x$$
$$y = 2x - 4 \qquad\qquad \frac{y}{2} = x - 2$$
$$y = 2x - 4$$

The equations are exactly the same; their graphs are the same line. The system has an infinite number of solutions.

(c) $x - 3y = 5$
$2x + y = 8$

In slope-intercept form, the equations are as follows.

$$x - 3y = 5 \qquad\qquad 2x + y = 8$$
$$-3y = -x + 5 \qquad\qquad y = -2x + 8$$
$$y = \frac{1}{3}x - \frac{5}{3}$$

The graphs of these equations are neither parallel lines nor the same line since the slopes are different. There will be exactly one solution to this system.   ✤

Examples 2–4 showed the three cases that may occur in a system of two equations with two unknowns.

Possible Types of Solutions

1. The graphs cross at exactly one point, which gives the solution of the system. The system is consistent and the equations are independent.

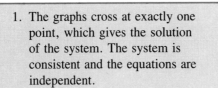

2. The graphs are parallel lines, so there is no solution. The system is inconsistent.

3. The graphs are the same line. There are an infinite number of solutions. The equations are dependent.

7.1 Exercises

Decide whether the given ordered pair is the solution of the given system. See Example 1.

1. $(2, -5)$
$$3x + y = 1$$
$$2x + 3y = -11$$

2. $(-1, 6)$
$$2x + y = 4$$
$$3x + 2y = 9$$

3. $(4, -2)$
$$x + y = 2$$
$$2x + 5y = 2$$

4. $(-6, 3)$
$$x + 2y = 0$$
$$3x + 5y = 3$$

5. $(2, 0)$
$$3x + 5y = 6$$
$$4x + 2y = 5$$

6. $(0, -4)$
$$2x - 5y = 20$$
$$3x + 6y = -20$$

7. $(5, 2)$
$$4x + 3y = 26$$
$$3x + 7y = 29$$

8. $(9, 1)$
$$2x + 5y = 23$$
$$3x + 2y = 29$$

9. $(6, -8)$
$$x + 2y + 10 = 0$$
$$2x - 3y + 30 = 0$$

10. $(-5, 2)$
$3x - 5y + 20 = 0$
$2x + 3y + 4 = 0$

11. $(5, -2)$
$x - 5 = 0$
$y + 2 = 0$

12. $(-8, 3)$
$x - 8 = 0$
$y - 3 = 0$

Solve each system of equations by graphing both equations on the same axes. See Examples 2 and 3.

13. $x + y = 6$
$x - y = 2$

14. $x + y = -1$
$x - y = 3$

15. $x + y = 4$
$y - x = 4$

16. $2x + 5y = 17$
$3x - 4y = -9$

17. $4x + 5y = 3$
$2x - 5y = 9$

18. $2x + y = 5$
$3x - 4y = 24$

19. $3x + 2y = -10$
$x - 2y = -6$

20. $4x + y = 5$
$3x - 2y = 12$

21. $3x - 4y = -24$
$5x + 2y = -14$

22. $x - 2y = -5$
$4x + 3y = -9$

23. $x - 2y = 5$
$4x + 3y = 20$

24. $-4x + 3y = 16$
$2x - 3y = -8$

25. $2x + 3y = 3$
$x + y = 3$

26. $x + 2y - 5 = 0$
$2x + y - 1 = 0$

27. $2x - y - 2 = 0$
$3x + y - 3 = 0$

28. $5x + y - 7 = 0$
$2x + 2y + 2 = 0$

29. $2x + 5y - 20 = 0$
$x + y - 10 = 0$

30. $x + y = 4$
$y = -2$

31. $2x + y = 1$
$y = 3$

32. $x = 2$
$3x - y = -1$

33. $x = 5$
$2x - y = 4$

34. $2x + 3y = 5$
$4x + 6y = 9$

35. $5x - 4y = 5$
$10x - 8y = 23$

36. $3x = y + 5$
$6x - 2y = 5$

37. $4y + 1 = x$
$2x - 3 = 8y$

38. $2x - y = 4$
$4x = 2y + 8$

39. $3x = 5 - y$
$6x + 2y = 10$

40. $\dfrac{3}{2}x - 2y = 4$

$\dfrac{4}{3}x + \dfrac{5}{3}y = -\dfrac{10}{3}$

41. $\dfrac{3}{5}x + \dfrac{2}{5}y = 2$

$2x - \dfrac{3}{2}y = -\dfrac{15}{2}$

42. $\dfrac{2}{3}x + \dfrac{5}{3}y = 4$

$\dfrac{1}{4}x - \dfrac{1}{2}y = -\dfrac{3}{4}$

Without graphing, answer the following questions for each linear system. See Example 4.

(a) Is the system inconsistent, are the equations dependent, or neither?

(b) Is the graph a pair of intersecting lines, a pair of parallel lines, or one line?

(c) Does the system have one solution, no solutions, or an infinite number of solutions?

43. $y - x = -5$
$x + y = 1$

44. $2x + y = 6$
$x - 3y = -4$

45. $x + 2y = 0$
$4y = -2x$

46. $4x + y = 2$
$2x - y = 4$

47. $3x - 2y = -3$
$3x + y = 6$

48. $y = 3x$
$y + 3 = 3x$

49. $5x + 4y = 7$
$2x - 3y = 12$

50. $2x + 3y = 12$
$2x - y = 4$

51. $x + y = 4$
$2x = 8 - 2y$

Review Exercises Add. See Section 3.4.

52. $2m + 3n$ $\quad\quad$ **53.** $6p + 2q$
$\quad\;\; 5m - \;\; n$ $\quad\quad\quad\quad\;\; 4p + 5q$

54. $\quad x - 2y$ $\quad\quad$ **55.** $3r - 2t$
$\quad\;\; -x + 3y$ $\quad\quad\quad\quad\;\; 5r + 2t$

56. $-2x + y$ and $-y + 5x$

57. $4a + 3b$ and $-3b - 2a$

58. $6z + 5w$ and $-5w - 6z$

59. $-3m + 2n$ and $-2n + 3m$

7.2 Solving Systems of Linear Equations by Addition

Objectives

1 Solve linear systems by addition.

2 Multiply one or both equations of a system so that the addition method can be used.

3 Put equations in the proper form to use the addition method.

4 Use the addition method to solve an inconsistent system.

5 Use the addition method to solve a system of dependent equations.

Graphing to solve a system of equations has a serious drawback: It is difficult to estimate a solution such as $(1/3, -5/6)$ accurately from a graph.

1 An algebraic method that depends on the addition property of equality can be used to solve systems. As mentioned earlier, adding the same quantity to each side of an equation results in equal sums.

$$\text{If} \quad A = B, \quad\quad \text{then} \quad A + C = B + C.$$

This addition can be taken a step further. Adding *equal* quantities, rather than the *same* quantity, to both sides of an equation also results in equal sums.

$$\text{If} \quad A = B \quad \text{and} \quad C = D, \quad\quad \text{then} \quad A + C = B + D.$$

The use of the addition property to solve systems is called the **addition method** for solving systems of equations. For most systems, this method is more efficient than graphing.

Example 1 Use the addition method to solve the system

$$x + y = 5$$
$$x - y = 3.$$

Each equation in this system is a statement of equality, so, as discussed above, the sum of the right-hand sides equals the sum of the left-hand sides. Adding in this way gives

$$(x + y) + (x - y) = 5 + 3.$$

Combine terms and simplify to get

$$2x = 8$$
$$x = 4.$$

This result, $x = 4$, gives the x-value of the solution of the given system. To find the y-value of the solution, substitute 4 for x in either of the two equations of the system. Choosing the first equation, $x + y = 5$, gives

$$x + y = 5$$
$$4 + y = 5$$
$$y = 1.$$

The solution, $(4, 1)$, can be checked by substituting 4 for x and 1 for y in both equations of the given system.

$$x + y = 5 \qquad\qquad x - y = 3$$
$$4 + 1 = 5 \qquad\qquad 4 - 1 = 3$$
$$5 = 5 \quad \text{True} \qquad\qquad 3 = 3 \quad \text{True}$$

Since both results are true, the solution of the given system is $(4, 1)$. ✦

Example 2 Solve the system

$$-2x + y = -11$$
$$5x - y = 26.$$

As above, add left-hand sides and right-hand sides. This may be done more easily by drawing a line under the second equation and adding vertically. (Like terms must be placed in columns first.)

$$-2x + y = -11$$
$$\underline{5x - y = \quad 26}$$
$$3x \qquad = \quad 15 \qquad \text{Add in columns}$$
$$x = 5$$

Substitute 5 for x in either of the original equations. Choosing the first gives

$$-2x + y = -11$$
$$-2(5) + y = -11 \qquad \text{Let } x = 5$$
$$-10 + y = -11$$
$$y = -1.$$

The solution is $(5, -1)$. Check the solution by substitution into both of the original equations. ✦

2 In both examples above, a variable was eliminated by the addition step. Sometimes it is necessary to multiply both sides of one or both equations in a system by some number before the addition step will eliminate a variable.

Example 3   Solve the system

$$x + 3y = 7 \qquad (1)$$
$$2x + 5y = 12. \qquad (2)$$

Adding the two equations gives $3x + 8y = 19$, which does not help to solve the system. However, if both sides of equation (1) are first multiplied by -2, the terms with the variable x will drop out after adding.

$$-2(x + 3y) = -2(7)$$
$$-2x - 6y = -14 \qquad (3)$$

Now add equations (3) and (2).

$$-2x - 6y = -14 \qquad (3)$$
$$\underline{2x + 5y = 12} \qquad (2)$$
$$-y = -2$$
$$y = 2$$

Substituting $y = 2$ into equation (1) gives

$$x + 3y = 7$$
$$x + 3(2) = 7 \qquad \text{Let } y = 2$$
$$x + 6 = 7$$
$$x = 1.$$

The solution of this system is (1, 2). Check that this ordered pair satisfies both of the original equations. ✤

Example 4   Solve the system

$$2x + 3y = -15 \qquad (1)$$
$$5x + 2y = 1. \qquad (2)$$

Here the multiplication property of equality must be used with both equations instead of just one, as in Example 3. Multiply by numbers that will cause the coefficients of x (or of y) in the two equations to be additive inverses of each other. For example, multiply both sides of equation (1) by 5, and both sides of equation (2) by -2.

$$10x + 15y = -75 \qquad \text{Multiply equation (1) by 5}$$
$$\underline{-10x - 4y = -2} \qquad \text{Multiply equation (2) by } -2$$
$$11y = -77 \qquad \text{Add}$$
$$y = -7$$

Substituting -7 for y in either (1) or (2) gives $x = 3$. Check that the solution of the system is (3, -7).

The same result would be obtained by multiplying both sides of equation (1) by 2 and both sides of equation (2) by -3 and then adding. This process would eliminate the y-terms so that the value of x would have been found first. ✛

3 Before a system can be solved by the addition method, the terms of the two equations must be in the same order. When this is not the case, the terms should first be rearranged, as the next example shows. This example also shows an alternate way to get the second number when solving a system.

Example 5 Solve the system

$$4x = 9 - 3y \qquad (1)$$
$$5x - 2y = 8. \qquad (2)$$

Rearrange the terms in equation (1) so that the like terms can be aligned in columns. Add $3y$ to both sides to get the system

$$4x + 3y = 9 \qquad (3)$$
$$5x - 2y = 8.$$

One way to proceed is to eliminate y by multiplying both sides of equation (3) by 2 and both sides of equation (2) by 3, and then adding.

$$\begin{array}{r} 8x + 6y = 18 \\ \underline{15x - 6y = 24} \\ 23x = 42 \\ x = \dfrac{42}{23} \end{array}$$

Substituting 42/23 for x in one of the given equations would give y, but the arithmetic involved would be messy. Instead, solve for y by starting again with the original equations and eliminating x. Multiply both sides of equation (3) by 5 and both sides of equation (2) by -4, and then add.

$$\begin{array}{r} 20x + 15y = 45 \\ \underline{-20x + 8y = -32} \\ 23y = 13 \\ y = \dfrac{13}{23} \end{array}$$

The solution is (42/23, 13/23). ✛

When the value of the first variable is a fraction, the method used in Example 5 avoids errors that often occur when working with fractions. Of course, this method could be used in solving any system of equations.

A summary of the steps used to solve a linear system of equations by the addition method is given on the next page.

Solving Linear Systems by Addition

Step 1 Write both equations of the system in the form $Ax + By = C$.

Step 2 Multiply one or both equations by appropriate numbers so that the coefficients of x (or y) are negatives of each other.

Step 3 Add the two equations to get an equation with only one variable.

Step 4 Solve the equation from Step 3.

Step 5 Substitute the solution from Step 4 into either of the original equations.

Step 6 Solve the resulting equation from Step 5 for the remaining variable.

Step 7 Check the answer.

4 In the previous section some of the systems had equations with graphs that were two parallel lines, while the equations of other systems had graphs that were the same line. This section shows how to solve these systems with the addition method. The next example illustrates the solution of an inconsistent system, where the graphs of the equations are parallel lines.

Example 6 Solve the following system by the addition method.

$$2x + 4y = 5$$
$$4x + 8y = -9$$

Multiply both sides of $2x + 4y = 5$ by -2 and then add $4x + 8y = -9$.

$$\begin{array}{r} -4x - 8y = -10 \\ 4x + 8y = -9 \\ \hline 0 = -19 \end{array} \qquad \text{False}$$

The false statement $0 = -19$ shows that the given system is self-contradictory. *It has no solution.* This means that the graphs of the equations of this system are parallel lines, as shown in Figure 7.5. Since this system has no solution, it is inconsistent. ✦

5 The next example shows the result of using the addition method when the equations of the system are dependent, with the graphs of the equations in the system the same line.

Example 7 Solve the following system by the addition method.

$$3x - y = 4$$
$$-9x + 3y = -12$$

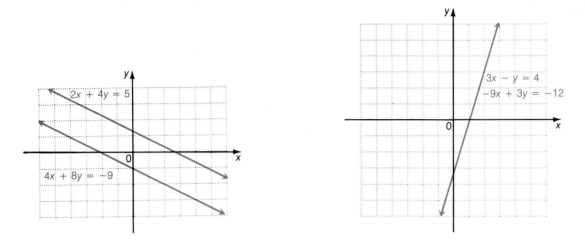

Figure 7.5   **Figure 7.6**

Multiply both sides of the first equation by 3 and then add the two equations to get

$$
\begin{array}{rcr}
9x - 3y &=& 12 \\
-9x + 3y &=& -12 \\
\hline
0 &=& 0. \quad \text{True}
\end{array}
$$

This result means that every solution of one equation is also a solution of the other, so the system has an infinite number of solutions: all the ordered pairs corresponding to points that lie on the common graph. As mentioned earlier, the equations of this system are dependent. In the answers at the back of this book, a solution of such a system of dependent equations is indicated by *infinite number of solutions*. A graph of the equations of this system is shown in Figure 7.6. ✤

One of three situations may occur when the addition method is used to solve a linear system of equations.
1. The result of the addition step is a statement such as $x = 2$ or $y = -3$. The solution will be exactly one ordered pair. The graphs of the equations of the system will intersect at exactly one point.
2. The result of the addition step is a false statement, such as $0 = 4$. In this case, the graphs are parallel lines and there is no solution for the system. The system is inconsistent.
3. The result of the addition step is a true statement, such as $0 = 0$. The graphs of the equations of the system are the same line, and an infinite number of ordered pairs are solutions. The equations are dependent.

7.2   Exercises

Solve each system by the addition method. Check your answers. See Examples 1 and 2.

1. $x - y = 3$
   $x + y = -1$

2. $x + y = 7$
   $x - y = -3$

3. $x + y = 2$
   $2x - y = 4$

4. $3x - y = 8$
   $x + y = 4$

5. $2x + y = 14$
   $x - y = 4$

6. $2x + y = 2$
   $-x - y = 1$

7. $3x + 2y = 6$
   $-3x - y = 0$

8. $5x - y = 9$
   $-5x + 2y = -8$

9. $6x - y = 1$
   $-6x + 5y = 7$

10. $6x + y = -2$
   $-6x + 3y = -14$

11. $2x - y = 5$
   $4x + y = 4$

12. $x - 4y = 13$
   $-x + 6y = -18$

13. $5x - y = 15$
   $7x + y = 21$

14. $x - 4y = 12$
   $-x + 6y = -18$

Solve each system by the addition method. Check your answers. See Example 3.

15. $2x - y = 7$
   $3x + 2y = 0$

16. $x + y = 7$
   $-3x + 3y = -9$

17. $x + 3y = 16$
   $2x - y = 4$

18. $4x - 3y = 8$
   $2x + y = 14$

19. $x + 4y = -18$
   $3x + 5y = -19$

20. $2x + y = 3$
   $5x - 2y = -15$

21. $3x - 2y = -6$
   $-5x + 4y = 16$

22. $-4x + 3y = 0$
   $5x - 6y = 9$

23. $2x - y = -8$
   $5x + 2y = -20$

24. $5x + 3y = -9$
   $7x + y = -3$

25. $2x + y = 5$
   $5x + 3y = 11$

26. $2x + 7y = -53$
   $4x + 3y = -7$

Solve each system by the addition method. Check your answers. See Examples 4 and 5.

27. $5x - 4y = -1$
   $-7x + 5y = 8$

28. $3x + 2y = 12$
   $5x - 3y = 1$

29. $3x + 5y = 33$
   $4x - 3y = 15$

30. $2x + 5y = 3$
   $5x = 3y + 23$

31. $3x + 7 = -5y$
   $5x + 4y = 10$

32. $3y + 11 = -2x$
   $5x + 2y = 22$

33. $2x + 3y = -12$
   $5x = 7y - 30$

34. $2x + 9y = 16$
   $5x = 6y + 40$

35. $4x + 7 = 3y$
   $6x + 5y = 18$

Use the addition method to solve each system. Check your answers. See Examples 6 and 7.

36. $2x - y = 1$
   $2x - y = 4$

37. $5x - 2y = 6$
   $10x - 4y = 10$

38. $3x - 5y = 2$
   $6x - 10y = 8$

39. $x + 3y = 5$
   $2x + 6y = 10$

40. $6x - 2y = 12$
   $-3x + y = -6$

41. $2x + 3y = 8$
   $4x + 6y = 12$

42. $4x + y = 6$
   $-8x - 2y = 21$

43. $3x + 5y = -19$
   $6x + 7y = -23$

44. $3x - 2y = -13$
   $9x + 5y = 16$

45. $24x + 12y = 19$
   $16x - 18y = -9$

46. $9x + 4y = -4$
   $6x + 6y = -11$

47. $9x - 5y = 7$
   $21x + 20y = 10$

48. $2x + \dfrac{3}{4}y = -1$

$4x + \dfrac{7}{3}y = \dfrac{4}{3}$

49. $\dfrac{6}{5}x + \dfrac{1}{5}y = \dfrac{33}{5}$

$-x + \dfrac{3}{2}y = \dfrac{19}{2}$

50. $\dfrac{5}{3}x - y = -1$

$\dfrac{7}{2}x + \dfrac{5}{2}y = -\dfrac{41}{2}$

51. $\dfrac{5}{2}x + y = -\dfrac{11}{2}$

$\dfrac{3}{4}x - \dfrac{3}{4}y = -\dfrac{15}{4}$

52. $\dfrac{3}{2}x - \dfrac{3}{4}y = \dfrac{15}{4}$

$\dfrac{4}{3}x - \dfrac{5}{3}y = \dfrac{19}{3}$

53. $\dfrac{2}{3}x - y = 0$

$\dfrac{4}{3}x + \dfrac{4}{3}y = 0$

54. $\dfrac{4}{3}x + \dfrac{1}{3}y = \dfrac{4}{3}$

$-\dfrac{8}{5}x - \dfrac{2}{5}y = -\dfrac{8}{5}$

55. $6.5x + 2.3y = 15$

$5.4x - 4.6y = 12$

56. $-2.2x + 7.1y = -4.8$

$3.8x + 14.2y = 6.8$

57. $.13x - .52y = -.39$

$.39x + .08y = -2.81$

58. $3x - 5y = 0$

$\dfrac{6}{7}x + \dfrac{10}{7}y = 0$

59. $\dfrac{2}{5}x - \dfrac{3}{5}y = 0$

$4x + 4y = 0$

Review Exercises Solve each equation. Check the solutions. See Sections 2.4 and 5.6.

60. $-2(y - 2) + 5y = 10$

61. $2m - 3(4 - m) = 8$

62. $p + 4(6 - 2p) = 3$

63. $4\left(\dfrac{3 - 2k}{2}\right) + 3k = -3$

64. $4x - 2\left(\dfrac{1 - 3x}{2}\right) = 6$

65. $a + 3\left(\dfrac{1 - a}{2}\right) = 5$

66. $4\left(\dfrac{-2 + y}{3}\right) - 2y = 2$

67. $3p + 2\left(\dfrac{2p - 1}{3}\right) = 4$

68. $4r + 5\left(\dfrac{2r + 7}{4}\right) = 1$

7.3 Solving Systems of Linear Equations by Substitution

Objectives

1 Solve linear systems by substitution.

2 Solve linear systems with fractions.

1 The graphical method and the addition method for solving systems of linear equations were discussed in the previous sections. A third method, the substitution method, is particularly useful for solving systems where one equation is solved, or can be solved easily, for one of the variables.

Example 1 Solve the system

$$3x + 5y = 26$$
$$y = 2x.$$

The second of these two equations says that $y = 2x$. Substituting $2x$ for y in the first equation gives

$$3x + 5y = 26$$
$$3x + 5(2x) = 26 \qquad \text{Let } y = 2x$$
$$3x + 10x = 26$$
$$13x = 26$$
$$x = 2.$$

Since $y = 2x$ and $x = 2$, $y = 2(2) = 4$. Check that the solution of the given system is $(2, 4)$. ✠

Example 2   Use substitution to solve the system

$$2x + 5y = 7$$
$$x = -1 - y.$$

The second equation is solved for x. Substitute $-1 - y$ for x in the first equation.

$$2x + 5y = 7$$
$$2(-1 - y) + 5y = 7 \qquad \text{Let } x = -1 - y$$
$$-2 - 2y + 5y = 7 \qquad \text{Distributive property}$$
$$-2 + 3y = 7$$
$$3y = 9$$
$$y = 3$$

Find x by letting $y = 3$ in $x = -1 - y$, to get $x = -1 - 3$, or $x = -4$. Check that the solution of the given system is $(-4, 3)$. ✠

Example 3   Use substitution to solve the system

$$x = 5 - 2y$$
$$2x + 4y = 6.$$

Substitute $5 - 2y$ for x in the second equation.

$$2x + 4y = 6$$
$$2(5 - 2y) + 4y = 6 \qquad \text{Let } x = 5 - 2y$$
$$10 - 4y + 4y = 6 \qquad \text{Distributive property}$$
$$10 = 6 \qquad \text{False}$$

As shown in the last section, this false result means that the equations of the system have graphs that are parallel lines. The system is inconsistent and has no solution. ✠

Example 4   Use substitution to solve the system

$$2x + 3y = 8$$
$$4x + 3y = 4.$$

The substitution method requires an equation solved for one of the variables. Choose the first equation of the system, $2x + 3y = 8$, and solve for x. Start by subtracting $3y$ on both sides.

$$2x + 3y = 8$$
$$2x = 8 - 3y$$

Now divide both sides of this equation by 2.

$$x = \frac{8 - 3y}{2}$$

Finally, substitute this value for x in the second equation of the system.

$$4x + 3y = 4$$

$$4\left(\frac{8 - 3y}{2}\right) + 3y = 4 \qquad \text{Let } x = \frac{8 - 3y}{2}$$

$$2(8 - 3y) + 3y = 4$$
$$16 - 6y + 3y = 4$$
$$-3y = -12$$
$$y = 4$$

Find x by letting $y = 4$ in $x = \frac{8 - 3y}{2}$.

$$x = \frac{8 - 3 \cdot 4}{2} = \frac{8 - 12}{2} = \frac{-4}{2} = -2$$

The solution of the given system is $(-2, 4)$. Check this solution in both equations. ✛

Example 5   Use substitution to solve the system

$$2x = 4 - y \qquad (1)$$
$$6 + 3y + 4x = 16 - x. \qquad (2)$$

Begin by simplifying the second equation by adding x and subtracting 6 on both sides. This gives the simplified system

$$2x = 4 - y \qquad (1)$$
$$5x + 3y = 10. \qquad (3)$$

For the substitution method, one of the equations must be solved for either x or y. Since the coefficient of y in equation (1) is -1, avoid fractions by solving this equation for y.

$$2x = 4 - y \qquad (1)$$
$$2x - 4 = -y$$
$$-2x + 4 = y$$

Now substitute $-2x + 4$ for y in equation (3).

$$5x + 3y = 10$$
$$5x + 3(-2x + 4) = 10$$
$$5x - 6x + 12 = 10$$
$$-x + 12 = 10$$
$$-x = -2$$
$$x = 2$$

Since $y = -2x + 4$ and $x = 2$,

$$y = -2(2) + 4 = 0,$$

and the solution is $(2, 0)$. Check the solution by substitution in both equations of the given system. ✜

2   When a system includes equations with fractions as coefficients, eliminate the fractions by multiplying both sides by a common denominator. Then solve the resulting system.

Example 6   Solve the system

$$3x + \frac{1}{4}y = 2 \qquad (1)$$

$$\frac{1}{2}x + \frac{3}{4}y = -\frac{5}{2} \qquad (2)$$

by any method.

Begin by eliminating fractions. Clear equation (1) of fractions by multiplying both sides by 4.

$$4\left(3x + \frac{1}{4}y\right) = 4(2)$$

$$4(3x) + 4\left(\frac{1}{4}y\right) = 4(2) \qquad \text{Distributive property}$$

$$12x + y = 8 \qquad (3)$$

Now clear equation (2) of fractions by multiplying both sides by the common denominator 4.

$$4\left(\frac{1}{2}x + \frac{3}{4}y\right) = 4\left(-\frac{5}{2}\right)$$

$$4\left(\frac{1}{2}x\right) + 4\left(\frac{3}{4}y\right) = 4\left(-\frac{5}{2}\right) \qquad \text{Distributive property}$$

$$2x + 3y = -10 \qquad (4)$$

The given system of equations has been simplified to

$$12x + y = 8 \qquad (3)$$
$$2x + 3y = -10. \qquad (4)$$

Solve the system by the substitution method. Equation (3) can be solved for y by subtracting $12x$ on each side.

$$12x + y = 8 \qquad (3)$$
$$y = -12x + 8$$

Now substitute the result for y in equation (4).

$$2x + 3(-12x + 8) = -10 \qquad \text{Let } y = -12x + 8$$
$$2x - 36x + 24 = -10 \qquad \text{Distributive property}$$
$$-34x = -34$$
$$x = 1$$

Using $y = -12x + 8$ and $x = 1$, gives $y = -4$. The solution is $(1, -4)$. Check by substituting 1 for x and -4 for y in both of the original equations. Verify that the same solution is found if the addition method is used to solve the system of equations (3) and (4).   ✚

7.3   Exercises

Solve each system by the substitution method. Check each solution. See Examples 1–4.

1. $x + y = 6$
$\qquad y = 2x$

2. $x + 3y = -11$
$\qquad y = -4x$

3. $3x + 2y = 26$
$\qquad x = y + 2$

4. $4x + 3y = -14$
$\qquad x = y - 7$

5. $x + 5y = 3$
$\qquad x = 2y + 10$

6. $5x + 2y = 14$
$\qquad y = 2x - 11$

7. $3x - y = 6$
$\qquad y = 3x - 5$

8. $4x - y = 4$
$\qquad y = 4x + 3$

9. $3x - 2y = 14$
$\qquad 2x + y = 0$

10. $2x - 5 = -y$
$\qquad x + 3y = 0$

11. $x + y = 6$
$\qquad x - y = 4$

12. $3x - 2y = 13$
$\qquad x + y = 6$

13. $6x - 8y = 4$
$\qquad 3x = 4y + 2$

14. $12x + 18y = 12$
$\qquad 2x = 2 - 3y$

15. $2x + 3y = 11$
$\qquad 4y - x = 0$

16. $4x + 5y = 5$
$\quad\ 2x + 3y = 1$

17. $6x + 5y = 13$
$\quad\ 3x + 2y = 4$

18. $5x - 2y = -3$
$\quad\ 3y + 2x = 11$

19. $3x + 4y = -10$
$\quad\ x + 6 = 0$

20. $4x + y = 5$
$\quad\ x - 2 = 0$

21. $5x + 2y = -19$
$\quad\ y - 3 = 0$

Solve each system by either the addition method or the substitution method. First simplify equations where necessary. Check each solution. See Example 5.

22. $5x - 4y = 42 - 8y - 2$
$\quad\ 2x + y = x + 1$

23. $4x - 2y + 8 = 3x + 4y - 1$
$\quad\ 3x + y = x + 8$

24. $5x - 4y - 8x - 2 = 6x + 3y - 3$
$\quad\ 4x - y = -2y - 8$

25. $2x - 8y + 3y + 2 = 5y + 16$
$\quad\ 8x - 2y = 4x + 28$

26. $7x - 9 + 2y - 8 = -3y + 4x + 13$
$\quad\ 4y - 8x = -8 + 9x + 32$

27. $\quad\ -2x + 3y = 12 + 2y$
$\quad\ 2x - 5y + 4 = -8 - 4y$

28. $\quad\ 2x + 5y = 7 + 4y - x$
$\quad\ 5x + 3y + 8 = 22 - x + y$

29. $\quad\ y + 9 = \quad 3x - 2y + 6$
$\quad\ 5 - 3x + 24 = -2x + 4y + 3$

30. $5x - 2y = 16 + 4x - 10$
$\quad\ 4x + 3y = 60 + 2x + y$

31. $4 + 4x - 3y = 34 + x$
$\quad\ 5y + 4x = 4y - 2 + 3x$

Solve each system by either the addition method or the substitution method. First clear all fractions. Check each solution. See Example 6.

32. $\quad \dfrac{5}{3}x + 2y = \dfrac{1}{3} + y$

$\quad 2x - 3 + \dfrac{y}{3} = -2 + x$

33. $\quad \dfrac{x}{6} + \dfrac{y}{6} = 1$

$\quad -\dfrac{1}{2}x - \dfrac{1}{3}y = -5$

34. $\quad \dfrac{x}{2} - \dfrac{y}{3} = \dfrac{5}{6}$

$\quad \dfrac{x}{5} - \dfrac{y}{4} = \dfrac{1}{10}$

35. $\dfrac{x}{3} - \dfrac{3y}{4} = -\dfrac{1}{2}$

$\quad \dfrac{2x}{3} + \dfrac{y}{2} = 3$

36. $\dfrac{x}{5} + 2y = \dfrac{8}{5}$

$\quad \dfrac{3x}{5} + \dfrac{y}{2} = \dfrac{-7}{10}$

37. $\dfrac{x}{2} + \dfrac{y}{3} = \dfrac{7}{6}$

$\quad \dfrac{x}{4} - \dfrac{3y}{2} = \dfrac{9}{4}$

38. $\dfrac{5x}{2} - \dfrac{y}{3} = \dfrac{5}{6}$

$\quad \dfrac{4x}{3} + y = \dfrac{19}{3}$

39. $\dfrac{2x}{3} + \dfrac{y}{6} = 9$

$\quad \dfrac{x}{4} - \dfrac{3y}{2} = -6$

40. $\dfrac{3x}{2} + \dfrac{y}{5} = 2$

$\quad \dfrac{5x}{2} - \dfrac{3y}{10} = \dfrac{13}{2}$

Review Exercises   *Solve each word problem. See Section 2.5.*

41. If three times a number is added to 6, the result is 69. Find the number.

42. The product of 5 and 1 more than a number is 35. Find the number.

43. The perimeter of a rectangle is 46 feet. The width is 7 feet less than the length. Find the width.

44. The area of a rectangle is numerically 20 more than the width, and the length is 6 centimeters. What is the width?

45. A cashier has ten-dollar bills and twenty-dollar bills. There are 6 more tens than twenties. If there are 32 bills altogether, how many of them are twenties?

46. Janet traveled for 2 hours at a constant speed. Because of road work, she reduced her speed by 7 miles per hour for the next 2 hours. If she traveled 206 miles, what was her speed on the first part of the trip?

7.4 Applications of Linear Systems

Objectives

Use linear systems to solve word problems about

1 numbers;

2 money;

3 mixtures;

4 rate or speed using the distance formula.

Many practical problems are more easily translated into equations if two variables are used. With two variables, a system of two equations is needed to find the desired solution. The examples in this section illustrate the method of solving word problems using two equations and two variables.

Recall from Chapter 3 the steps used in solving word problems. The steps presented there can be modified as follows to allow for two variables and two equations.

Solving Word Problems with Two Variables	*Step 1* Choose two variables, one to represent each of the unknown values that must be found. *Write down what each variable is to represent.*
	Step 2 Translate the problem into two equations using both variables.
	Step 3 Solve the system of two equations.
	Step 4 Answer the question or questions asked in the problem.
	Step 5 Check your solution by using the words of the original problem.

1 The first example shows how to use two variables to solve a problem about two unknown numbers.

Example 1 The sum of two numbers is 63. Their difference is 19. Find the two numbers.

Step 1 Let x represent one number and y the other.

Step 2 From the information in the problem, set up a system of equations.

$$x + y = 63 \qquad \text{The sum is 63}$$
$$x - y = 19 \qquad \text{The difference is 19}$$

Step 3   Solve the system from Step 2. Here the addition method is used. Adding gives

$$\begin{aligned} x + y &= 63 \\ \underline{x - y} &= \underline{19} \\ 2x \quad\;\; &= 82. \end{aligned}$$

From this last equation, $x = 41$. Substitute 41 for x in either equation to find $y = 22$.

Step 4   The numbers required in the problem are 41 and 22.

Step 5   The sum of 41 and 22 is 63, and their difference is $41 - 22 = 19$. The solution satisfies the conditions of the problem.   ✤

2   The next example shows how to solve a common type of word problem that involves two quantities and their costs.

Example 2   An office supply store sold 454 rolls of tape for total receipts of $2528. Large rolls were priced at $6 and smaller rolls were priced at $2. The manager needs to know how many rolls of each kind of tape were sold.

Step 1   Let $x =$ the number of large rolls of tape sold;
          $y =$ the number of small rolls of tape sold.

Step 2   The information given in the problem is summarized in the chart. The entries in the receipts column were found by multiplying the number of rolls of each size of tape sold by the appropriate price per roll.

Size of Roll	Number Sold	Price (in dollars)	Receipts (in dollars)
large	x	6	$6x$
small	y	2	$2y$

The total number of rolls of tape sold was 454, so

$$x + y = 454.$$

Since the total receipts were $2528, the right-hand column of the chart indicates that

$$6x + 2y = 2528.$$

Step 3   The information in the problem has produced the system of equations

$$\begin{aligned} x + y &= 454 &\quad (1) \\ 6x + 2y &= 2528. &\quad (2) \end{aligned}$$

Solve the system by multiplying both sides of equation (1) by -2 and then adding the result to equation (2).

$$\begin{array}{rcr} -2x - 2y & = & -908 \\ 6x + 2y & = & 2528 \\ \hline 4x \quad\quad & = & 1620 \\ x & = & 405 \end{array}$$

Substitute 405 for x in equation 1 to find $y = 49$.

Step 4   There were 405 large and 49 small rolls of tape sold.

Step 5   The receipts from the sale of 405 large and 49 small rolls of tape should be

$$405(6) + 49(2) = 2528,$$

which agrees with the given information.   ✛

Try to identify each step in the solution of the problems in the rest of the examples.

3   Mixture problems occur in many fields of application of mathematics. One important type of mixture problem occurs in the study of chemistry. This kind of problem can be solved as shown in the next example.

Example 3   A pharmacist needs 100 liters of 50% alcohol solution. She has on hand 30% alcohol solution and 80% alcohol solution, which she can mix. How many liters of each will be required to make the 100 liters of 50% alcohol solution?

Let $x =$ the number of liters of 30% alcohol needed;

$y =$ the number of liters of 80% alcohol needed.

The information given in the problem is summarized in the table. (Each percent has been changed to its decimal form in the third column.)

Liters of Solution	Percent	Liters of Pure Alcohol
x	30	$.30x$
y	80	$.80y$
100	50	$.50(100)$

The pharmacist will have $.30x$ liters of alcohol from the x liters of 30% solution and $.80y$ liters of alcohol from the y liters of 80% solution, for a total of $.30x + .80y$ liters of pure alcohol. In the mixture, she wants 100 liters of 50% solution.

This 100 liters would contain .50(100) = 50 liters of pure alcohol. Since the amounts of pure alcohol must be equal,

$$.30x + .80y = 50.$$

The total number of liters is 100, or

$$x + y = 100.$$

These two equations give the system

$$.30x + .80y = 50$$
$$x + y = 100.$$

Solve this system by the substitution method. Solving the second equation of the system for x gives $x = 100 - y$. Substitute $100 - y$ for x in the first equation to get

$$.30 (100 - y) + .80y = 50 \qquad \text{Let } x = 100 - y$$
$$30 - .30y + .80y = 50$$
$$.50y = 20$$
$$y = 40.$$

Then $x = 100 - y = 100 - 40 = 60$. The pharmacist should use 60 liters of the 30% solution and 40 liters of the 80% solution. Since .30(60) + (.80)40 = 50, this mix will give 100 liters of 50% solution, as required in the original problem. ✛

4 Word problems that use the distance formula, relating distance, rate, and time, often result in a system of two linear equations.

Example 4 Two executives in cities 400 miles apart drive to a business meeting at a location on the line between their cities. They meet after 4 hours. Find the speed of each car if one car travels 20 miles per hour faster than the other.

Let $\quad x =$ the speed of the faster car;

$\quad\quad y =$ the speed of the slower car.

Use the formula that relates distance, rate, and time, $d = rt$. Since each car travels for 4 hours, the time, t, for each car is 4. This information is shown in the chart. The distance is found by using the formula $d = rt$ and the expressions already entered in the chart.

	r	t	d
Faster car	x	4	$4x$
Slower car	y	4	$4y$

Since the total distance traveled by both cars is 400 miles,

$$4x + 4y = 400.$$

The faster car goes 20 miles per hour more than the slower, giving

$$x = 20 + y.$$

This system of equations,

$$4x + 4y = 400$$
$$x = 20 + y,$$

can be solved by substitution. Replace x with $20 + y$ in the first equation of the system and solve for y.

$$4(20 + y) + 4y = 400$$
$$80 + 4y + 4y = 400$$
$$80 + 8y = 400$$
$$8y = 320$$
$$y = 40$$

Since $x = 20 + y$ and $y = 40$, $x = 20 + 40 = 60$. The speeds of the two cars were 40 miles per hour and 60 miles per hour. If each car travels for 4 hours, the total distance traveled is

$$4(40) + 4(60) = 160 + 240 = 400$$

miles, as required. ✛

The problems in this section also could be solved using only one variable, but for most of them the solution is simpler with two variables. Be careful: do not forget that two variables require two equations.

7.4 Exercises

Write a system of equations for each problem. Then solve the system. Formulas are given inside the front cover of the book. See Example 1.

1. The sum of two numbers is 52, and their difference is 34. Find the numbers.
2. Find two numbers whose sum is 56 and whose difference is 18.
3. A certain number is three times as large as a second number. Their sum is 96. What are the two numbers?
4. One number is five times as large as another. The difference of the numbers is 48. Find the numbers.
5. Two angles have a sum of 90°. Their difference is 20°. Find the angles.
6. The sum of two angles is 180°. One angle is 30° less than twice the other. Find the angles.

Write a system of equations for each problem. Then solve the system. See Example 2.

7. The cashier at the Evergreen Ranch has some ten-dollar bills and some twenty-dollar bills. The total value of the money is $1480. If there is a total of 85 bills, how many of each type are there?

Kind of Bill	Number of Bills	Total Value
$10	x	$10x$
$20	y	
Totals	85	$1480

8. A bank teller has 154 bills of $1 and $5 denominations. How many of each type of bill does he have if the total value of the money is $466?

9. A club secretary bought 16¢ and 20¢ pieces of candy to give to the members. She spent a total of $31.04. If she bought 170 pieces of candy, how many of each kind did she buy?

10. There were 311 tickets sold for a basketball game, some for students and some for nonstudents. Student tickets cost $1.25 each and nonstudent tickets cost $3.75 each. The total receipts were $543.75. How many of each type of ticket were sold?

11. A bank clerk has a total of 124 bills, both fives and tens. The total value of the money is $840. How many of each type of bill does he have?

12. A library buys a total of 54 books. Some cost $32 each and some cost $44 each. The total cost of the books is $1968. How many of each type of book does the library buy?

13. An artist bought some large canvases at $7 each and some small ones at $4 each, paying $219 in total. Altogether, the artist bought 39 canvases. How many of each size were bought?

14. A hospital bought a total of 146 bottles of glucose solution. Small bottles cost $2 each, and large bottles are $3 each. The total cost was $336. How many of each size bottle were bought?

Write a system of equations for each problem. Then solve the system. See Example 3.

15. A 90% antifreeze solution is to be mixed with a 75% solution to make 20 liters of a 78% solution. How many liters of 90% and 75% solutions should be used? See the chart at the top of the next page.

Liters of solution	Percent	Liters of pure antifreeze
x	90	$.90x$
y	75	$.75y$
20	78	

16. A 40% potassium iodide solution is to be mixed with a 70% solution to get 60 liters of a 50% solution. How many liters of the 40% and 70% solutions will be needed?

17. How many liters of a 25% indicator solution should be mixed with a 55% solution to get 12 liters of a 45% solution?

18. A 60% solution of salt is to be mixed with an 80% solution to get 40 liters of a 65% solution. How many liters of the 60% and the 80% solutions will be needed?

19. How many barrels of olives worth $40 per barrel must be mixed with olives worth $60 per barrel to get 50 barrels of a mixture worth $48 per barrel?

20. A glue merchant wishes to mix some glue worth $70 per barrel with some glue worth $90 per barrel to get 80 barrels of a mixture worth $77.50 per barrel. How many barrels of each type should be used?

Write a system of equations for each problem. Then solve the system. See Example 4.

21. A boat takes 3 hours to go 24 miles upstream, and it can go 36 miles downstream in the same time. Find the speed of the current and the speed of the boat in still water. Let x = the speed of the boat in still water and y = the speed of the current.

	d	r	t
Downstream	36	$x + y$	3
Upstream	24	$x - y$	3

22. It takes a boat 1 1/2 hours to go 12 miles downstream, and 6 hours to return. Find the speed of the boat in still water and the speed of the current.

23. If a plane can travel 400 miles per hour into the wind and 540 miles per hour with the wind, find the speed of the wind and the speed of the plane in still air.

24. A small plane travels 100 miles per hour with the wind and 60 miles per hour against it. Find the speed of the wind and the speed of the plane in still air.

Write a system of equations for each problem. Then solve the system.

25. Ms. Sullivan has $10,000 to invest, part at 10% and part at 14%. She wants the income from simple interest on the two investments to total $1100 yearly. How much should she invest at each rate?

26. A grocer wishes to blend candy selling for $1.20 a pound with candy selling for $1.80 a pound to get a mixture that will be sold for $1.40 a pound. How many pounds of the $1.20 and the $1.80 candy should be used to get 30 pounds of the mixture?

27. A rectangle is twice as long as it is wide. Its perimeter is 60 inches. Find the dimensions of the rectangle.

28. Mr. Emerson has twice as much money invested at 14% as he has at 16%. If his yearly income from investments is $880, how much does he have invested at each rate?

29. A merchant wishes to mix coffee worth $6 per pound with coffee worth $3 per pound to get 90 pounds of a mixture worth $4 per pound. How many pounds of the $6 and the $3 coffee will be needed?

30. The perimeter of a triangle is 21 inches. If two sides are of equal length, and the third side is 3 inches longer than one of the equal sides, find the length of the three sides.

31. At the beginning of a walk for charity, John and Harriet are 30 miles apart. If they leave at the same time and walk in the same direction, John overtakes Harriet in 60 hours. If they walk toward each other, they meet in 5 hours. What are their speeds?

32. Mr. Anderson left Farmersville in a plane at noon to travel to Exeter. Mr. Bentley left Exeter in his automobile at 2 P.M. to travel to Farmersville. It is 400 miles from Exeter to Farmersville. If the sum of their speeds is 120 miles per hour, and if they met at 4 P.M., find the speed of each.

33. The Smith family is coming to visit, and no one knows how many children they have. Janet, one of the girls, says she has as many brothers as sisters; her brother Steve says he has twice as many sisters as brothers. How many boys and how many girls are in the family?

34. In the Lopez family, the number of boys is one more than half the number of girls. One of the Lopez boys, Rico, says that he has one more sister than brothers. How many boys and girls are in the family?

Review Exercises   *Graph each linear inequality. See Section 6.5.*

35. $x + y \leq 5$

36. $2x - y \geq 7$

37. $3x + 2y > 6$

38. $x + 3y < 9$

39. $5x + 4y > 0$

40. $3x - 4y < 0$

7.5   *Solving Systems of Linear Inequalities*

Objective

1   Solve systems of linear inequalities by graphing.

Graphing the solution of a linear inequality was discussed in Section 6.5. For example, to graph the solution of $x + 3y > 12$, first graph the line $x + 3y = 12$ by finding a few ordered pairs that satisfy the equation. Because the points on the line do not satisfy the inequality, use a dashed line. To decide which side of the line should be shaded, choose any test point not on the line, such as $(0, 0)$. Substitute 0 for x and 0 for y in the given inequality.

$$x + 3y > 12$$
$$0 + 0 > 12$$
$$0 > 12 \qquad \text{False}$$

Since the test point does not satisfy the inequality, shade the region on the side of the line that does not include $(0, 0)$, as in Figure 7.7.

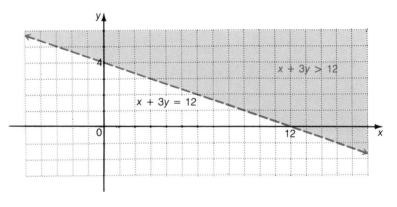

Figure 7.7

1   The same method is used to find the solution of a system of two linear inequalities, as shown in Examples 1 and 2. A **system of linear inequalities** consists of two or more linear inequalities. The solution of a system of linear inequalities includes all points that make all inequalities of the system true at the same time.

Example 1   Graph the solution of the linear system

$$3x + 2y \leq 6$$
$$2x - 5y \geq 10.$$

First graph the inequality $3x + 2y \le 6$, using the steps described above. Make the line solid rather than dashed, because the points on the line are also part of the solution. Then, on the same axes, graph the second inequality, $2x - 5y \ge 10$. The solution of the system is given by the overlap of the regions of the two graphs. This solution is the darkest shaded region in Figure 7.8 and includes portions of the two boundary lines. ✥

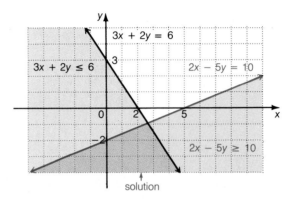

Figure 7.8

Example 2   Graph the solution of the system

$$x - y > 5$$
$$2x + y < 2.$$

Figure 7.9 shows the graphs of both $x - y > 5$ and $2x + y < 2$. Dashed lines show that the graphs of the inequalities do not include their boundary lines. The solution of the system is the darkest shaded region in the figure. The solution does not include either boundary line. ✥

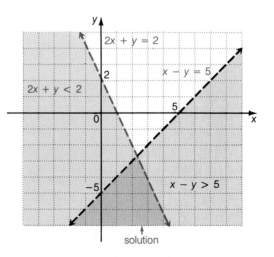

Figure 7.9

Example 3   Graph the solution of the system

$$4x - 3y \le 8$$
$$x \ge 2.$$

Recall that $x = 2$ is a vertical line through the point $(2, 0)$. The graph of the solution is the darkest shaded region in Figure 7.10.   ⬥

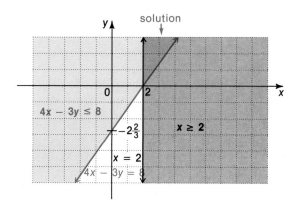

Figure 7.10

7.5   Exercises

Graph the solution of each system of linear inequalities. See Examples 1–3.

1. $x + y \le 6$
$ x - y \le 1$

2. $x + y \ge 2$
$ x - y \le 3$

3. $2x - 3y \le 6$
$ x + y \ge -1$

4. $4x + 5y \le 20$
$ x - 2y \le 5$

5. $ x + 4y \le 8$
$2x - y \le 4$

6. $3x + y \le 6$
$2x - y \le 8$

7. $x - 4y \le 3$
$ x \ge 2y$

8. $2x + 3y \le 6$
$ x - y \ge 5$

9. $x + 2y \le 4$
$ x + 1 \ge y$

10. $ y \le 2x - 5$
$x - 3y \le 2$

11. $4x + 3y \le 6$
$ x - 2y \ge 4$

12. $ 3x - y \le 4$
$-6x + 2y \le -10$

13. $ x - 2y > 6$
$2x + y > 4$

14. $3x + y < 4$
$ x + 2y > 2$

15. $ x < 2y + 3$
$x + y > 0$

16. $2x + 3y < 6$
$4x + 6y > 18$

17. $x - 3y \le 6$
$ x \ge -1$

18. $2x + 5y \ge 10$
$ x \le 4$

19. $3x - 2y \ge 9$
$ y \le 3$

20. $x \ge -1$
$ y \ge 4$

21. $x \ge 2$
$ y \le 3$

22. $4x + 5y < 8$
$ y > -2$
$ x > -4$

23. $x + y \ge -3$
$ x - y \le 3$
$ y \le 3$

24. $3x - 2y \ge 6$
$ x + y \le 4$
$ x \ge 0$
$ y \ge -4$

Review Exercises *Find each value. See Section 1.3.*

25. 5^2 **26.** 7^2 **27.** 11^2 **28.** 15^2

29. 6^3 **30.** 3^3 **31.** 2^4 **32.** 3^5

Chapter 7 *Summary*

Solving Linear Systems by Addition	*Step 1* Write both equations of the system in the form $Ax + By = C$.
	Step 2 Multiply one or both equations by appropriate numbers so that the coefficients of x (or y) are negatives of each other.
	Step 3 Add the two equations to get an equation with only one variable.
	Step 4 Solve the equation from Step 3.
	Step 5 Substitute the solution from Step 4 into either of the original equations.
	Step 6 Solve the resulting equation from Step 5 for the remaining variable.
	Step 7 Check the answer.

One of three situations may occur when the addition method is used to solve a linear system of equations.

1. The result of the addition step is a statement such as $x = 2$ or $y = -3$. The solution will be exactly one ordered pair. The graphs of the equations of the system will intersect at exactly one point.
2. The result of the addition step is a false statement, such as $0 = 4$. In this case, the graphs are parallel lines and there is no solution for the system. The system is inconsistent.
3. The result of the addition step is a true statement, such as $0 = 0$. The graphs of the equations of the system are the same line, and an infinite number of ordered pairs are solutions. The equations are dependent.

Solving Word Problems with Two Variables	*Step 1* Choose two variables, one to represent each of the unknown values that must be found. Write down what each variable is to represent.
	Step 2 Translate the problem into two equations using both variables.
	Step 3 Solve the system of two equations.
	Step 4 Answer the question or questions asked in the problem.
	Step 5 Check your solution by using the words of the original problem.

Chapter 7 *Review Exercises*

❶ [7.1] *Decide whether the given ordered pair is a solution of the given system.*

1. $(3, 4)$
$$4x - 2y = 4$$
$$5x + y = 17$$

2. $(-2, 1)$
$$5x + 3y = -7$$
$$2x - 3y = -7$$

3. $(-5, 2)$
$$x - 4y = -10$$
$$2x + 3y = -4$$

4. $(6, 3)$
$$3x + 8y = 42$$
$$4x - 3y = 15$$

5. $(-1, -3)$
$$x + 2y = -7$$
$$-x + 3y = -8$$

6. $(2, 6)$
$$3x - y = 0$$
$$4x + 2y = 20$$

Solve each system by graphing.

7. $x + y = 3$
$2x - y = 3$

8. $x - 2y = -6$
$2x + y = -2$

9. $2x + 3y = 1$
$4x - y = -5$

10. $x = 2y + 2$
$2x - 4y = 4$

11. $y + 2 = 0$
$3y = 6x$

12. $2x + 5y = 10$
$2x + 5y = 8$

[7.2] *Solve each system by the addition method. Identify any inconsistent systems or systems with dependent equations.*

13. $x + y = 6$
$2x + y = 8$

14. $3x - y = 13$
$x - 2y = 1$

15. $5x + 4y = 7$
$3x - 4y = 17$

16. $-4x + 3y = -7$
$6x - 5y = 11$

17. $3x - 4y = 7$
$6x - 8y = 14$

18. $2x + y = 5$
$2x + y = 8$

19. $3x - 2y = 14$
$5x - 4y = 24$

20. $2x + 6y = 18$
$3x + 5y = 19$

21. $3x - 4y = -1$
$-6x + 8y = 2$

[7.3] *Solve each system by the substitution method.*

22. $2x - 5y = 5$
$y = x + 2$

23. $3x + y = 14$
$x = 2y$

24. $5x + y = -6$
$x = 3y + 2$

25. $4x + 5y = 35$
$x = 2y - 1$

26. $6x + 5y = 9$
$2x - 3y = 17$

27. $2x + 3y = 5$
$3x + 4y = 8$

[7.4] *Solve each word problem by any method. Use two variables.*

28. One number is 2 more than twice as large as another. Their sum is 17. Find the numbers.

29. The perimeter of a rectangle is 40 meters. Its length is 1 1/2 times its width. Find the length and width of the rectangle.

30. A cashier has 20 bills, all of which are ten-dollar or twenty-dollar bills. The total value of the money is $250. How many of each type does he have?

31. Ms. Branson has $18,000 to invest. She wants the total annual income from the money to be $2600. She can invest part of it at 12% and the rest at 16%. How much should she invest at each rate?

32. A certain plane flying with the wind travels 540 miles in 2 hours. Later, flying against the same wind, the plane travels 690 miles in 3 hours. Find the speed of the plane in still air and the speed of the wind.

33. A 40% antifreeze solution is to be mixed with a 70% solution to get 60 liters of a 50% solution. How many liters of the 40% and 70% solutions will be needed?

[7.5] *Graph each system of linear inequalities.*

34. $x + y \geq 4$
$x - y \leq 2$

35. $x + y \leq 3$
$2x \geq y$

36. $y + 2 < 2x$
$x > 3$

37. $y \geq 3x$
$2x + 3y \leq 4$

❷ *Solve by any method.*

38. $\dfrac{3x}{4} - \dfrac{y}{3} = \dfrac{7}{6}$

$\dfrac{x}{2} + \dfrac{2y}{3} = \dfrac{5}{3}$

39. $2x + y - x = 3y + 5$

$y + 2 = x - 5$

40. Candy that sells for $1.30 a pound is to be mixed with candy selling for $.90 a pound to get 50 pounds of a mix that will sell for $1 per pound. How much of each type should be used?

41. $5x + 4y = 3$

$7x + 5y = 3$

42. $\dfrac{5x}{3} - \dfrac{y}{2} = -\dfrac{31}{3}$

$2x + \dfrac{y}{3} = \dfrac{26}{3}$

43. $2x + y < 6$

$y - 2x < 6$

44. $3x + 5y = -1$

$5x + 4y = 7$

45. $\dfrac{2x}{5} - \dfrac{y}{2} = \dfrac{7}{10}$

$\dfrac{x}{3} + y = 2$

46. $4x + y = 5$

$-12x - 3y = -15$

47. The sum of two numbers is 42, and their difference is 14. Find the numbers.

48. $x \geq 2$

$y \leq 5$

49. $\dfrac{x}{2} + \dfrac{y}{3} = \dfrac{-8}{3}$

$\dfrac{x}{4} + 2y = \dfrac{1}{2}$

50. $5x - 3 + y = 4y + 8$

$2y + 1 = x - 3$

Chapter 7   *Test*

Solve each system by graphing.

1. $2x + y = 5$

$3x - y = 15$

2. $3x + 2y = 8$

$5x + 4y = 10$

3. $x + 2y = 6$

$2x - y = 7$

Solve each system by the addition method.

4. $2x - 5y = -13$

$3x + 5y = 43$

5. $4x + 3y = 25$

$5x + 4y = 32$

6. $6x + 5y = -13$

$-3x - \dfrac{5y}{2} = \dfrac{13}{2}$

7. $2x - y = 5$

$4x + 3y = 0$

8. $4x + 5y = 8$

$-8x - 10y = -6$

9. $\dfrac{6}{5}x - y = \dfrac{1}{5}$

$-\dfrac{2}{3}x + \dfrac{1}{6}y = \dfrac{1}{3}$

Solve each system by substitution.

10. $2x + y = 1$
$\qquad x = 8 + y$

11. $4x + 3y = 0$
$\qquad 2x + \ y = -4$

Solve each system by any method.

12. $8 + 3x - 4y = 14 - 3y$
$\qquad 3x + y + 12 = 9x - y$

13. $\dfrac{x}{2} - \dfrac{y}{4} = -4$

$\qquad \dfrac{2x}{3} + \dfrac{5y}{4} = 1$

Solve the following word problems.

14. The sum of two numbers is 39. If one number is doubled, the result is 3 less than the other. Find the numbers.

15. The local record shop is having a sale. Some records cost \$5 and some cost \$7.50. Joe has exactly \$40 to spend and wants to buy 6 records. How many can he buy at each price?

16. A 40% solution of acid is to be mixed with a 60% solution to get 100 liters of a 45% solution. How many liters of each solution should be used?

17. Two cars leave from the same place and travel in the same direction. One car travels one and one third times as fast as the other. After 3 hours, they are 45 miles apart. What is the speed of each car?

Graph the solution of each system of inequalities.

18. $2x + 7y \le 14$
$\qquad x - \ y \ge 1$

19. $\ \ 2x - y \le 6$
$\qquad 4y + 12 \ge -3x$

20. $3x - 5y < 15$
$\qquad\qquad y < 2$

Roots and Radicals

Objectives

1 Find square roots.

2 Decide whether a given root is rational or irrational.

3 Find decimal approximations for irrational square roots.

4 Use the Pythagorean formula.

5 Find higher roots.

The square of a number is found by multiplying it by itself.

If $a = 7$, then $a^2 = 7 \cdot 7 = 49$.

If $a = 10$, then $a^2 = 10 \cdot 10 = 100$.

If $a = -5$, then $a^2 = (-5) \cdot (-5) = 25$.

If $a = -\dfrac{1}{2}$, then $a^2 = \left(-\dfrac{1}{2}\right) \cdot \left(-\dfrac{1}{2}\right) = \dfrac{1}{4}$.

In this chapter, the opposite problem is considered.

$$\text{If } a^2 = 49, \text{ then } a = ?$$
$$\text{If } a^2 = 100, \text{ then } a = ?$$
$$\text{If } a^2 = 25, \text{ then } a = ?$$
$$\text{If } a^2 = \frac{1}{4}, \text{ then } a = ?$$

1   In these examples, finding a requires a number whose square is the given number. The number a is called the **square root** of the number a^2.

Example 1   If $a^2 = 49$, find a.

Find a square root of 49 by thinking of a number whose square is 49. One square root of 49 is 7, since $7 \cdot 7 = 49$. Another square root of 49 is -7, since $(-7)(-7) = 49$. Therefore, 49 has two square roots, 7 and -7. One is positive and one is negative.   ✤

All numbers with rational square roots are called **perfect squares.** The first 100 integer perfect squares are listed in Table 3 inside the back cover of this book.

The positive square root of a number is written with the symbol $\sqrt{}$. For example, the positive square root of 121 is 11, written

$$\sqrt{121} = 11.$$

The symbol $-\sqrt{}$ is used for the negative square root of a number. For example, the negative square root of 121 is -11, written

$$-\sqrt{121} = -11.$$

If a is a nonnegative real number,

$$\sqrt{a} \text{ is the positive square root of } a,$$
$$-\sqrt{a} \text{ is the negative square root of } a.$$

Also, for nonnegative a,

$$\sqrt{a} \cdot \sqrt{a} = (\sqrt{a})^2 = a \text{ and } -\sqrt{a} \cdot -\sqrt{a} = (-\sqrt{a})^2 = a.$$

The symbol $\sqrt{}$ is called a **radical sign** and always represents the nonnegative square root. The number inside the radical sign is called the **radicand** and the entire expression, radical sign and radicand, is called a **radical.** An algebraic expression containing a radical is called a **radical expression.**

Example 2   Find each square root.

(a) $\sqrt{144}$

The radical $\sqrt{144}$ represents the positive square root of 144. Think of a positive number whose square is 144. The number is 12 (see Table 3) and

$$\sqrt{144} = 12.$$

(b) $-\sqrt{1024}$

This symbol represents the negative square root of 1024. From Table 3,

$$-\sqrt{1024} = -32.$$

(c) $\sqrt{256} = 16$ $\qquad\qquad$ **(d)** $-\sqrt{900} = -30$

(e) $\sqrt{\dfrac{4}{9}} = \dfrac{2}{3}$ $\qquad\qquad$ **(f)** $-\sqrt{\dfrac{16}{49}} = -\dfrac{4}{7}$ ✜

2 A number that is not a perfect square has a square root that is not a rational number. For example, $\sqrt{5}$ is not a rational number, because it cannot be written as the ratio of two integers. However, $\sqrt{5}$ is a real number and corresponds to a point on the number line. A real number that is not rational is called an **irrational number.** The number $\sqrt{5}$ is irrational. Many square roots of integers are irrational.

> If a is a positive integer that is not a perfect square, then $\sqrt{a}$ is irrational.

Not every number has a *real number* square root. For example, there is no real number that can be squared to get -36. (The square of a real number can never be negative.) Because of this $\sqrt{-36}$ is not a real number.

> If a is a negative number, then $\sqrt{a}$ is not a real number.

Example 3   Tell whether each square root is rational or irrational.

(a) $\sqrt{17}$

Since 17 is not a perfect square (see Table 3 or check with a calculator), $\sqrt{17}$ is irrational.

(b) $\sqrt{64}$

The number 64 is a perfect square, 8^2, so $\sqrt{64} = 8$, a rational number.

(c) $\sqrt{85}$ is irrational.

(d) $\sqrt{81}$ is rational ($\sqrt{81} = 9$). ✜

Not all irrational numbers are square roots of integers. For example, π (approximately 3.14159) is an irrational number that is not a square root of any integer.

3 Even if a number is irrational, a decimal that approximates the number can be found with a table or certain calculators. For square roots, Table 3 can be used, as shown in Example 4.

Example 4   Find a decimal approximation for each square root. Round answers to the nearest thousandth.

(a) $\sqrt{11}$

Using the square root key of a calculator gives $3.31662479 \approx 3.317$. To use the table of powers and roots, find 11 at the left. The approximate square root is given in the column having $\sqrt{n}$ at the top. You should find that

$$\sqrt{11} \approx 3.317,$$

where $\approx$ means approximately equal to.

(b) $\sqrt{39} \approx 6.245$ Use Table 3 or a calculator.

(c) $\sqrt{740}$

There is no 740 in the "n" column of Table 3. However, $740 = 74 \times 10$, so $\sqrt{740}$ can be found in the "$\sqrt{10n}$" column. Using the table as shown or a calculator,

$$\sqrt{740} \approx 27.203.$$

(d) $\sqrt{180} \approx 13.416$

Look in the "n" column for 18; then read across to the "$\sqrt{10n}$" column, or use a calculator.   ✚

4   One application of square roots comes from using the Pythagorean formula. Recall from Section 4.6 that this formula says that if c is the length of the hypotenuse of a right triangle, and a and b are the lengths of the two legs, then

$$a^2 + b^2 = c^2.$$

Example 5   Find the third side of the given right triangle with sides a, b, and c, where c is the hypotenuse.

(a) $a = 3, b = 4$

Use the formula to find c^2 first.

$$c^2 = a^2 + b^2$$
$$c^2 = 3^2 + 4^2$$
$$= 9 + 16 = 25$$

Now find the positive square root of 25 to get c.

$$c = \sqrt{25} = 5$$

(Although -5 is also a square root of 25, the length of a side of a triangle must be a positive number.)

(b) $c = 9, b = 5$

Substitute the given values in the formula $c^2 = a^2 + b^2$. Then solve for a^2.

$$9^2 = a^2 + 5^2$$
$$81 = a^2 + 25$$
$$56 = a^2$$

Use Table 3 or a calculator to find $a = \sqrt{56} \approx 7.483$.   ✚

Be careful not to make the common mistake of thinking that $\sqrt{a^2 + b^2}$ equals $a + b$. As Example 5(a) shows,

$$\sqrt{9 + 16} = \sqrt{25} = 5 \neq \sqrt{9} + \sqrt{16} = 3 + 4.$$

so that, in general,

$$\sqrt{a^2 + b^2} \neq a + b.$$

5   Finding the square root of a number is the inverse of squaring a number. In a similar way, there are inverses to finding the cube of a number, or finding the fourth or higher power of a number. These inverses are called finding the **cube root**, written $\sqrt[3]{a}$, the **fourth root**, written $\sqrt[4]{a}$, and so on. In $\sqrt[n]{a}$, the number n is the **index** or **order** of the radical. It would be possible to write $\sqrt[2]{a}$ instead of $\sqrt{a}$, but the simpler symbol $\sqrt{a}$ is customary since the square root is the most commonly used root. A table of selected powers that may help you find roots is given inside the back cover of this book. Some calculators also can be used to find these roots.

Example 6   Find each cube root.

(a) $\sqrt[3]{8}$

Look for a number that can be cubed to give 8. Since $2^3 = 8$, then $\sqrt[3]{8} = 2$.

(b) $\sqrt[3]{-8}$

$\sqrt[3]{-8} = -2$ because $(-2)^3 = -8$.   ✦

As these examples suggest, the cube root of a positive number is positive, and the cube root of a negative number is negative.

There is only one real number cube root for each real number.

When the index of the radical is even (square root, fourth root, and so on), the radicand must be nonnegative to get a real number root. Also, for even indexes the symbols $\sqrt{}$, $\sqrt[4]{}$, $\sqrt[6]{}$, and so on are used for the *nonnegative* roots.

Example 7   Find each root.

(a) $\sqrt[4]{16}$

$\sqrt[4]{16} = 2$ because 2 is positive and $2^4 = 16$. Also $-\sqrt[4]{16} = -2$, but there is no real number that equals $\sqrt[4]{-16}$.

(b) $\sqrt[3]{64}$

$$\sqrt[3]{64} = 4 \text{ since } 4^3 = 64.$$

(c) $-\sqrt[5]{32}$

First find $\sqrt[5]{32}$. Since 2 is the number whose fifth power is 32,

$$\sqrt[5]{32} = 2.$$

If $\sqrt[5]{32} = 2$, then

$$-\sqrt[5]{32} = -2.   ✦$$

8.1 Exercises

Find all square roots of each number. See Example 1.

1. 9   **2.** 16   **3.** 121   **4.** $\dfrac{196}{25}$

5. $\dfrac{400}{81}$ **6.** $\dfrac{900}{49}$ **7.** 625   **8.** 961

9. 1521   **10.** 2209   **11.** 3969   **12.** 4624

Find each root that exists. See Example 2.

13. $\sqrt{4}$ **14.** $\sqrt{9}$ **15.** $\sqrt{25}$ **16.** $\sqrt{36}$

17. $-\sqrt{64}$ **18.** $-\sqrt{100}$ **19.** $\sqrt{\dfrac{169}{49}}$ **20.** $\sqrt{\dfrac{196}{25}}$

21. $-\sqrt{\dfrac{49}{9}}$ **22.** $-\sqrt{\dfrac{81}{16}}$ **23.** $\sqrt{-9}$ **24.** $\sqrt{-25}$

Write rational *or* irrational *for each number. If a number is rational, give its exact value. If a number is irrational, give a decimal approximation for the square root. See Examples 3 and 4.*

25. $\sqrt{16}$ **26.** $\sqrt{81}$ **27.** $\sqrt{15}$ **28.** $\sqrt{31}$

29. $\sqrt{47}$ **30.** $\sqrt{53}$ **31.** $\sqrt{68}$ **32.** $\sqrt{72}$

33. $-\sqrt{121}$ **34.** $-\sqrt{144}$ **35.** $\sqrt{110}$ **36.** $\sqrt{170}$

37. $\sqrt{400}$ **38.** $\sqrt{900}$ **39.** $-\sqrt{200}$ **40.** $-\sqrt{260}$

Use a calculator with a square root key to find the following. Round to the nearest thousandth.

41. $\sqrt{571}$ **42.** $\sqrt{693}$ **43.** $\sqrt{798}$ **44.** $\sqrt{453}$

45. $\sqrt{3.94}$ **46.** $\sqrt{1.03}$ **47.** $\sqrt{.00895}$ **48.** $\sqrt{.000402}$

Find the missing side of each right triangle with sides a, b, and c, where c is the longest side. See Example 5.

49. $a = 6, b = 8$ **50.** $a = 5, b = 12$ **51.** $c = 17, a = 8$ **52.** $c = 26, b = 10$

53. $a = 10, b = 8$ **54.** $a = 9, b = 7$ **55.** $c = 12, b = 7$ **56.** $c = 8, a = 3$

Find each of the following roots that are real numbers. Use Table 2 inside the back cover if necessary. See Examples 6 and 7.

57. $\sqrt[3]{1000}$ **58.** $\sqrt[3]{8}$ **59.** $\sqrt[3]{125}$ **60.** $\sqrt[3]{216}$

61. $-\sqrt[3]{8}$ **62.** $-\sqrt[3]{64}$ **63.** $\sqrt[3]{-8}$ **64.** $\sqrt[3]{-27}$

65. $\sqrt[4]{1}$ **66.** $-\sqrt[4]{16}$ **67.** $\sqrt[4]{-16}$ **68.** $-\sqrt[4]{-625}$

69. $-\sqrt[4]{81}$ **70.** $-\sqrt[4]{256}$ **71.** $\sqrt[5]{1}$ **72.** $\sqrt[5]{-32}$

*Use the square root key on your calculator twice to find approximations for the
following roots. Round answers to the nearest thousandth.*

73. $-\sqrt[4]{32}$ **74.** $\sqrt[4]{125}$ **75.** $\sqrt[4]{27}$ **76.** $-\sqrt[4]{28}$

77. $\sqrt[4]{1.42}$ **78.** $\sqrt[4]{2.04}$ **79.** $\sqrt[4]{265.3}$ **80.** $-\sqrt[4]{57.68}$

Use Table 3 to find each root that exists in Exercises 81–88.

81. $-\sqrt{16.81}$ **82.** $-\sqrt{21.16}$ **83.** $\sqrt{26.01}$ **84.** $\sqrt{3025}$

85. $\sqrt{.0049}$ **86.** $\sqrt{.0016}$ **87.** $\sqrt{.0121}$ **88.** $\sqrt{.0841}$

89. The hypotenuse of a triangle measures 9 inches and one leg measures 3
inches. What is the measure of the other leg? Give the answer to the nearest
thousandth.

90. Two sides of a rectangle measure 8 centimeters and 15 centimeters. How long
are the diagonals of the figure?

Review Exercises   *Find all positive integer factors of each number. See Section 4.1.*

91. 24   **92.** 70   **93.** 300   **94.** 48

95. 72   **96.** 75   **97.** 150   **98.** 120

8.2 Products and Quotients of Radicals

Objectives

1 Multiply radicals.

2 Simplify radicals using the product rule.

3 Simplify radical quotients using the quotient rule.

4 Use the product and quotient rules to simplify higher roots.

1 Several useful rules for finding products and quotients of radicals are developed in this section. To get a rule for products, recall that

$$\sqrt{4} \cdot \sqrt{9} = 2 \cdot 3 = 6 \quad \text{and} \quad \sqrt{4 \cdot 9} = \sqrt{36} = 6,$$

showing that

$$\sqrt{4} \cdot \sqrt{9} = \sqrt{4 \cdot 9}.$$

This result is an example of the **product rule for radicals:**

Product Rule for Radicals	For nonnegative real numbers x and y, $\sqrt{x} \cdot \sqrt{y} = \sqrt{x \cdot y}$.

(Recall that the square root of a negative number is not a real number.) That is,
the product of two radicals is the radical of the product.

Example 1   Use the product rule for radicals to find each product.

(a) $\sqrt{2} \cdot \sqrt{3} = \sqrt{2 \cdot 3} = \sqrt{6}$

(b) $\sqrt{7} \cdot \sqrt{5} = \sqrt{35}$

(c) $\sqrt{11} \cdot \sqrt{a} = \sqrt{11a}$ Assume $a > 0$ ✛

2   An important use of the product rule is in simplifying radicals. A radical is **simplified** when no perfect square factor remains under the radical sign. Example 2 shows how this is done.

Example 2   Simplify each radical.

(a) $\sqrt{20}$

Since 20 has a perfect square factor of 4,

$$\sqrt{20} = \sqrt{4 \cdot 5} \qquad \text{4 is a perfect square}$$
$$= \sqrt{4} \cdot \sqrt{5} \qquad \text{Product rule}$$
$$= 2\sqrt{5}. \qquad \sqrt{4} = 2$$

Thus, $\sqrt{20} = 2\sqrt{5}$. Since 5 has no perfect square factor, other than 1, $2\sqrt{5}$ is called the **simplifed form** of $\sqrt{20}$.

(b) $\sqrt{72}$

Look down the list of perfect squares in the table of powers and roots. Find the largest of these numbers that is a factor of 72. The largest is 36, so

$$\sqrt{72} = \sqrt{36 \cdot 2} \qquad \text{36 is a perfect square}$$
$$= \sqrt{36} \cdot \sqrt{2}$$
$$= 6\sqrt{2}.$$

(c) $\sqrt{300} = \sqrt{100 \cdot 3}$ 100 is a perfect square
$$= \sqrt{100} \cdot \sqrt{3}$$
$$= 10\sqrt{3}$$

(d) $\sqrt{15}$

The number 15 has no perfect square factors, so $\sqrt{15}$ cannot be simplified further.   ✛

Sometimes the product rule can be used to simplify an answer, as Example 3 shows.

Example 3   Find each product and simplify.

(a) $\sqrt{9} \cdot \sqrt{75} = 3\sqrt{75}$
$$= 3\sqrt{25 \cdot 3}$$
$$= 3\sqrt{25} \cdot \sqrt{3}$$
$$= 3 \cdot 5\sqrt{3}$$
$$= 15\sqrt{3}$$

(b) $\sqrt{8} \cdot \sqrt{12} = \sqrt{8 \cdot 12}$
$= \sqrt{4 \cdot 2 \cdot 4 \cdot 3}$
$= \sqrt{4} \cdot \sqrt{4} \cdot \sqrt{2 \cdot 3}$
$= 2 \cdot 2 \cdot \sqrt{6}$
$= 4\sqrt{6}$ ✜

3 The **quotient rule for radicals** is very similar to the product rule:

Quotient Rule for Radicals	If x and y are nonnegative real numbers and $y \neq 0$, $$\frac{\sqrt{x}}{\sqrt{y}} = \sqrt{\frac{x}{y}}.$$

That is, the quotient of the radicals is the radical of the quotient.
The next example shows the quotient rule used to simplify radicals.

Example 4 Use the quotient rule to simplify each radical.

(a) $\sqrt{\dfrac{25}{9}} = \dfrac{\sqrt{25}}{\sqrt{9}} = \dfrac{5}{3}$

(b) $\sqrt{\dfrac{144}{49}} = \dfrac{\sqrt{144}}{\sqrt{49}} = \dfrac{12}{7}$

(c) $\sqrt{\dfrac{3}{4}} = \dfrac{\sqrt{3}}{\sqrt{4}} = \dfrac{\sqrt{3}}{2}$ ✜

Example 5 Use the quotient rule to divide $27\sqrt{15}$ by $9\sqrt{3}$.
Work as follows.

$$\frac{27\sqrt{15}}{9\sqrt{3}} = \frac{27}{9} \cdot \frac{\sqrt{15}}{\sqrt{3}} = 3\sqrt{\frac{15}{3}} = 3\sqrt{5}$$ ✜

Some problems require both the product and the quotient rules, as Example 6 shows.

Example 6 Simplify $\sqrt{\dfrac{3}{5}} \cdot \sqrt{\dfrac{4}{5}}$.

Use the quotient rule, and then the product rule.

$$\sqrt{\frac{3}{5}} \cdot \sqrt{\frac{4}{5}} = \frac{\sqrt{3}}{\sqrt{5}} \cdot \frac{\sqrt{4}}{\sqrt{5}} = \frac{\sqrt{3} \cdot \sqrt{4}}{\sqrt{5} \cdot \sqrt{5}}$$
$$= \frac{\sqrt{3} \cdot 2}{\sqrt{25}} = \frac{2\sqrt{3}}{5}$$ ✜

The properties of this section also apply when variables appear under the radical sign, as long as all the variables represent only nonnegative numbers. For example, $\sqrt{5^2} = 5$, but $\sqrt{(-5)^2} \neq -5$. That is,

$$\sqrt{a^2} = a \text{ only if } a \text{ is nonnegative.}$$

Example 7   Simplify each radical. Assume all variables represent positive real numbers.

(a) $\sqrt{25m^4} = \sqrt{25} \cdot \sqrt{m^4} = 5m^2$

(b) $\sqrt{64p^{10}} = 8p^5$

(c) $\sqrt{r^9} = \sqrt{r^8 \cdot r} = \sqrt{r^8} \cdot \sqrt{r} = r^4\sqrt{r}$

(d) $\sqrt{\dfrac{5}{x^2}} = \dfrac{\sqrt{5}}{\sqrt{x^2}} = \dfrac{\sqrt{5}}{x}$ ✚

4   The product rule and the quotient rule for radicals also work for other roots, as shown in Example 8. To simplify cube roots, look for factors that are *perfect cubes*. A **perfect cube** is a number with a rational cube root. For example, $\sqrt[3]{64} = 4$, and since 4 is a rational number, 64 is a perfect cube. Higher roots are handled in a similar manner.

For all real numbers where the indicated roots exist,

$$\sqrt[n]{x} \cdot \sqrt[n]{y} = \sqrt[n]{xy}$$

and

$$\dfrac{\sqrt[n]{x}}{\sqrt[n]{y}} = \sqrt[n]{\dfrac{x}{y}}.$$

Example 8   Simplify each radical.

(a) $\sqrt[3]{32} = \sqrt[3]{8 \cdot 4} = \sqrt[3]{8} \cdot \sqrt[3]{4} = 2\sqrt[3]{4}$

(b) $\sqrt[3]{108} = \sqrt[3]{27 \cdot 4} = \sqrt[3]{27} \cdot \sqrt[3]{4} = 3\sqrt[3]{4}$

(c) $\sqrt[3]{\dfrac{8}{125}} = \dfrac{\sqrt[3]{8}}{\sqrt[3]{125}} = \dfrac{2}{5}$

(d) $\sqrt[4]{32} = \sqrt[4]{16} \cdot \sqrt[4]{2} = 2\sqrt[4]{2}$

(e) $\sqrt[4]{\dfrac{16}{625}} = \dfrac{\sqrt[4]{16}}{\sqrt[4]{625}} = \dfrac{2}{5}$ ✚

8.2   Exercises

Use the product rule to simplify each expression. See Examples 1–3.

1. $\sqrt{8} \cdot \sqrt{2}$

2. $\sqrt{27} \cdot \sqrt{3}$

3. $\sqrt{6} \cdot \sqrt{6}$

4. $\sqrt{11} \cdot \sqrt{11}$

5. $\sqrt{21} \cdot \sqrt{21}$

6. $\sqrt{17} \cdot \sqrt{17}$

7. $\sqrt{3} \cdot \sqrt{7}$

8. $\sqrt{2} \cdot \sqrt{5}$

9. $\sqrt{27}$ **10.** $\sqrt{45}$ **11.** $\sqrt{18}$ **12.** $\sqrt{75}$

13. $\sqrt{48}$ **14.** $\sqrt{80}$ **15.** $\sqrt{150}$ **16.** $\sqrt{700}$

17. $10\sqrt{27}$ **18.** $4\sqrt{8}$ **19.** $2\sqrt{20}$ **20.** $5\sqrt{80}$

21. $\sqrt{27} \cdot \sqrt{48}$ **22.** $\sqrt{75} \cdot \sqrt{27}$ **23.** $\sqrt{50} \cdot \sqrt{72}$ **24.** $\sqrt{98} \cdot \sqrt{8}$

25. $\sqrt{80} \cdot \sqrt{15}$ **26.** $\sqrt{60} \cdot \sqrt{12}$ **27.** $\sqrt{50} \cdot \sqrt{20}$ **28.** $\sqrt{72} \cdot \sqrt{12}$

Use the quotient rule and the product rule, as necessary, to simplify each expression. See Examples 4–6.

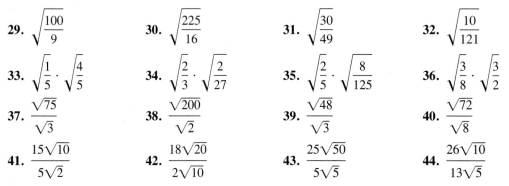

29. $\sqrt{\dfrac{100}{9}}$ **30.** $\sqrt{\dfrac{225}{16}}$ **31.** $\sqrt{\dfrac{30}{49}}$ **32.** $\sqrt{\dfrac{10}{121}}$

33. $\sqrt{\dfrac{1}{5}} \cdot \sqrt{\dfrac{4}{5}}$ **34.** $\sqrt{\dfrac{2}{3}} \cdot \sqrt{\dfrac{2}{27}}$ **35.** $\sqrt{\dfrac{2}{5}} \cdot \sqrt{\dfrac{8}{125}}$ **36.** $\sqrt{\dfrac{3}{8}} \cdot \sqrt{\dfrac{3}{2}}$

37. $\dfrac{\sqrt{75}}{\sqrt{3}}$ **38.** $\dfrac{\sqrt{200}}{\sqrt{2}}$ **39.** $\dfrac{\sqrt{48}}{\sqrt{3}}$ **40.** $\dfrac{\sqrt{72}}{\sqrt{8}}$

41. $\dfrac{15\sqrt{10}}{5\sqrt{2}}$ **42.** $\dfrac{18\sqrt{20}}{2\sqrt{10}}$ **43.** $\dfrac{25\sqrt{50}}{5\sqrt{5}}$ **44.** $\dfrac{26\sqrt{10}}{13\sqrt{5}}$

Simplify each expression. Assume that all variables represent positive real numbers. See Example 7.

45. $\sqrt{y} \cdot \sqrt{y}$ **46.** $\sqrt{m} \cdot \sqrt{m}$ **47.** $\sqrt{x} \cdot \sqrt{z}$ **48.** $\sqrt{p} \cdot \sqrt{q}$

49. $\sqrt{x^2}$ **50.** $\sqrt{y^2}$ **51.** $\sqrt{x^4}$ **52.** $\sqrt{y^4}$

53. $\sqrt{x^2 y^4}$ **54.** $\sqrt{x^4 y^8}$ **55.** $\sqrt{x^3}$ **56.** $\sqrt{y^5}$

57. $\sqrt{\dfrac{16}{x^2}}$ **58.** $\sqrt{\dfrac{100}{m^4}}$ **59.** $\sqrt{\dfrac{11}{r^4}}$ **60.** $\sqrt{\dfrac{23}{y^6}}$

61. $\sqrt{28m^3}$ **62.** $\sqrt{40p^4}$ **63.** $\sqrt{125x^2 y}$ **64.** $\sqrt{1100z^2 t}$

Simplify each radical. See Example 8.

65. $\sqrt[3]{40}$ **66.** $\sqrt[3]{48}$ **67.** $\sqrt[3]{54}$ **68.** $\sqrt[3]{135}$

69. $\sqrt[3]{128}$ **70.** $\sqrt[3]{192}$ **71.** $\sqrt[4]{80}$ **72.** $\sqrt[4]{243}$

73. $\sqrt[3]{\dfrac{8}{27}}$ **74.** $\sqrt[4]{\dfrac{1}{256}}$ **75.** $\sqrt[4]{\dfrac{10,000}{81}}$ **76.** $\sqrt[3]{\dfrac{216}{125}}$

77. $\sqrt[3]{2} \cdot \sqrt[3]{4}$ **78.** $\sqrt[3]{9} \cdot \sqrt[3]{3}$ **79.** $\sqrt[4]{4} \cdot \sqrt[4]{4}$ **80.** $\sqrt[4]{3} \cdot \sqrt[4]{27}$

81. $\sqrt[3]{4x} \cdot \sqrt[3]{8x^2}$ **82.** $\sqrt[3]{25m} \cdot \sqrt[3]{125m^3}$ **83.** $\sqrt[3]{\dfrac{3m}{8n^3}}$ **84.** $\sqrt[4]{\dfrac{4k^2}{81p^4}}$

The volume of a cube is found with the formula $V = s^3$, *where s is the length of an edge of the cube.*

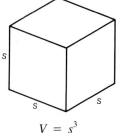

$$V = s^3$$

85. A container in the shape of a cube has a volume of 216 cubic centimeters. What is the depth of the container?

86. A cube-shaped box must be constructed to contain 128 cubic feet. What should the dimensions (height, width, and length) of the box be?

Review Exercises   *Rewrite each fraction with the given denominator. See Section 5.3.*

87. $\dfrac{9}{13} = \dfrac{}{39}$

88. $\dfrac{5}{11} = \dfrac{}{66}$

89. $\dfrac{2}{3} = \dfrac{}{15}$

90. $\dfrac{3}{14} = \dfrac{}{56}$

91. $\dfrac{2}{x} = \dfrac{}{3x}$

92. $\dfrac{5}{2y} = \dfrac{}{2y^2}$

93. $\dfrac{5}{\sqrt{3}} = \dfrac{}{2\sqrt{3}}$

94. $\dfrac{8}{\sqrt{5}} = \dfrac{}{3\sqrt{5}}$

8.3   Addition and Subtraction of Radicals

Objectives

1 Add and subtract radicals.

2 Simplify radical sums and differences.

3 Simplify radical expressions involving multiplication.

1 Add or subtract radicals with the distributive property. For example,

$$8\sqrt{3} + 6\sqrt{3} = (8 + 6)\sqrt{3} \qquad \text{Distributive property}$$
$$= 14\sqrt{3}.$$

Also,

$$2\sqrt{11} - 7\sqrt{11} = -5\sqrt{11}.$$

Only **like radicals,** those that are multiples of the same root of the same number, can be combined this way.

Example 1   Add or subtract, as indicated.

(a) $3\sqrt{6} + 5\sqrt{6} = (3 + 5)\sqrt{6} = 8\sqrt{6}$

(b) $5\sqrt{10} - 7\sqrt{10} = (5 - 7)\sqrt{10} = -2\sqrt{10}$

(c) $\sqrt[3]{5} + \sqrt[3]{5} = 1\sqrt[3]{5} + 1\sqrt[3]{5} = (1 + 1)\sqrt[3]{5} = 2\sqrt[3]{5}$

(d) $\sqrt[4]{7} + 2\sqrt[4]{7} = 1\sqrt[4]{7} + 2\sqrt[4]{7} = 3\sqrt[4]{7}$

(e) $\sqrt{3} + \sqrt{7}$ cannot be simplified further.   ✤

2   Sometimes each radical in a sum or difference must be simplified first. Doing this may result in like radicals, which then can be added or subtracted.

Example 2   Simplify as much as possible.

(a) $3\sqrt{2} + \sqrt{8} = 3\sqrt{2} + \sqrt{4 \cdot 2}$ Simplify $\sqrt{8}$
$= 3\sqrt{2} + \sqrt{4} \cdot \sqrt{2}$
$= 3\sqrt{2} + 2\sqrt{2}$
$= 5\sqrt{2}$

(b) $\sqrt{18} - \sqrt{27} = \sqrt{9 \cdot 2} - \sqrt{9 \cdot 3}$ Simplify $\sqrt{18}$ and $\sqrt{27}$
$= \sqrt{9} \cdot \sqrt{2} - \sqrt{9} \cdot \sqrt{3}$
$= 3\sqrt{2} - 3\sqrt{3}$

Since $\sqrt{2}$ and $\sqrt{3}$ are unlike radicals, this difference cannot be simplified further.

(c) $2\sqrt{12} + 3\sqrt{75} = 2(\sqrt{4} \cdot \sqrt{3}) + 3(\sqrt{25} \cdot \sqrt{3})$
$= 2(2\sqrt{3}) + 3(5\sqrt{3})$
$= 4\sqrt{3} + 15\sqrt{3}$
$= 19\sqrt{3}$

(d) $3\sqrt[3]{16} + 5\sqrt[3]{2} = 3(\sqrt[3]{8} \cdot \sqrt[3]{2}) + 5\sqrt[3]{2}$
$= 3(2\sqrt[3]{2}) + 5\sqrt[3]{2}$
$= 6\sqrt[3]{2} + 5\sqrt[3]{2}$
$= 11\sqrt[3]{2}$ ✤

3   Some radical expressions require both multiplication and addition (or subtraction). The order of operations presented earlier still applies.

Example 3   Simplify each expression. Assume that all variables represent nonnegative real numbers.

(a) $\sqrt{5} \cdot \sqrt{15} + 4\sqrt{3} = \sqrt{5 \cdot 15} + 4\sqrt{3}$
$= \sqrt{75} + 4\sqrt{3}$
$= \sqrt{25 \cdot 3} + 4\sqrt{3}$
$= \sqrt{25} \cdot \sqrt{3} + 4\sqrt{3}$
$= 5\sqrt{3} + 4\sqrt{3}$
$= 9\sqrt{3}$

$$\textbf{(b)}\ \sqrt{2} \cdot \sqrt{6k} + \sqrt{27k} = \sqrt{12k} + \sqrt{27k}$$
$$= \sqrt{4 \cdot 3k} + \sqrt{9 \cdot 3k}$$
$$= \sqrt{4} \cdot \sqrt{3k} + \sqrt{9} \cdot \sqrt{3k}$$
$$= 2\sqrt{3k} + 3\sqrt{3k}$$
$$= 5\sqrt{3k}$$

$$\textbf{(c)}\ \sqrt[3]{2} \cdot \sqrt[3]{16m^3} - \sqrt[3]{108m^3} = \sqrt[3]{32m^3} - \sqrt[3]{108m^3}$$
$$= \sqrt[3]{(8m^3)4} - \sqrt[3]{(27m^3)4}$$
$$= 2m\sqrt[3]{4} - 3m\sqrt[3]{4}$$
$$= -m\sqrt[3]{4} \quad \clubsuit$$

Remember that a sum or difference of radicals can be simplified only if the radicals are like radicals.

For example, $\sqrt{5} + 3\sqrt{5} = 4\sqrt{5}$, but $\sqrt{5} + 5\sqrt{3}$ cannot be simplified further.

8.3   Exercises

Simplify and combine terms wherever possible. See Examples 1 and 2.

1. $2\sqrt{3} + 5\sqrt{3}$ **2.** $6\sqrt{5} + 8\sqrt{5}$ **3.** $4\sqrt{7} - 9\sqrt{7}$ **4.** $6\sqrt{2} - 8\sqrt{2}$

5. $\sqrt{6} + \sqrt{6}$ **6.** $\sqrt{11} + \sqrt{11}$ **7.** $\sqrt{17} + 2\sqrt{17}$ **8.** $3\sqrt{19} + \sqrt{19}$

9. $5\sqrt{7} - \sqrt{7}$ **10.** $3\sqrt{27} - \sqrt{27}$ **11.** $3\sqrt{18} + \sqrt{2}$ **12.** $2\sqrt{48} - \sqrt{3}$

13. $-\sqrt{12} + \sqrt{75}$ **14.** $2\sqrt{27} - \sqrt{300}$ **15.** $5\sqrt{72} - 2\sqrt{50}$ **16.** $6\sqrt{18} - 4\sqrt{32}$

17. $5\sqrt{7} - 2\sqrt{28} + 6\sqrt{63}$ **18.** $3\sqrt{11} + 5\sqrt{44} - 3\sqrt{99}$ **19.** $9\sqrt{24} - 2\sqrt{54} + 3\sqrt{20}$

20. $2\sqrt{8} - 5\sqrt{32} + 2\sqrt{48}$ **21.** $5\sqrt{72} - 3\sqrt{48} - 4\sqrt{128}$ **22.** $4\sqrt{50} + 3\sqrt{12} + 5\sqrt{45}$

23. $\dfrac{1}{4}\sqrt{288} - \dfrac{1}{6}\sqrt{72}$ **24.** $\dfrac{2}{3}\sqrt{27} - \dfrac{3}{4}\sqrt{48}$ **25.** $\dfrac{3}{5}\sqrt{75} - \dfrac{2}{3}\sqrt{45}$

26. $\dfrac{5}{8}\sqrt{128} - \dfrac{3}{4}\sqrt{160}$ **27.** $\sqrt{3} \cdot \sqrt{7} + 2\sqrt{21}$ **28.** $\sqrt{13} \cdot \sqrt{2} + 3\sqrt{26}$

29. $\sqrt{6} \cdot \sqrt{2} + 3\sqrt{3}$ **30.** $4\sqrt{15} \cdot \sqrt{3} - 2\sqrt{5}$ **31.** $4\sqrt[3]{16} - 3\sqrt[3]{54}$

32. $5\sqrt[3]{128} + 3\sqrt[3]{250}$ **33.** $3\sqrt[3]{24} + 6\sqrt[3]{81}$ **34.** $2\sqrt[4]{48} - \sqrt[4]{243}$

35. $5\sqrt[4]{32} + 2\sqrt[4]{32} \cdot \sqrt[4]{4}$ **36.** $8\sqrt[3]{48} + 10\sqrt[3]{3} \cdot \sqrt[3]{18}$

Simplify each expression in Exercises 37–57. Assume that all variables represent nonnegative real numbers. See Example 3.

37. $\sqrt{9x} + \sqrt{49x} - \sqrt{16x}$ **38.** $\sqrt{4a} - \sqrt{16a} + \sqrt{9a}$ **39.** $\sqrt{4a} + 6\sqrt{a} + \sqrt{25a}$

40. $\sqrt{6x^2} + x\sqrt{54}$ **41.** $\sqrt{75x^2} + x\sqrt{300}$ **42.** $\sqrt{20y^2} - 3y\sqrt{5}$

43. $3\sqrt{8x^2} - 4x\sqrt{2}$ **44.** $\sqrt{2b^2} + 3b\sqrt{18}$ **45.** $5\sqrt{75p^2} - 4\sqrt{27p^2}$

46. $-3\sqrt{32k} + 6\sqrt{8k}$ **47.** $2\sqrt{125x^2z} + 8x\sqrt{80z}$ **48.** $4p\sqrt{14m} - 6\sqrt{28mp^2}$

49. $3k\sqrt{24k^2h^2} + 9h\sqrt{54k^3}$ **50.** $6r\sqrt{27r^2s} + 3r^2\sqrt{3s}$ **51.** $6\sqrt[3]{8p^2} - 2\sqrt[3]{27p^2}$

52. $5\sqrt[4]{m^3} + 8\sqrt[4]{16m^3}$ **53.** $5\sqrt[4]{m^4} + 3\sqrt[4]{81m^4}$ **54.** $2\sqrt[4]{p^5} - 5p\sqrt[4]{16p}$

55. $8k\sqrt[3]{54k} + 6\sqrt[3]{16k^4}$ **56.** $-5\sqrt[3]{256z^4} - 2z\sqrt[3]{32z}$ **57.** $10\sqrt[3]{4m^4} - 3m\sqrt[3]{32m}$

58. A rectangular room has a width of $\sqrt{125}$ feet and a length of $\sqrt{245}$ feet. What is the perimeter?

59. Find the perimeter of a triangular lot with sides measuring $3\sqrt{27}$, $5\sqrt{12}$, and $2\sqrt{48}$ meters.

Review Exercises   *Find each product. See Section 3.6.*

60. $(m + 3)(m + 5)$ **61.** $(p - 2)(p + 7)$ **62.** $(2k - 3)(3k - 4)$

63. $(4z + 2)(5z - 8)$ **64.** $(3x - 1)^2$ **65.** $(2y + 5)^2$

66. $(a + 3)(a - 3)$ **67.** $(2n + 3)(2n - 3)$ **68.** $(4k + 5)(4k - 5)$

8.4   Rationalizing the Denominator

Objectives

1 Rationalize denominators with square roots.

2 Simplify expressions by rationalizing the denominator.

3 Rationalize denominators with cube roots.

1 Decimal approximations for radicals were found in the first section of this chapter. For more complicated radical expressions it is easier to find these decimals if the denominators do not contain radicals. For example, the radical in the denominator of

$$\frac{\sqrt{3}}{\sqrt{2}}$$

can be eliminated by multiplying the numerator and the denominator by $\sqrt{2}$.

$$\frac{\sqrt{3}}{\sqrt{2}} = \frac{\sqrt{3} \cdot \sqrt{2}}{\sqrt{2} \cdot \sqrt{2}} = \frac{\sqrt{6}}{2} \qquad \text{Since } \sqrt{2} \cdot \sqrt{2} = 2$$

This process of changing the denominator from a radical (irrational number) to a rational number is called **rationalizing the denominator.** The value of the number is not changed; only the form of the number is changed.

Example 1   Rationalize each denominator.

(a) $\dfrac{9}{\sqrt{6}}$

Multiply both numerator and denominator by $\sqrt{6}$.

$$\frac{9}{\sqrt{6}} = \frac{9 \cdot \sqrt{6}}{\sqrt{6} \cdot \sqrt{6}} = \frac{9\sqrt{6}}{6} = \frac{3\sqrt{6}}{2} \qquad \text{Since } \sqrt{6} \cdot \sqrt{6} = 6$$

(b) $\dfrac{12}{\sqrt{8}}$

The denominator here could be rationalized by multiplying by $\sqrt{8}$. However, the result can be found with less work by first simplifying the denominator.

$$\sqrt{8} = \sqrt{4} \cdot \sqrt{2} = 2\sqrt{2}$$

Then multiply numerator and denominator by $\sqrt{2}$.

$$\frac{12}{\sqrt{8}} = \frac{12}{2\sqrt{2}} = \frac{12 \cdot \sqrt{2}}{2\sqrt{2} \cdot \sqrt{2}} = \frac{12\sqrt{2}}{2\sqrt{4}} = \frac{12\sqrt{2}}{2 \cdot 2} = \frac{12\sqrt{2}}{4} = 3\sqrt{2} \quad \clubsuit$$

2   Radicals are considered simplified only if any denominators are rationalized, as shown in Examples 2–4.

Example 2   Simplify $\sqrt{\dfrac{27}{5}}$ by rationalizing the denominator.

First, use the quotient rule for radicals.

$$\sqrt{\frac{27}{5}} = \frac{\sqrt{27}}{\sqrt{5}}$$

Now rationalize the denominator by multiplying both numerator and denominator by $\sqrt{5}$.

$$\frac{\sqrt{27}}{\sqrt{5}} = \frac{\sqrt{27} \cdot \sqrt{5}}{\sqrt{5} \cdot \sqrt{5}}$$

$$= \frac{\sqrt{9 \cdot 3} \cdot \sqrt{5}}{5} = \frac{\sqrt{9} \cdot \sqrt{3} \cdot \sqrt{5}}{5}$$

$$= \frac{3 \cdot \sqrt{3 \cdot 5}}{5} = \frac{3\sqrt{15}}{5} \quad \clubsuit$$

Example 3   Simplify $\sqrt{\dfrac{5}{8}} \cdot \sqrt{\dfrac{1}{6}}$.

Use both the product rule and the quotient rule.

$$\sqrt{\frac{5}{8}} \cdot \sqrt{\frac{1}{6}} = \sqrt{\frac{5}{8} \cdot \frac{1}{6}} = \sqrt{\frac{5}{48}} = \frac{\sqrt{5}}{\sqrt{48}} = \frac{\sqrt{5}}{\sqrt{16} \cdot \sqrt{3}} = \frac{\sqrt{5}}{4\sqrt{3}}$$

Now rationalize the denominator by multiplying both numerator and denominator by $\sqrt{3}$.

$$\frac{\sqrt{5}}{4\sqrt{3}} = \frac{\sqrt{5} \cdot \sqrt{3}}{4\sqrt{3} \cdot \sqrt{3}} = \frac{\sqrt{15}}{4\sqrt{9}} = \frac{\sqrt{15}}{4 \cdot 3} = \frac{\sqrt{15}}{12} \quad \clubsuit$$

Example 4 Rationalize the denominator of $\dfrac{\sqrt{4x}}{\sqrt{y}}$. Assume that x and y represent positive real numbers.

Multiply numerator and denominator by $\sqrt{y}$.

$$\frac{\sqrt{4x}}{\sqrt{y}} = \frac{\sqrt{4x} \cdot \sqrt{y}}{\sqrt{y} \cdot \sqrt{y}} = \frac{\sqrt{4xy}}{y} = \frac{2\sqrt{xy}}{y} \quad \text{✦}$$

3 A denominator with a cube root is rationalized by changing the radicand in the denominator to a perfect cube, as shown in the next example.

Example 5 Rationalize each denominator.

(a) $\sqrt[3]{\dfrac{3}{2}}$

Multiply the numerator and the denominator by enough factors of 2 to make the denominator a perfect cube. This will eliminate the radical in the denominator. Here, multiply by $\sqrt[3]{2^2}$.

$$\sqrt[3]{\frac{3}{2}} = \frac{\sqrt[3]{3}}{\sqrt[3]{2}} = \frac{\sqrt[3]{3} \cdot \sqrt[3]{2^2}}{\sqrt[3]{2} \cdot \sqrt[3]{2^2}} = \frac{\sqrt[3]{3 \cdot 2^2}}{\sqrt[3]{2^3}} = \frac{\sqrt[3]{12}}{2} \qquad \text{Since } \sqrt[3]{2^3} = \sqrt[3]{8} = 2$$

(b) $\dfrac{\sqrt[3]{3}}{\sqrt[3]{4}}$

Since $4 \cdot 2 = 2^2 \cdot 2 = 2^3$, multiply numerator and denominator by $\sqrt[3]{2}$.

$$\frac{\sqrt[3]{3}}{\sqrt[3]{4}} = \frac{\sqrt[3]{3} \cdot \sqrt[3]{2}}{\sqrt[3]{4} \cdot \sqrt[3]{2}} = \frac{\sqrt[3]{6}}{\sqrt[3]{8}} = \frac{\sqrt[3]{6}}{2} \quad \text{✦}$$

8.4 Exercises

Perform the indicated operations, and write all answers in simplest form. See Examples 1–3.

1. $\dfrac{6}{\sqrt{5}}$

2. $\dfrac{4}{\sqrt{2}}$

3. $\dfrac{5}{\sqrt{5}}$

4. $\dfrac{15}{\sqrt{15}}$

5. $\dfrac{3}{\sqrt{7}}$

6. $\dfrac{12}{\sqrt{10}}$

7. $\dfrac{8\sqrt{3}}{\sqrt{5}}$

8. $\dfrac{9\sqrt{6}}{\sqrt{5}}$

9. $\dfrac{12\sqrt{10}}{8\sqrt{3}}$

10. $\dfrac{9\sqrt{15}}{6\sqrt{2}}$

11. $\dfrac{8}{\sqrt{27}}$

12. $\dfrac{12}{\sqrt{18}}$

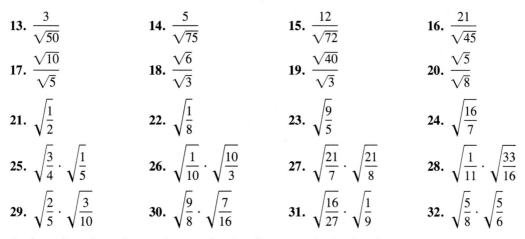

13. $\dfrac{3}{\sqrt{50}}$

14. $\dfrac{5}{\sqrt{75}}$

15. $\dfrac{12}{\sqrt{72}}$

16. $\dfrac{21}{\sqrt{45}}$

17. $\dfrac{\sqrt{10}}{\sqrt{5}}$

18. $\dfrac{\sqrt{6}}{\sqrt{3}}$

19. $\dfrac{\sqrt{40}}{\sqrt{3}}$

20. $\dfrac{\sqrt{5}}{\sqrt{8}}$

21. $\sqrt{\dfrac{1}{2}}$

22. $\sqrt{\dfrac{1}{8}}$

23. $\sqrt{\dfrac{9}{5}}$

24. $\sqrt{\dfrac{16}{7}}$

25. $\sqrt{\dfrac{3}{4}} \cdot \sqrt{\dfrac{1}{5}}$

26. $\sqrt{\dfrac{1}{10}} \cdot \sqrt{\dfrac{10}{3}}$

27. $\sqrt{\dfrac{21}{7}} \cdot \sqrt{\dfrac{21}{8}}$

28. $\sqrt{\dfrac{1}{11}} \cdot \sqrt{\dfrac{33}{16}}$

29. $\sqrt{\dfrac{2}{5}} \cdot \sqrt{\dfrac{3}{10}}$

30. $\sqrt{\dfrac{9}{8}} \cdot \sqrt{\dfrac{7}{16}}$

31. $\sqrt{\dfrac{16}{27}} \cdot \sqrt{\dfrac{1}{9}}$

32. $\sqrt{\dfrac{5}{8}} \cdot \sqrt{\dfrac{5}{6}}$

Perform the indicated operations, and write all answers in simplest form. Rationalize all denominators. Assume that all variables represent positive real numbers. See Example 4.

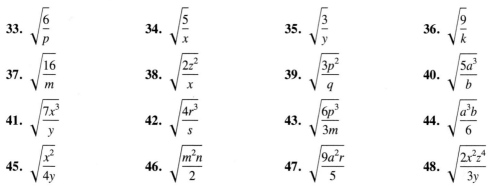

33. $\sqrt{\dfrac{6}{p}}$

34. $\sqrt{\dfrac{5}{x}}$

35. $\sqrt{\dfrac{3}{y}}$

36. $\sqrt{\dfrac{9}{k}}$

37. $\sqrt{\dfrac{16}{m}}$

38. $\sqrt{\dfrac{2z^2}{x}}$

39. $\sqrt{\dfrac{3p^2}{q}}$

40. $\sqrt{\dfrac{5a^3}{b}}$

41. $\sqrt{\dfrac{7x^3}{y}}$

42. $\sqrt{\dfrac{4r^3}{s}}$

43. $\sqrt{\dfrac{6p^3}{3m}}$

44. $\sqrt{\dfrac{a^3b}{6}}$

45. $\sqrt{\dfrac{x^2}{4y}}$

46. $\sqrt{\dfrac{m^2n}{2}}$

47. $\sqrt{\dfrac{9a^2r}{5}}$

48. $\sqrt{\dfrac{2x^2z^4}{3y}}$

Rationalize the denominators of the cube roots in Exercises 49–60. See Example 5.

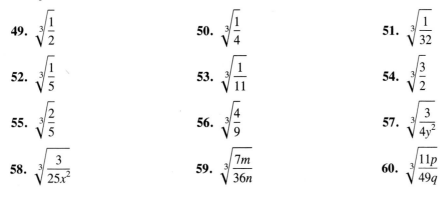

49. $\sqrt[3]{\dfrac{1}{2}}$

50. $\sqrt[3]{\dfrac{1}{4}}$

51. $\sqrt[3]{\dfrac{1}{32}}$

52. $\sqrt[3]{\dfrac{1}{5}}$

53. $\sqrt[3]{\dfrac{1}{11}}$

54. $\sqrt[3]{\dfrac{3}{2}}$

55. $\sqrt[3]{\dfrac{2}{5}}$

56. $\sqrt[3]{\dfrac{4}{9}}$

57. $\sqrt[3]{\dfrac{3}{4y^2}}$

58. $\sqrt[3]{\dfrac{3}{25x^2}}$

59. $\sqrt[3]{\dfrac{7m}{36n}}$

60. $\sqrt[3]{\dfrac{11p}{49q}}$

61. The period p of a pendulum is given by

$$p = k \cdot \sqrt{\frac{L}{g}},$$

where L is the length of the pendulum, g is the acceleration due to gravity, and k is a constant. Find the period when $k = 6$, $L = 9$ feet, and $g = 32$ feet per second per second. Leave the answer in the form of a simplified radical expression.

62. The velocity v of a meteorite approaching the earth is given by

$$v = \frac{k}{\sqrt{d}},$$

where d is its distance from the center of the earth and k is a constant. What is the velocity of a meteorite that is 6000 kilometers away from the center of the earth, if $k = 450$?

Review Exercises *Combine like terms. See Section 2.1.*

63. $2x + 3x - x$

64. $-5a + 2a + 7a$

65. $m + 2 - 3m - 5$

66. $-2p + 4p + 7 - 9 - 3p$

67. $4y + 3z + 5z - y$

68. $5w + 2t - w + 3t$

8.5 Simplifying Radical Expressions

Objectives

1 Simplify radical expressions with sums.

2 Simplify radical expressions with products.

3 Simplify radical expressions with quotients.

4 Write radical expressions with quotients in lowest terms.

It can be difficult to decide on the "simplest" form of a radical. In this book, a radical expression is simplified when the following five conditions are satisfied. Although the conditions are illustrated with square roots, they apply to higher roots as well.

Simplifying Radicals

1. If a radical represents a rational number, then that rational number should be used in place of the radical.

   For example, $\sqrt{49}$ is simplified by writing 7; $\sqrt{64}$ by writing 8; $\sqrt{\dfrac{169}{9}}$ by writing $\dfrac{13}{3}$.

2. If a radical expression contains products of radicals, the product rule for radicals, $\sqrt{x} \cdot \sqrt{y} = \sqrt{xy}$, should be used to get a single radical.

   For example, $\sqrt{3} \cdot \sqrt{2}$ is simplified to $\sqrt{6}$; $\sqrt{5} \cdot \sqrt{x}$ to $\sqrt{5x}$.

3. If a radicand has a factor that is a perfect square, the radical should be expressed as the product of the positive square root of the perfect square and the remaining radical factor.

   For example, $\sqrt{20}$ is simplified to $\sqrt{20} = \sqrt{4 \cdot 5} = \sqrt{4} \cdot \sqrt{5} = 2\sqrt{5}$; $\sqrt{75}$ to $5\sqrt{3}$.

4. If a radical expression contains sums or differences of radicals, the distributive property should be used to combine like terms.

   For example, $3\sqrt{2} + 4\sqrt{2}$ is combined as $7\sqrt{2}$, but $3\sqrt{2} + 4\sqrt{3}$ cannot be further combined.

5. Any radical in a denominator should be changed to a rational number.

   For example, $\dfrac{5}{\sqrt{3}}$ is rationalized as

   $$\frac{5}{\sqrt{3}} = \frac{5\sqrt{3}}{\sqrt{3} \cdot \sqrt{3}} = \frac{5\sqrt{3}}{3};$$

   $\sqrt{\dfrac{3}{2}}$ is rationalized as

   $$\sqrt{\frac{3}{2}} = \frac{\sqrt{3}}{\sqrt{2}} = \frac{\sqrt{3}}{\sqrt{2}} \cdot \frac{\sqrt{2}}{\sqrt{2}} = \frac{\sqrt{6}}{2}.$$

1 The first example shows how radical expressions involving sums may be simplified.

Example 1 Simplify each of the following.

(a) $\sqrt{16} + \sqrt{9}$

Here $\sqrt{16} + \sqrt{9} = 4 + 3 = 7$.

(b) $5\sqrt{2} + 2\sqrt{18}$

First simplify $\sqrt{18}$.

$$
\begin{aligned}
5\sqrt{2} + \sqrt{18} &= 5\sqrt{2} + 2(\sqrt{9} \cdot \sqrt{2}) \\
&= 5\sqrt{2} + 2(3\sqrt{2}) \\
&= 5\sqrt{2} + 6\sqrt{2} \\
&= 11\sqrt{2} \quad \clubsuit
\end{aligned}
$$

2 The next examples show how to simplify radical expressions involving products.

Example 2   Simplify $\sqrt{5}(\sqrt{8} - \sqrt{32})$.

Start by simplifying $\sqrt{8}$ and $\sqrt{32}$.

$$\sqrt{8} = 2\sqrt{2} \qquad \text{and} \qquad \sqrt{32} = 4\sqrt{2}$$

Now simplify inside the parentheses.

$$
\begin{aligned}
\sqrt{5}(\sqrt{8} - \sqrt{32}) &= \sqrt{5}(2\sqrt{2} - 4\sqrt{2}) \\
&= \sqrt{5}(-2\sqrt{2}) \\
&= -2\sqrt{5 \cdot 2} \\
&= -2\sqrt{10} \quad \clubsuit
\end{aligned}
$$

Example 3   Simplify each product.

(a) $(\sqrt{3} + 2\sqrt{5})(\sqrt{3} - 4\sqrt{5})$

The products of these sums of radicals can be found in the same way as the product of binomials in Chapter 3. The pattern of multiplication is the same, using the FOIL method.

$$
\begin{aligned}
&(\sqrt{3} + 2\sqrt{5})(\sqrt{3} - 4\sqrt{5}) \\
&= \sqrt{3} \cdot \sqrt{3} + \sqrt{3}(-4\sqrt{5}) + 2\sqrt{5} \cdot \sqrt{3} + 2\sqrt{5}(-4\sqrt{5}) \\
&= 3 - 4\sqrt{15} + 2\sqrt{15} - 8 \cdot 5 \\
&= 3 - 2\sqrt{15} - 40 \\
&= -37 - 2\sqrt{15}
\end{aligned}
$$

(b) $(\sqrt{3} + \sqrt{21})(\sqrt{3} - \sqrt{7})$

$$
\begin{aligned}
&(\sqrt{3} + \sqrt{21})(\sqrt{3} - \sqrt{7}) \\
&= \sqrt{3}(\sqrt{3}) + \sqrt{3}(-\sqrt{7}) + \sqrt{21}(\sqrt{3}) + \sqrt{21}(-\sqrt{7}) \\
&= 3 - \sqrt{21} + \sqrt{63} - \sqrt{147} \\
&= 3 - \sqrt{21} + \sqrt{9} \cdot \sqrt{7} - \sqrt{49} \cdot \sqrt{3} \\
&= 3 - \sqrt{21} + 3\sqrt{7} - 7\sqrt{3}
\end{aligned}
$$

Since there are no like radicals, no terms may be combined. $\clubsuit$

Since radicals represent real numbers, the special products of binomials discussed in Chapter 3 can be used to find products of radicals. Example 4 uses the rule

$$(a + b)(a - b) = a^2 - b^2.$$

Example 4   Find each product.

(a) $(4 + \sqrt{3})(4 - \sqrt{3})$

Follow the pattern given above. Let $a = 4$ and $b = \sqrt{3}$.

$$(4 + \sqrt{3})(4 - \sqrt{3}) = (4)^2 - (\sqrt{3})^2$$
$$= 16 - 3 = 13$$

(b) $(\sqrt{12} - \sqrt{6})(\sqrt{12} + \sqrt{6}) = (\sqrt{12})^2 - (\sqrt{6})^2$
$$= 12 - 6 = 6 \quad \clubsuit$$

Both products in Example 4 resulted in rational numbers. The expressions in these products, such as $4 + \sqrt{3}$ and $4 - \sqrt{3}$, are called **conjugates** of each other.

3   Products similar to those in Example 4 can be used to rationalize the denominators in more complicated quotients, such as

$$\frac{2}{4 - \sqrt{3}}.$$

By Example 4(a), if this denominator, $4 - \sqrt{3}$, is multiplied by $4 + \sqrt{3}$, then the product $(4 - \sqrt{3})(4 + \sqrt{3})$ is the rational number 13. Multiplying numerator and denominator by $4 + \sqrt{3}$ gives

$$\frac{2}{4 - \sqrt{3}} = \frac{2(4 + \sqrt{3})}{(4 - \sqrt{3})(4 + \sqrt{3})} = \frac{2(4 + \sqrt{3})}{13}.$$

The denominator now has been rationalized; it contains no radical signs.

Example 5   Rationalize the denominator in the quotient

$$\frac{4}{3 + \sqrt{5}}.$$

The radical in the denominator can be eliminated by multiplying numerator and denominator by $3 - \sqrt{5}$, the conjugate of $3 + \sqrt{5}$.

$$\frac{4}{3 + \sqrt{5}} = \frac{4(3 - \sqrt{5})}{(3 + \sqrt{5})(3 - \sqrt{5})}$$
$$= \frac{4(3 - \sqrt{5})}{3^2 - (\sqrt{5})^2} = \frac{4(3 - \sqrt{5})}{9 - 5}$$
$$= \frac{4(3 - \sqrt{5})}{4} = 3 - \sqrt{5}$$

The last step gives the result in lowest terms. $\clubsuit$

Example 6   Simplify $\dfrac{6 + \sqrt{2}}{\sqrt{2} - 5}$.

Multiply numerator and denominator by the conjugate $\sqrt{2} + 5$.

$$\frac{6 + \sqrt{2}}{\sqrt{2} - 5} = \frac{(6 + \sqrt{2})(\sqrt{2} + 5)}{(\sqrt{2} - 5)(\sqrt{2} + 5)}$$

$$= \frac{6\sqrt{2} + 30 + 2 + 5\sqrt{2}}{2 - 25}$$

$$= \frac{11\sqrt{2} + 32}{-23} = -\frac{11\sqrt{2} + 32}{23} \quad \blacklozenge$$

4   The final example shows how to write quotients with radicals in lowest terms.

Example 7   Write $\dfrac{3\sqrt{3} + 15}{12}$ in lowest terms.

Factor the numerator, and then divide numerator and denominator by any common factors.

$$\frac{3\sqrt{3} + 15}{12} = \frac{3(\sqrt{3} + 5)}{12} = \frac{\sqrt{3} + 5}{4} \quad \blacklozenge$$

8.5   Exercises

Simplify each expression. Use the five rules given in the text. Assume all variables represent nonnegative real numbers. See Examples 1–4.

1. $3\sqrt{5} + 8\sqrt{45}$ **2.** $6\sqrt{2} + 4\sqrt{18}$ **3.** $9\sqrt{50} - 4\sqrt{72}$

4. $3\sqrt{80} - 5\sqrt{45}$ **5.** $\sqrt{2}(\sqrt{8} - \sqrt{32})$ **6.** $\sqrt{3}(\sqrt{27} - \sqrt{3})$

7. $\sqrt{5}(\sqrt{3} + \sqrt{7})$ **8.** $\sqrt{7}(\sqrt{10} - \sqrt{3})$ **9.** $2\sqrt{5}(\sqrt{2} + \sqrt{5})$

10. $3\sqrt{7}(2\sqrt{7} - 4\sqrt{5})$ **11.** $-\sqrt{14} \cdot \sqrt{2} - \sqrt{28}$ **12.** $\sqrt{6} \cdot \sqrt{3} - 2\sqrt{50}$

13. $(2\sqrt{6} + 3)(3\sqrt{6} - 5)$ **14.** $(4\sqrt{5} - 2)(2\sqrt{5} + 3)$

15. $(5\sqrt{7} - 2\sqrt{3})(3\sqrt{7} + 3\sqrt{3})$ **16.** $(2\sqrt{10} + 5\sqrt{2})(3\sqrt{10} - 4\sqrt{2})$

17. $(3\sqrt{2} + 4)(3\sqrt{2} + 4)$ **18.** $(4\sqrt{5} - 1)(4\sqrt{5} - 1)$

19. $(2\sqrt{7} - 3)^2$ **20.** $(3\sqrt{5} + 5)^2$

21. $(\sqrt{2} + \sqrt{3})^2$ **22.** $(\sqrt{6} + \sqrt{2})^2$

23. $(3 - \sqrt{2})(3 + \sqrt{2})$ **24.** $(7 - \sqrt{5})(7 + \sqrt{5})$

25. $(2 + \sqrt{8})(2 - \sqrt{8})$ **26.** $(3 + \sqrt{11})(3 - \sqrt{11})$

27. $(\sqrt{6} - \sqrt{5})(\sqrt{6} + \sqrt{5})$ **28.** $(\sqrt{11} + \sqrt{10})(\sqrt{11} - \sqrt{10})$

29. $(\sqrt{8} - \sqrt{2})(\sqrt{2} + \sqrt{4})$ **30.** $(\sqrt{6} - \sqrt{3})(\sqrt{3} + \sqrt{18})$

31. $(\sqrt{5x} + \sqrt{30})(\sqrt{6x} + \sqrt{3})$

32. $(\sqrt{10y} - \sqrt{20})(\sqrt{2y} - \sqrt{5})$

33. $(3\sqrt{t} + \sqrt{7})(2\sqrt{t} - \sqrt{14})$

34. $(2\sqrt{z} - \sqrt{3})(\sqrt{z} - \sqrt{5})$

35. $(\sqrt{3m} + \sqrt{2n})(\sqrt{5m} - \sqrt{5n})$

36. $(\sqrt{4p} - \sqrt{3k})(\sqrt{2p} + \sqrt{9k})$

Write each quotient in lowest terms. See Example 7.

37. $\dfrac{5\sqrt{7} - 10}{5}$

38. $\dfrac{6\sqrt{5} - 9}{3}$

39. $\dfrac{2\sqrt{3} + 10}{8}$

40. $\dfrac{4\sqrt{6} + 6}{10}$

41. $\dfrac{12 - 2\sqrt{10}}{4}$

42. $\dfrac{9 - 6\sqrt{2}}{12}$

43. $\dfrac{16 + 8\sqrt{2}}{24}$

44. $\dfrac{25 + 5\sqrt{3}}{10}$

Rationalize the denominators. See Examples 5 and 6.

45. $\dfrac{5}{2 + \sqrt{5}}$

46. $\dfrac{6}{3 + \sqrt{7}}$

47. $\dfrac{7}{2 - \sqrt{11}}$

48. $\dfrac{38}{5 - \sqrt{6}}$

49. $\dfrac{\sqrt{12}}{\sqrt{3} + 1}$

50. $\dfrac{\sqrt{18}}{\sqrt{2} - 1}$

51. $\dfrac{2\sqrt{3}}{\sqrt{3} + 5}$

52. $\dfrac{\sqrt{12}}{2 - \sqrt{10}}$

53. $\dfrac{\sqrt{2} + 3}{\sqrt{3} - 1}$

54. $\dfrac{\sqrt{5} + 2}{2 - \sqrt{3}}$

55. $\dfrac{6 - \sqrt{5}}{\sqrt{2} + 2}$

56. $\dfrac{3 + \sqrt{2}}{\sqrt{2} + 1}$

57. $\dfrac{2\sqrt{6} + 1}{\sqrt{2} + 5}$

58. $\dfrac{3\sqrt{2} - 4}{\sqrt{3} + 2}$

59. $\dfrac{\sqrt{7} + \sqrt{2}}{\sqrt{3} - \sqrt{2}}$

60. $\dfrac{\sqrt{6} + \sqrt{5}}{\sqrt{3} + \sqrt{5}}$

61. $\dfrac{3 + \sqrt{3}}{\sqrt{2}}$

62. $\dfrac{2 - \sqrt{5}}{\sqrt{3}}$

63. $\dfrac{\sqrt{6} + \sqrt{2}}{\sqrt{2}}$

64. $\dfrac{\sqrt{7} + \sqrt{5}}{\sqrt{5}}$

65. $\dfrac{\sqrt{8} + \sqrt{3}}{\sqrt{2}}$

66. $\dfrac{\sqrt{12} + \sqrt{10}}{\sqrt{2}}$

Simplify the radical expressions in Exercises 67–74.

67. $\sqrt[3]{4}(\sqrt[3]{2} - 3)$

68. $\sqrt[3]{5}(4\sqrt[3]{5} - \sqrt[3]{25})$

69. $2\sqrt[4]{2}(3\sqrt[4]{8} + 5\sqrt[4]{4})$

70. $6\sqrt[4]{9}(2\sqrt[4]{9} - \sqrt[4]{27})$

71. $(\sqrt[3]{2} - 1)(\sqrt[3]{4} + 3)$

72. $(\sqrt[3]{9} + 5)(\sqrt[3]{3} - 4)$

73. $(\sqrt[3]{5} - \sqrt[3]{4})(\sqrt[3]{25} + \sqrt[3]{20} + \sqrt[3]{16})$

74. $(\sqrt[3]{4} + \sqrt[3]{2})(\sqrt[3]{16} - \sqrt[3]{8} + \sqrt[3]{4})$

75. The radius of the circular top or bottom of a tin can with a surface area A and a height h is given by

$$r = \frac{-h + \sqrt{h^2 + .64A}}{2}.$$

What radius should be used to make a can with a height of 12 centimeters and a surface area of 100 square centimeters? Leave the answer as a simplified radical expression.

76. If an investment of P dollars grows to A dollars in two years, the annual rate of return on the investment is given by

$$r = \frac{\sqrt{A} - \sqrt{P}}{\sqrt{P}}.$$

First rationalize this expression, then find the annual rate of return (in percent) if $50,000 increases to $58,320.

Review Exercises   *Solve each equation. See Section 4.5.*

77. $y^2 - 4y + 3 = 0$ **78.** $x^2 - x - 20 = 0$ **79.** $2m^2 + m = 15$

80. $3a^2 = 14a + 5$ **81.** $k - 1 = (k - 1)^2$ **82.** $(t + 2)^2 = -(t + 2)$

8.6   Equations with Radicals

Objectives

1 Solve equations with radicals.

2 Identify equations with no solutions.

3 Solve equations by squaring a binomial.

1   The addition and multiplication properties are not enough to solve an equation with radicals such as

$$\sqrt{x + 1} = 3.$$

Solving equations that have square roots requires a new property, the **squaring property.**

Squaring Property of Equality	If both sides of a given equation are squared, all solutions of the original equation are *among* the solutions of the squared equation.

Be very careful with the squaring property: Although this property usually produces an equation without radicals that can be solved, the new equation may have more solutions than the original equation. For example, starting with the equation $y = 4$ and squaring both sides gives

$$y^2 = 4^2 \quad \text{or} \quad y^2 = 16.$$

This last equation, $y^2 = 16$, has *two* solutions, 4 or -4, while the original equation, $y = 4$, has only *one* solution, 4. Because of this possibility,

all proposed solutions from the squared equation must be checked in the original equation.

Example 1   Solve the equation $\sqrt{p + 1} = 3$.
Use the squaring property of equality to square both sides of the equation and then solve this new equation.

$$(\sqrt{p + 1})^2 = 3^2$$
$$p + 1 = 9$$
$$p = 8$$

Now check this answer in the original equation.

$$\sqrt{p + 1} = 3$$
$$\sqrt{8 + 1} = 3$$
$$\sqrt{9} = 3$$
$$3 = 3 \quad \text{True}$$

Since this statement is true, the solution of $\sqrt{p + 1} = 3$ is the number 8. In this case the squared equation had just one solution, which also satisfied the original equation.   ✦

Example 2   Solve $3\sqrt{x} = \sqrt{x + 8}$.
Squaring both sides gives

$$(3\sqrt{x})^2 = (\sqrt{x + 8})^2$$
$$3^2(\sqrt{x})^2 = (\sqrt{x + 8})^2 \qquad \text{Power rule (b) for exponents}$$
$$9x = x + 8$$
$$8x = 8$$
$$x = 1.$$

Check this proposed solution.

$$3\sqrt{x} = \sqrt{x + 8}$$
$$3\sqrt{1} = \sqrt{1 + 8} \qquad \text{Let } x = 8$$
$$3(1) = \sqrt{9}$$
$$3 = 3 \qquad \text{True}$$

The check shows that the solution of the given equation is 1.   ✦

2   Not all equations with radicals have a solution, as shown by the equations in Examples 3 and 4.

Example 3   Solve the equation $\sqrt{y} = -3$.
Square both sides.

$$(\sqrt{y})^2 = (-3)^2$$
$$y = 9$$

Check this proposed answer in the original equation.

$$\sqrt{y} = -3$$
$$\sqrt{9} = -3 \qquad \text{Let } y = 9$$
$$3 = -3 \qquad \text{False}$$

Since the statement $3 = -3$ is false, the number 9 is not a solution of the given equation, and $\sqrt{y} = -3$ has no solution. Noticing that $\sqrt{y}$ represents the *nonnegative* square root would have shown immediately that there is no solution. ✤

Example 4   Solve $a = \sqrt{a^2 + 5a + 10}$.

Square both sides.

$$(a)^2 = (\sqrt{a^2 + 5a + 10})^2$$
$$a^2 = a^2 + 5a + 10$$
$$0 = 5a + 10$$
$$a = -2$$

Check the proposed solution $a = -2$ in the original equation.

$$a = \sqrt{a^2 + 5a + 10}$$
$$-2 = \sqrt{(-2)^2 + 5(-2) + 10} \qquad \text{Let } a = -2$$
$$-2 = \sqrt{4 - 10 + 10}$$
$$-2 = 2 \qquad\qquad\qquad \text{False}$$

Since $a = -2$ leads to a false result, there is no solution. In this example, it was not possible to tell from the given equation that there would be no solution. ✤

3   The next examples use the facts that

$$(a + b)^2 = a^2 + 2ab + b^2 \qquad \text{and} \qquad (a - b)^2 = a^2 - 2ab + b^2.$$

By the second pattern, for example,

$$(y - 3)^2 = y^2 - 2(y)(3) + (3)^2$$
$$= y^2 - 6y + 9.$$

Example 5   Solve the equation $\sqrt{2y - 3} = y - 3$.

From above,

$$(y - 3)^2 = y^2 - 6y + 9.$$

Square both sides of the given equation.

$$(\sqrt{2y - 3})^2 = (y - 3)^2$$
$$2y - 3 = y^2 - 6y + 9$$

This equation is quadratic, because of the y^2 term. Solve the equation by first getting one side equal to 0. Subtract $2y$ and add 3, getting

$$0 = y^2 - 8y + 12.$$

This equation can be solved by factoring.

$$0 = (y - 6)(y - 2)$$

Set each factor equal to 0.

$$y - 6 = 0 \quad \text{or} \quad y - 2 = 0$$
$$y = 6 \quad \text{or} \quad y = 2$$

Check both of these proposed solutions in the original equation.

If $y = 6$,

$$\sqrt{2y - 3} = y - 3$$
$$\sqrt{2(6) - 3} = 6 - 3$$
$$\sqrt{12 - 3} = 3$$
$$\sqrt{9} = 3$$
$$3 = 3. \qquad \text{True}$$

If $y = 2$,

$$\sqrt{2y - 3} = y - 3$$
$$\sqrt{2(2) - 3} = 2 - 3$$
$$\sqrt{4 - 3} = -1$$
$$\sqrt{1} = -1$$
$$1 = -1. \qquad \text{False}$$

Only 6 is a solution of the equation. ✛

Sometimes it is necessary to write an equation in a different form before squaring both sides. The next example shows why.

Example 6   Solve the equation $\sqrt{x} + 1 = 2x$.
Squaring both sides gives

$$(\sqrt{x} + 1)^2 = (2x)^2$$
$$x + 2\sqrt{x} + 1 = 4x^2,$$

an equation that is more complicated, and still contains a radical. It would be better instead to rewrite the original equation so that the radical is alone on one side of the equals sign. Do this by subtracting 1 from both sides to get

$$\sqrt{x} = 2x - 1.$$

Now square both sides.

$$(\sqrt{x})^2 = (2x - 1)^2$$
$$x = 4x^2 - 4x + 1$$

Subtract x from both sides.

$$0 = 4x^2 - 5x + 1$$

This equation is a quadratic equation that can be solved by factoring.

$$0 = (4x - 1)(x - 1)$$
$$4x - 1 = 0 \quad \text{or} \quad x - 1 = 0$$
$$x = \frac{1}{4} \quad \text{or} \quad x = 1$$

Both of these proposed solutions must be checked in the original equation.

If $x = \frac{1}{4}$,

$$\sqrt{x} + 1 = 2x$$
$$\sqrt{\frac{1}{4}} + 1 = 2\left(\frac{1}{4}\right)$$
$$\frac{1}{2} + 1 = \frac{1}{2}. \qquad \text{False}$$

If $x = 1$,

$$\sqrt{x} + 1 = 2x$$
$$\sqrt{1} + 1 = 2(1)$$
$$2 = 2. \qquad \text{True}$$

The only solution to the original equation is 1. ✦

Some equations with radicals require squaring twice, as in the next example.

Example 7   Solve $\sqrt{21 + x} = 3 + \sqrt{x}$.
Square both sides.

$$(\sqrt{21 + x})^2 = (3 + \sqrt{x})^2$$
$$21 + x = 9 + 6\sqrt{x} + x$$

Combine terms and simplify.

$$12 = 6\sqrt{x}$$
$$2 = \sqrt{x}$$

Square both sides a second time.

$$4 = x$$

Check the proposed solution.

If $x = 4$,

$$\sqrt{21 + x} = 3 + \sqrt{x}$$
$$\sqrt{21 + 4} = 3 + \sqrt{4}$$
$$5 = 5. \qquad \text{True}$$

The solution is 4. ✦

Here is a summary of the steps to use when solving an equation with radicals.

Solving an Equation with Radicals	*Step 1* Arrange the terms so that there is no more than one radical on each side of the equation. *Step 2* Square both sides. *Step 3* Combine like terms. *Step 4* If there is still a term with a radical, repeat Steps 1–3. *Step 5* Solve the equation from Step 3. *Step 6* Check all solutions from Step 5 in the original equation.

8.6   *Exercises*

Find all solutions for each equation. See Examples 1–4.

1. $\sqrt{x} = 2$

2. $\sqrt{m} = 5$

3. $\sqrt{y + 3} = 2$

4. $\sqrt{z + 1} = 5$

5. $\sqrt{t - 3} = 2$

6. $\sqrt{r + 5} = 4$

7. $\sqrt{n + 8} = 1$

8. $\sqrt{k + 10} = 2$

9. $\sqrt{m + 5} = 0$

10. $\sqrt{y - 4} = 0$

11. $\sqrt{z + 5} = -2$

12. $\sqrt{t - 3} = -2$

13. $\sqrt{k} - 2 = 5$

14. $\sqrt{p} - 3 = 7$

15. $\sqrt{y} + 4 = 2$

16. $\sqrt{m} + 6 = 5$

17. $\sqrt{5t - 9} = 2\sqrt{t}$

18. $\sqrt{3n + 4} = 2\sqrt{n}$

19. $3\sqrt{r} = \sqrt{8r + 16}$

20. $2\sqrt{r} = \sqrt{3r + 9}$

21. $\sqrt{5y - 5} = \sqrt{4y + 1}$

22. $\sqrt{2x + 2} = \sqrt{3x - 5}$

23. $\sqrt{x + 2} = \sqrt{2x - 5}$

24. $\sqrt{3m + 3} = \sqrt{5m - 1}$

25. $p = \sqrt{p^2 - 3p - 12}$

26. $k = \sqrt{k^2 - 2k + 10}$

27. $2r = \sqrt{4r^2 + 5r - 30}$

Find all solutions for each equation. Remember that $(a + b)^2 = a^2 + 2ab + b^2$ and $(\sqrt{a})^2 = a$. See Examples 5 and 6.

28. $\sqrt{5x + 11} = x + 3$

29. $\sqrt{2x + 1} = x - 7$

30. $\sqrt{5x + 1} = x + 1$

31. $\sqrt{3x + 10} = 2x - 5$

32. $\sqrt{4x + 13} = 2x - 1$

33. $\sqrt{x + 1} - 1 = x$

34. $\sqrt{3x + 3} + 5 = x$

35. $\sqrt{4x + 5} - 2 = 2x - 7$

36. $\sqrt{6x + 7} - 1 = x + 1$

37. $3\sqrt{x + 13} = x + 9$

38. $2\sqrt{x + 7} = x - 1$

39. $\sqrt{4x} - x + 3 = 0$

40. $\sqrt{2x} - x + 4 = 0$

41. $\sqrt{3x} - 4 = x - 10$

42. $\sqrt{x} + 9 = x + 3$

In the following two exercises, it is necessary to square both sides twice. See Example 7.

43. $\sqrt{x} = \sqrt{x - 5} + 1$

44. $\sqrt{2x} = \sqrt{x + 7} - 1$

Solve each word problem.

45. The square root of the sum of a number and 4 is 5. Find the number.

46. A certain number is the same as the square root of the product of 8 and the number. Find the number.

47. Three times the square root of 2 equals the square root of the sum of some number and 10. Find the number.

48. The negative square root of a number equals that number decreased by 2. Find the number.

49. To estimate the speed at which a car was traveling at the time of an accident, police sometimes use the following procedure. A police officer drives the car involved in the accident under conditions similar to those during which the accident took place, and then skids to a stop. If the car is driven at 30 miles per hour, the speed at the time of the accident is given by

$$s = 30\sqrt{\frac{a}{p}},$$

where a is the length of the skid marks left at the time of the accident and p is the length of the skid marks in the police test. Find s if
(a) $a = 900$ feet and $p = 100$ feet;   (b) $a = 400$ feet and $p = 25$ feet;
(c) $a = 80$ feet and $p = 20$ feet;   (d) $a = 120$ feet and $p = 30$ feet.

Review Exercises   *Use the rules for exponents to simplify each expression. Write each answer in exponential form with only positive exponents. See Section 3.2.*

50. $(5^2)^3$

51. $3^{-4} \cdot 3^{-1}$

52. $\dfrac{a^{-2}a^3}{a^4}$

53. $(2x^3)^{-1}$

54. $\left(\dfrac{p}{3}\right)^{-2}$

55. $\left(\dfrac{2y^3}{y^{-1}}\right)^{-2}$

56. $\dfrac{(c^3)^2 c^4}{(c^{-1})^3}$

8.7 Fractional Exponents

Objectives
Use fractional exponents of the form

1 $a^{1/n}$;

2 $a^{m/n}$.

3 Use rules for exponents with fractional exponents.

4 Use fractional exponents to simplify radicals.

This section introduces exponential expressions with fractional exponents, such as $5^{1/2}$, $16^{3/4}$, and $8^{-2/3}$.

1 How should $5^{1/2}$ be defined? This expression should be defined so that all the rules for exponents developed earlier in this book still hold. In particular, power rule (a) should work, so that

$$(5^{1/2})^2 = 5^{2/2} = 5^1 = 5.$$

By definition,

$$(\sqrt{5})^2 = 5.$$

Since both $(5^{1/2})^2$ and $(\sqrt{5})^2$ equal 5,

$$5^{1/2} \text{ should equal } \sqrt{5}.$$

Also, by power rule (a) for exponents,

$$(5^{1/3})^3 = 5 \quad \text{and} \quad (\sqrt[3]{5})^3 = 5,$$

so

$$5^{1/3} \text{ should equal } \sqrt[3]{5}.$$

These examples suggest the following definition.

$a^{1/n}$

> If a is a nonnegative number and n is a positive integer,
> $$a^{1/n} = \sqrt[n]{a}.$$

Example 1 Simplify each expression by first writing it in radical form.
(a) $16^{1/2}$

By definition,

$$16^{1/2} = \sqrt{16} = 4.$$

(b) $27^{1/3} = \sqrt[3]{27} = 3$

(c) $64^{1/3} = \sqrt[3]{64} = 4$

(d) $64^{1/6} = \sqrt[6]{64} = 2$ ✤

2 Now a more general exponential expression like $16^{3/4}$ can be defined. By power rule (a) again,

$$16^{3/4} = (16^{1/4})^3 = (\sqrt[4]{16})^3 = 2^3 = 8.$$

However, $16^{3/4}$ could also be written as

$$16^{3/4} = (16^3)^{1/4} = (4096)^{1/4} = \sqrt[4]{4096} = 8.$$

The expression can be evaluated either way to get the same answer. As the example suggests, taking the root first involves smaller numbers and is often easier. This example suggests the following definition for $a^{m/n}$.

$a^{m/n}$

> If a is a nonnegative number and m and n are integers, with $n > 0$,
> $$a^{m/n} = \sqrt[n]{a^m} = (\sqrt[n]{a})^m.$$

Example 2   Evaluate each expression.

(a) $9^{3/2}$

Use the definition to write

$$9^{3/2} = (9^{1/2})^3 = 3^3 = 27.$$

(b) $64^{2/3} = (64^{1/3})^2 = 4^2 = 16$

(c) $(-64)^{2/3} = [(-64)^{1/3}]^2 = (-4)^2 = 16$ ✚

Earlier, a^{-n} was defined as

$$a^{-n} = \frac{1}{a^n}$$

for nonzero numbers a and integers n. This same result applies for negative fractional exponents.

$a^{-m/n}$

If a is a positive number and m and n are integers, with $n > 0$, $$a^{-m/n} = \frac{1}{a^{m/n}}.$$

Example 3   Write each expression with a positive exponent and then evaluate.

(a) $32^{-3/5} = \dfrac{1}{32^{3/5}} = \dfrac{1}{(32^{1/5})^3} = \dfrac{1}{2^3} = \dfrac{1}{8}$

(b) $27^{-4/3} = \dfrac{1}{27^{4/3}} = \dfrac{1}{3^4} = \dfrac{1}{81}$ ✚

3   All the rules for exponents given earlier still hold when the exponents are fractions. The next examples show how to use these rules to simplify expressions with fractional exponents.

Example 4   Simplify each expression. Write each answer in exponential form with only positive exponents.

(a) $3^{2/3} \cdot 3^{5/3} = 3^{2/3+5/3} = 3^{7/3}$

(b) $\dfrac{5^{1/4}}{5^{3/4}} = 5^{1/4-3/4} = 5^{-2/4} = 5^{-1/2} = \dfrac{1}{5^{1/2}}$

(c) $(9^{1/4})^2 = 9^{2(1/4)} = 9^{2/4} = 9^{1/2} = \sqrt{9} = 3$

(d) $\dfrac{2^{1/2} \cdot 2^{-1}}{2^{-3/2}} = \dfrac{2^{1/2+(-1)}}{2^{-3/2}} = \dfrac{2^{-1/2}}{2^{-3/2}} = 2^{-1/2-(-3/2)} = 2^{2/2} = 2^1 = 2$

(e) $\left(\dfrac{9}{4}\right)^{5/2} = \dfrac{9^{5/2}}{4^{5/2}} = \dfrac{(9^{1/2})^5}{(4^{1/2})^5} = \dfrac{(\sqrt{9})^5}{(\sqrt{4})^5} = \dfrac{3^5}{2^5}$ ✦

Example 5   Simplify each expression. Write each answer in exponential form with only positive exponents. Assume that all variables represent positive numbers.

(a) $m^{1/5} \cdot m^{3/5} = m^{1/5 + 3/5} = m^{4/5}$

(b) $\dfrac{p^{5/3}}{p^{4/3}} = p^{5/3 - 4/3} = p^{1/3}$

(c) $(x^2 y^{1/2})^4 = (x^2)^4 (y^{1/2})^4 = x^8 y^2$

(d) $\left(\dfrac{z^{1/4}}{w^{1/3}}\right)^5 = \dfrac{(z^{1/4})^5}{(w^{1/3})^5} = \dfrac{z^{5/4}}{w^{5/3}}$

(e) $\dfrac{k^{2/3} \cdot k^{-1/3}}{k^{5/3}} = k^{2/3 + (-1/3) - 5/3} = k^{-4/3} = \dfrac{1}{k^{4/3}}$ ✦

4   Sometimes a radical expression can be simplified by writing it in exponential form. The next example shows how this is done.

Example 6   Simplify each radical by writing it in exponential form.

(a) $\sqrt[6]{9^3} = (9^3)^{1/6} = 9^{3/6} = 9^{1/2} = \sqrt{9} = 3$

(b) $(\sqrt[4]{m})^2 = (m^{1/4})^2 = m^{2/4} = m^{1/2} = \sqrt{m}$

Here it is assumed that $m \geq 0$. ✦

8.7   *Exercises*

Simplify each expression by first writing it in radical form. See Examples 1 and 2.

1. $16^{1/2}$ **2.** $64^{1/2}$ **3.** $25^{1/2}$ **4.** $49^{1/2}$

5. $8^{1/3}$ **6.** $27^{1/3}$ **7.** $64^{1/3}$ **8.** $125^{1/3}$

9. $16^{1/4}$ **10.** $81^{1/4}$ **11.** $32^{1/5}$ **12.** $243^{1/5}$

13. $4^{3/2}$ **14.** $9^{5/2}$ **15.** $27^{2/3}$ **16.** $8^{5/3}$

17. $16^{3/4}$ **18.** $64^{5/4}$ **19.** $32^{2/5}$ **20.** $144^{3/2}$

21. $(-8)^{2/3}$ **22.** $(-27)^{5/3}$ **23.** $(-64)^{1/3}$ **24.** $(-125)^{5/3}$

Simplify each expression. Write each answer in exponential form with only positive exponents. Assume that all variables represent positive numbers. See Examples 3–5.

25. $2^{1/2} \cdot 2^{5/2}$ **26.** $5^{2/3} \cdot 5^{4/3}$ **27.** $6^{1/4} \cdot 6^{-3/4}$ **28.** $12^{2/5} \cdot 12^{-1/5}$

29. $\dfrac{15^{3/4}}{15^{5/4}}$

30. $\dfrac{7^{3/5}}{7^{-1/5}}$

31. $\dfrac{11^{-2/7}}{11^{-3/7}}$

32. $\dfrac{4^{-2/3}}{4^{1/3}}$

33. $(8^{3/2})^2$

34. $(5^{2/5})^{10}$

35. $(6^{1/3})^{3/2}$

36. $(7^{2/5})^{5/3}$

37. $(9^{1/4})^{3/2}$

38. $(13^{2/3})^{4/5}$

39. $\left(\dfrac{25}{4}\right)^{3/2}$

40. $\left(\dfrac{9}{16}\right)^{5/2}$

41. $\left(\dfrac{8}{27}\right)^{2/3}$

42. $\left(\dfrac{64}{27}\right)^{2/3}$

43. $\dfrac{2^{2/5} \cdot 2^{-3/5}}{2^{7/5}}$

44. $\dfrac{3^{-3/4} \cdot 3^{5/4}}{3^{-1/4}}$

45. $\dfrac{6^{-2/9}}{6^{1/9} \cdot 6^{-5/9}}$

46. $\dfrac{8^{6/7}}{8^{2/7} \cdot 8^{-1/7}}$

47. $p^{2/3} \cdot p^{7/3}$

48. $k^{-1/4} \cdot k^{5/4}$

49. $\dfrac{z^{2/3}}{z^{-1/3}}$

50. $\dfrac{r^{5/4}}{r^{3/4}}$

51. $(m^3 n^{1/4})^{2/3}$

52. $(p^4 \cdot q^{1/2})^{4/3}$

53. $(x^{1/2} y^{2/3})^{3/4}$

54. $(z^{2/5} w^{1/5})^{4/3}$

55. $\left(\dfrac{a^{1/2}}{b^{1/3}}\right)^{4/3}$

56. $\left(\dfrac{m^{2/3}}{n^{3/4}}\right)^{1/2}$

57. $\dfrac{c^{2/3} \cdot c^{-1/3}}{c^{5/3}}$

58. $\dfrac{d^{3/4} \cdot d^{5/4}}{d^{1/4}}$

59. $\left(\dfrac{k^{1/4} \cdot k^{-3/4}}{k^{5/4}}\right)^2$

60. $\left(\dfrac{p^{3/8} \cdot p^{-2}}{p^{1/8}}\right)^4$

Simplify each radical in Exercises 61–68 by writing it in exponential form. Give final answers in radical form. Assume that all variables represent positive numbers. See Example 6.

61. $\sqrt[6]{4^3}$

62. $\sqrt[9]{8^3}$

63. $\sqrt[8]{16^2}$

64. $\sqrt[9]{27^3}$

65. $\sqrt[4]{a^2}$

66. $\sqrt[9]{b^3}$

67. $\sqrt[6]{k^4}$

68. $\sqrt[8]{m^4}$

69. A formula for calculating the distance, d, one can see from an airplane to the horizon on a clear day is

$$d = 1.22x^{1/2},$$

where x is the altitude of the plane in feet and d is given in miles. How far can one see to the horizon in a plane flying at the following altitudes? (a) 20,000 feet (b) 30,000 feet

70. A biologist has shown that the number of different plant species S on a Galápagos Island is related to the area of the island, A, by

$$S = 28.6A^{1/3}.$$

How many plant species would exist on such an island with the following areas? (a) 8 square miles (b) 27,000 squares miles

Review Exercises   *Find all square roots of each number. Simplify where possible. See Section 8.1*

71. 25

72. 49

73. 14

74. 29

75. 18

76. 48

77. 80

78. 75

Chapter 8   *Summary*

| **Product Rule for Radicals** | For nonnegative real numbers x and y, $\sqrt{x} \cdot \sqrt{y} = \sqrt{x \cdot y}$. |

| **Quotient Rule for Radicals** | If x and y are nonnegative real numbers and $y \neq 0$, $$\frac{\sqrt{x}}{\sqrt{y}} = \sqrt{\frac{x}{y}}.$$ |

| **Simplifying Radicals** | 1. If a radical represents a rational number, then that rational number should be used in place of that radical.
2. If a radical expression contains products of radicals, the product rule for radicals, $\sqrt{x} \cdot \sqrt{y} = \sqrt{xy}$, should be used to get a single radical.
3. If a radicand has a factor that is a perfect square, the radical should be expressed as the product of the positive square root of the perfect square and the remaining radical factor.
4. If a radical expression contains sums or differences of radicals, the distributive property should be used to combine like terms.
5. Any radical in the denominator should be changed to a rational number. |

| $a^{m/n}$ | If a is a nonnegative number and m and n are integers, with $n > 0$, $$a^{m/n} = \sqrt[n]{a^m} = (\sqrt[n]{a})^m.$$ |

Chapter 8   *Review Exercises*

❶ [8.1] *Find all square roots of each number.*

1. 49   **2.** 81   **3.** 225   **4.** 729

Find each root that exists.

5. $\sqrt{16}$ **6.** $-\sqrt{4225}$ **7.** $\sqrt{-8100}$

8. $\sqrt[3]{-64}$ **9.** $\sqrt[4]{16}$ **10.** $\sqrt[5]{32}$

Write rational *or* irrational *for each number. If a number is rational, give its exact value. If a number is irrational, give a decimal approximation for the number. Round approximations to the nearest thousandth.*

11. $\sqrt{15}$ **12.** $\sqrt{64}$ **13.** $-\sqrt{169}$ **14.** $-\sqrt{170}$

[8.2] *Use the product rule to simplify each expression.*

15. $\sqrt{5} \cdot \sqrt{15}$ **16.** $\sqrt{160}$ **17.** $\sqrt{98} \cdot \sqrt{50}$

Use the quotient rule and the product rule, as necessary, to simplify each expression.

18. $\sqrt{\dfrac{9}{4}}$

19. $\sqrt{\dfrac{10}{169}}$

20. $\dfrac{24\sqrt{12}}{16\sqrt{3}}$

Simplify each expression. Assume that all variables represent nonnegative real numbers.

21. $\sqrt{p} \cdot \sqrt{p}$

22. $\sqrt{m^2 p^4}$

23. $\sqrt{x^3}$

Simplify each radical. Assume that all variables represent nonnegative real numbers.

24. $\sqrt[3]{\dfrac{5}{8}}$

25. $\sqrt[3]{\dfrac{6}{125}}$

26. $\sqrt[3]{375x^4}$

[8.3] *Simplify and combine terms wherever possible.*

27. $3\sqrt{2} + 5\sqrt{2}$

28. $\dfrac{1}{3}\sqrt{18} + \dfrac{1}{4}\sqrt{32}$

29. $\dfrac{2}{5}\sqrt{75} - \dfrac{3}{4}\sqrt{160}$

30. $\sqrt{5} \cdot \sqrt{3} + 4\sqrt{15} - \sqrt{3}$ **31.** $3\sqrt[3]{54} - 2\sqrt[3]{16}$

32. $4\sqrt[4]{16} + 3\sqrt[4]{32} \cdot \sqrt[4]{2}$

Simplify each expression. Assume that all variables represent nonnegative real numbers.

33. $\sqrt{16p} + 3\sqrt{p} - \sqrt{49p}$ **34.** $\sqrt{20m^2} - m\sqrt{45}$

35. $3k\sqrt{8k^2n} + 5k^2\sqrt{2n}$

[8.4] *Perform the indicated operations, and write all answers in simplest form. Rationalize all denominators.*

36. $\dfrac{10}{\sqrt{3}}$

37. $\dfrac{3\sqrt{2}}{\sqrt{5}}$

38. $\sqrt{\dfrac{3}{5}}$

39. $\sqrt{\dfrac{2}{3}} \cdot \sqrt{\dfrac{1}{5}}$

40. $\sqrt[3]{\dfrac{1}{3}}$

41. $\sqrt[3]{\dfrac{5}{4}}$

Perform the indicated operations, and write all answers in simplest form. Rationalize all denominators. Assume that all variables represent positive real numbers.

42. $\sqrt{\dfrac{7}{x}}$

43. $\sqrt{\dfrac{r^3 t}{5w^2}}$

44. $\sqrt[3]{\dfrac{k^4}{4n}}$

[8.5] *Simplify each expression.*

45. $3\sqrt{2}(\sqrt{3} + 2\sqrt{2})$

46. $(2\sqrt{3} - 4)(5\sqrt{3} + 2)$

47. $(5\sqrt{7} + 2)^2$

48. $(\sqrt{5} - \sqrt{7})(\sqrt{5} + \sqrt{7})$

Rationalize the denominators.

49. $\dfrac{\sqrt{3}}{1 + \sqrt{3}}$

50. $\dfrac{\sqrt{5} - 1}{\sqrt{2} + 3}$

51. $\dfrac{2 + \sqrt{6}}{\sqrt{3} - 1}$

[8.6] *Find all solutions for each equation.*

52. $\sqrt{p} = -2$

53. $\sqrt{k + 1} = 10$

54. $\sqrt{x} + 2 = 1$

55. $\sqrt{-2k - 4} = k + 2$

56. $\sqrt{2 - x} + 3 = x + 7$

[8.7] *Simplify each expression. Assume that all variables represent positive numbers.*

57. $81^{1/2}$

58. $125^{1/3}$

59. $7^{2/3} \cdot 7^{5/3}$

60. $\dfrac{12^{2/5}}{12^{-2/5}}$

61. $\dfrac{x^{1/4} \cdot x^{5/4}}{x^{3/4}}$

62. $\sqrt[8]{49^4}$

❷ *Simplify each expression.*

63. $\sqrt[3]{2} \cdot \sqrt[3]{8}$

64. $4\sqrt{24} - 3\sqrt{54} + \sqrt{6}$

65. $\sqrt{\dfrac{36}{p^2}}$

66. $\dfrac{1}{2 + \sqrt{5}}$

67. $\sqrt{\dfrac{1}{6}} \cdot \sqrt{\dfrac{18}{5}}$

68. $\sqrt[3]{54x^2}$

69. $\sqrt[12]{125^4}$

70. $-\sqrt{3}(\sqrt{5} - \sqrt{27})$

71. $\sqrt{\dfrac{9m^3}{2p}}$

72. $\sqrt[3]{27}$

73. $3\sqrt{75} + 2\sqrt{27}$

74. $\dfrac{(7^{5/2})^{1/5}}{7^{3/2}}$

75. $(2\sqrt{3} + 5)(2\sqrt{3} - 5)$

76. $-\sqrt{36}$

77. $\sqrt{48}$

78. $\dfrac{12}{\sqrt{24}}$

79. $\sqrt{\dfrac{2}{5}} \cdot \sqrt{\dfrac{2}{45}}$

80. $\sqrt{15} \cdot \sqrt{2} + 5\sqrt{30}$

Chapter 8   *Test*

In this test, assume that all variables represent positive numbers. In Exercises 1–5, find the indicated root. Use Table 3 if necessary.

1. $\sqrt{100}$

2. $\sqrt{77}$

3. $-\sqrt{190}$

4. $\sqrt[3]{-27}$

5. $\sqrt[4]{625}$

Simplify where possible.

6. $\sqrt{27}$

7. $\sqrt{128}$

8. $\sqrt[3]{-32}$

9. $3\sqrt{6} + \sqrt{14}$

10. $3\sqrt{28} + \sqrt{63}$

11. $m\sqrt{20} - m\sqrt{45}$

12. $3\sqrt{27x} - 4\sqrt{48x} + 2\sqrt{3x}$

13. $\sqrt[3]{32x^2y^3}$

14. $\sqrt[4]{32m^3n^4p^6}$

15. $(6 - \sqrt{5})(6 + \sqrt{5})$

16. $(2 - \sqrt{7})(3\sqrt{2} + 1)$

17. $(\sqrt{5} + \sqrt{6})^2$

Rationalize each denominator.

18. $\dfrac{3\sqrt{2}}{\sqrt{6}}$

19. $\dfrac{4p}{\sqrt{k}}$

20. $\sqrt[3]{\dfrac{5}{9}}$

21. $\dfrac{-3}{4 - \sqrt{3}}$

22. $\dfrac{\sqrt{2} + 1}{3 - \sqrt{7}}$

Solve each equation.

23. $\sqrt{k + 2} + 3 = -2$

24. $\sqrt{2y + 8} = 2\sqrt{y}$

25. $6\sqrt{k} - 3 = k + 2$

26. $\sqrt{y + 3} = \sqrt{y} + 1$

Simplify each expression. Write answers with only positive exponents.

27. $8^{4/3}$

28. $5^{3/4} \cdot 5^{7/4}$

29. $\dfrac{(3^{1/4})^3}{3^{7/4}}$

30. $\dfrac{m^{2/3} \cdot m^{5/6}}{m^{4/3}}$

Quadratic Equations

9.1 Solving Quadratic Equations by the Square Root Method

Objectives

1 Solve equations of the form $x^2 =$ a number.

2 Solve equations of the form $(ax + b)^2 =$ a number.

Recall that a *quadratic equation* is an equation that can be written in the form

$$ax^2 + bx + c = 0$$

for real numbers a, b, and c, with $a \neq 0$. In Chapter 4, these equations were solved by factoring. However, not all quadratic equations can be solved by factoring. Other ways to solve quadratic equations are shown in this chapter. For example, the quadratic equation

$$(x - 3)^2 = 15,$$

in which the square of a binomial is equal to some number, can be solved with square roots.

1 The **square root property of equations** justifies taking square roots of both sides of an equation.

Square Root Property	If b is a positive number and if $a^2 = b$, then $a = \sqrt{b}$ or $a = -\sqrt{b}$.

Example 1   Solve each equation. Write radicals in simplified form.

(a) $x^2 = 16$

By the square root property, since $x^2 = 16$, then

$$x = \sqrt{16} = 4 \qquad \text{or} \qquad x = -\sqrt{16} = -4.$$

Check each solution in the original equation.

(b) $z^2 = 5$

The solutions are $z = \sqrt{5}$ or $z = -\sqrt{5}$.

(c) $m^2 = 8$

Use the square root property to get $m = \sqrt{8}$ or $m = -\sqrt{8}$. Simplify $\sqrt{8}$ as $\sqrt{8} = 2\sqrt{2}$, so

$$m = 2\sqrt{2} \qquad \text{or} \qquad m = -2\sqrt{2}.$$

(d) $y^2 = -4$

Since -4 is a negative number and since the square of a real number cannot be negative, there is no real number solution for this equation. (The square root property cannot be used because of the requirement that b must be positive.)

(e) $3x^2 + 5 = 11$

First solve the equation for x^2.

$$3x^2 + 5 = 11$$
$$3x^2 = 6 \qquad \text{Subtract 5}$$
$$x^2 = 2 \qquad \text{Divide by 3}$$

Now use the square root property to get

$$x = \sqrt{2} \qquad \text{or} \qquad x = -\sqrt{2}. \quad \clubsuit$$

2   The equation $(x - 3)^2 = 16$ also can be solved with the square root property. If $(x - 3)^2 = 16$, then

$$x - 3 = 4 \qquad \text{or} \qquad x - 3 = -4.$$

Solve each of the last two equations to get

$$x - 3 = 4 \qquad \text{or} \qquad x - 3 = -4$$
$$x = 7 \qquad \text{or} \qquad x = -1.$$

Check both answers in the original equation.

$$(7 - 3)^2 = 4^2 = 16 \qquad \text{and} \qquad (-1 - 3)^2 = (-4)^2 = 16$$

Both 7 and -1 are solutions.

Example 2 Solve $(x - 1)^2 = 6$.

By the square root property,

$$x - 1 = \sqrt{6} \qquad \text{or} \qquad x - 1 = -\sqrt{6}$$
$$x = 1 + \sqrt{6} \qquad \text{or} \qquad x = 1 - \sqrt{6}.$$

Check: $\qquad (1 + \sqrt{6} - 1)^2 = (\sqrt{6})^2 = 6;$
$\qquad\qquad (1 - \sqrt{6} - 1)^2 = (-\sqrt{6})^2 = 6.$

The solutions are $1 + \sqrt{6}$ and $1 - \sqrt{6}$. ✤

Example 3 Solve the equation $(3r - 2)^2 = 27$.

The square root property gives

$$3r - 2 = \sqrt{27} \qquad \text{or} \qquad 3r - 2 = -\sqrt{27}.$$

Now simplify the radical: $\sqrt{27} = \sqrt{9 \cdot 3} = \sqrt{9} \cdot \sqrt{3} = 3\sqrt{3}$, so

$$3r - 2 = 3\sqrt{3} \qquad \text{or} \qquad 3r - 2 = -3\sqrt{3}$$
$$3r = 2 + 3\sqrt{3} \qquad \text{or} \qquad 3r = 2 - 3\sqrt{3}$$
$$r = \frac{2 + 3\sqrt{3}}{3} \qquad \text{or} \qquad r = \frac{2 - 3\sqrt{3}}{3}.$$

The solutions are

$$\frac{2 + 3\sqrt{3}}{3} \qquad \text{and} \qquad \frac{2 - 3\sqrt{3}}{3}.$$

These fractions cannot be reduced since 3 is *not* a common factor in the numerator. ✤

Example 4 Solve $(x + 3)^2 = -9$.

The square root of -9 is not a real number. There is no real number solution for this equation. ✤

9.1 Exercises

Solve each equation by using the square root property. Express all radicals in simplest form. See Example 1.

1. $x^2 = 25$	**2.** $y^2 = 100$	**3.** $x^2 = 64$	**4.** $z^2 = 81$
5. $m^2 = 13$	**6.** $x^2 = 7$	**7.** $y^2 = -15$	**8.** $p^2 = -10$
9. $3p^2 = 6$	**10.** $2q^2 = 12$	**11.** $4k^2 = 5$	**12.** $5k^2 = 8$
13. $3x^2 - 8 = 64$	**14.** $2t^2 + 7 = 61$	**15.** $5a^2 + 4 = 8$	**16.** $4p^2 - 3 = 7$
17. $k^2 = 2.56$	**18.** $z^2 = 9.61$	**19.** $r^2 = 77.44$	**20.** $y^2 = 43.56$

Solve each equation by using the square root property. Express all radicals in simplest form. Round answers to the nearest hundredth in Exercises 41–42. See Examples 2–4.

21. $(x - 2)^2 = 16$ **22.** $(r + 4)^2 = 25$ **23.** $(a + 4)^2 = 10$ **24.** $(r - 3)^2 = 15$

25. $(m - 1)^2 = -4$ **26.** $(t + 2)^2 = -8$ **27.** $(x - 1)^2 = 32$ **28.** $(y + 5)^2 = 28$

29. $(2m - 1)^2 = 9$ **30.** $(3y - 7)^2 = 4$ **31.** $(6m - 2)^2 = 121$ **32.** $(7m - 10)^2 = 144$

33. $(2a - 5)^2 = 30$ **34.** $(2y + 3)^2 = 45$ **35.** $(3p - 1)^2 = 18$ **36.** $(5r - 6)^2 = 75$

37. $(2k - 5)^2 = 98$ **38.** $(4x - 1)^2 = 48$ **39.** $(3m + 4)^2 = 8$ **40.** $(5y - 3)^2 = 50$

41. $(2.11p + 3.42)^2 = 9.58$ **42.** $(1.71m - 6.20)^2 = 5.41$

Solve the following word problems.

43. One expert at marksmanship can hold a silver dollar at forehead level, drop it, draw his gun, and shoot the coin as it passes waist level. The distance traveled by a falling object is given by

$$d = 16t^2,$$

where d is the distance the object falls in t seconds. If the coin falls about 4 feet, use the formula to estimate the time that elapses between the dropping of the coin and the shot.

44. The illumination produced by a light source depends on the distance from the source. For a particular light source, this relationship can be expressed as

$$d^2 = \frac{4050}{I},$$

where d is the distance from the source (in feet) and I is the amount of illumination in footcandles. How far from the source is the illumination equal to 50 footcandles?

45. The amount A that P dollars invested at a rate of interest r will amount to in 2 years is

$$A = P(1 + r)^2.$$

At what interest rate will \$1 grow to \$1.21 in two years?

46. The price p, in dollars, of a new product depends on the demand d, in thousands, for the product, according to the expression

$$p = (d - 2)^2.$$

What demand produces a price of \$5?

Review Exercises   Simplify all radicals and combine terms. See Sections 8.2 and 8.4.

47. $\dfrac{3}{2} + \sqrt{\dfrac{27}{4}}$ **48.** $-\dfrac{1}{4} + \sqrt{\dfrac{5}{16}}$ **49.** $6 + \sqrt{\dfrac{2}{3}}$ **50.** $5 - \sqrt{\dfrac{7}{2}}$

Factor each perfect square trinomial. See Section 4.4.

51. $x^2 + 8x + 16$ **52.** $m^2 - 10m + 25$ **53.** $p^2 - 5p + \dfrac{25}{4}$ **54.** $z^2 + 3z + \dfrac{9}{4}$

9.2   Solving Quadratic Equations by Completing the Square

Objectives

1   Solve quadratic equations by completing the square when the coefficient of the squared term is 1.

2   Solve quadratic equations by completing the square when the coefficient of the squared term is not 1.

1   The properties and methods studied so far are not enough to solve the equation

$$x^2 + 6x + 7 = 0.$$

For a method of solving this equation, recall the method from the preceding section for solving equations of the type

$$(x + 3)^2 = 2.$$

If the equation $x^2 + 6x + 7 = 0$ could be rewritten in the form $(x + 3)^2 = 2$, it could be solved by using the square root property. The following example shows how to rewrite the equation in this form.

Example 1   Solve $x^2 + 6x + 7 = 0$.

Start by subtracting 7 from both sides of the equation to get $x^2 + 6x = -7$. If $x^2 + 6x = -7$ is to be written in the form $(x + 3)^2 = 2$, the quantity on the left-hand side of $x^2 + 6x = -7$ must be made into a perfect square trinomial. The expression $x^2 + 6x + 9$ is a perfect square, since

$$x^2 + 6x + 9 = (x + 3)^2.$$

Therefore, if 9 is added to both sides of $x^2 + 6x = -7$, the equation will have a perfect square trinomial on one side, as needed.

$$x^2 + 6x + 9 = -7 + 9 \qquad \text{Add 9 on both sides}$$
$$(x + 3)^2 = 2 \qquad \text{Simplify}$$

Now use the square root property to complete the solution.

$$x + 3 = \sqrt{2} \qquad \text{or} \qquad x + 3 = -\sqrt{2}$$
$$x = -3 + \sqrt{2} \qquad \text{or} \qquad x = -3 - \sqrt{2}$$

The solutions of the original equation are $-3 + \sqrt{2}$ and $-3 - \sqrt{2}$. Check this by substituting $-3 + \sqrt{2}$ and $-3 - \sqrt{2}$ for x in the original equation.   ✦

The process of changing the form of the equation in Example 1 from

$$x^2 + 6x + 7 = 0 \qquad \text{to} \qquad (x + 3)^2 = 2$$

is called **completing the square.** Completing the square changes only the form of the equation. To see this, multiply out $(x + 3)^2 = 2$ and combine terms; the result will be $x^2 + 6x + 7 = 0$.

Example 2 Solve the quadratic equation $m^2 - 8m = 5$.

A suitable number must be added to both sides to make one side a perfect square. Find this number as follows: recall from Chapter 3 that

$$(m + a)^2 = m^2 + 2am + a^2.$$

In the equation $m^2 - 8m = 5$, the value of $2am$ is $-8m$ and a^2 must be found. Set $2am$ equal to $-8m$ to find a.

$$2am = -8m$$
$$a = -4$$

Squaring -4 gives 16, the number to be added to both sides.

$$m^2 - 8m + 16 = 5 + 16 \qquad (1)$$

The trinomial $m^2 - 8m + 16$ is a perfect square trinomial. Factor this trinomial to get

$$m^2 - 8m + 16 = (m - 4)^2.$$

Equation (1) becomes

$$(m - 4)^2 = 21.$$

Now use the square root property.

$$m - 4 = \sqrt{21} \qquad \text{or} \qquad m - 4 = -\sqrt{21}$$
$$m = 4 + \sqrt{21} \qquad \text{or} \qquad m = 4 - \sqrt{21}$$

Check that the solutions are

$$4 + \sqrt{21} \qquad \text{and} \qquad 4 - \sqrt{21}. \quad \clubsuit$$

As illustrated by Example 2, the number to be added to both sides of the quadratic equation $x^2 + 2ax = b$ to complete the square is found by taking 1/2 of $2a$, the coefficient of the x term, and squaring it. This gives $(1/2)(2a) = a$, with a^2 then added to both sides of the equation as shown in Example 2.

2 The process of completing the square just described requires the coefficient of the squared variable to be 1. For an equation of the form $ax^2 + bx + c = 0$, get a coefficient of 1 on x^2 by first dividing both sides of the equation by a. The next example shows this approach.

Example 3 Solve $4y^2 + 24y = 13$.

Before completing the square, get the coefficient of the squared term equal to 1 by dividing both sides of the equation by 4.

$$4y^2 + 24y = 13$$
$$y^2 + 6y = \frac{13}{4}$$

Now take half the coefficient of y, or $(1/2)(6) = 3$, and square the result: $3^2 = 9$. Add 9 to both sides of the equation and perform the addition on the right-hand side.

$$y^2 + 6y + 9 = \frac{13}{4} + 9$$

$$y^2 + 6y + 9 = \frac{49}{4}$$

Factor on the left.

$$(y + 3)^2 = \frac{49}{4}$$

Use the square root property and solve for y.

$$y + 3 = \frac{7}{2} \qquad \text{or} \qquad y + 3 = -\frac{7}{2}$$

$$y = -3 + \frac{7}{2} \qquad \text{or} \qquad y = -3 - \frac{7}{2}$$

$$y = \frac{1}{2} \qquad \text{or} \qquad y = -\frac{13}{2}$$

The two solutions are $1/2$ and $-13/2$. Check by substitution into the original equation. ✦

Example 4 Solve $2x^2 - 7x = 9$.

Divide both sides of the equation by 2 to get a coefficient of 1 for the x^2 term.

$$x^2 - \frac{7}{2}x = \frac{9}{2}$$

Now take half the coefficient of x and square it. Half of $-7/2$ is $-7/4$, and $-7/4$ squared is $49/16$. Add $49/16$ to both sides of the equation, write the left side as a perfect square, and perform the addition on the right.

$$x^2 - \frac{7}{2}x + \frac{49}{16} = \frac{9}{2} + \frac{49}{16}$$

$$\left(x - \frac{7}{4}\right)^2 = \frac{121}{16}$$

Apply the square root property.

$$x - \frac{7}{4} = \sqrt{\frac{121}{16}} \qquad \text{or} \qquad x - \frac{7}{4} = -\sqrt{\frac{121}{16}}$$

Since $\sqrt{\dfrac{121}{16}} = \dfrac{11}{4}$,

$$x - \frac{7}{4} = \frac{11}{4} \quad \text{or} \quad x - \frac{7}{4} = -\frac{11}{4}$$

$$x = \frac{18}{4} \quad \text{or} \quad x = -\frac{4}{4}$$

$$x = \frac{9}{2} \quad \text{or} \quad x = -1.$$

The solutions are 9/2 and -1. ✤

Example 5 Solve $4p^2 + 8p + 5 = 0$.

First divide both sides by 4 to get the coefficient 1 for the p^2 term. The result is

$$p^2 + 2p + \frac{5}{4} = 0.$$

Subtract 5/4 from both sides, which gives

$$p^2 + 2p = -\frac{5}{4}.$$

The coefficient of p is 2. Take half of 2, square the result, and add it to both sides. The left-hand side can then be written as a perfect square.

$$p^2 + 2p + 1 = -\frac{5}{4} + 1$$

$$(p + 1)^2 = -\frac{1}{4}$$

The square root of $-1/4$ is not a real number so the square root property does not apply. This equation has no real number solution. ✤

The steps in solving a quadratic equation by completing the square are summarized below.

Solving a Quadratic Equation by Completing the Square	
Step 1	If the coefficient of the squared term is 1, proceed to Step 2. If the coefficient of the squared term is not 1 but some other nonzero number a, divide both sides of the equation by a. This gives an equation that has 1 as coefficient of the squared term.
Step 2	Make sure that all terms with variables are on one side of the equals sign and that all numbers are on the other side.
Step 3	Take half the coefficient of x and square the result. Add the square to both sides of the equation. The side containing the variables now can be written as a perfect square.
Step 4	Apply the square root property.

9.2 Exercises

Find the number that should be added to each expression to make it a perfect square.

1. $x^2 + 2x$ **2.** $y^2 - 4y$ **3.** $x^2 + 18x$ **4.** $m^2 - 3m$

5. $z^2 + 9z$ **6.** $p^2 + 22p$ **7.** $y^2 + 5y$ **8.** $r^2 + 7r$

Solve each equation by completing the square. You may have to simplify first. Round answers to the nearest thousandth in Exercises 29–32. See Examples 2–5.

9. $x^2 + 4x = -3$ **10.** $y^2 - 4y = 0$ **11.** $a^2 + 2a = 5$

12. $m^2 + 4m = -1$ **13.** $z^2 + 6z = -8$ **14.** $q^2 - 8q = -16$

15. $x^2 - 6x + 1 = 0$ **16.** $b^2 - 2b - 2 = 0$ **17.** $c^2 + 3c = 2$

18. $k^2 + 5k - 3 = 0$ **19.** $2m^2 + 4m = -7$ **20.** $3y^2 - 9y + 5 = 0$

21. $6q^2 - 8q + 3 = 0$ **22.** $4y^2 + 4y - 3 = 0$ **23.** $-x^2 + 6x = 4$

24. $-x^2 + 4 = 2x$ **25.** $3x^2 - 2x = 1$ **26.** $-x^2 - 4 = 2x$

27. $m^2 - 4m + 8 = 6m$ **28.** $2z^2 = 8z + 5 - 4z^2$ **29.** $3r^2 - 2 = 6r + 3$

30. $4p - 3 = p^2 + 2p$ **31.** $(x + 1)(x + 3) = 2$ **32.** $(x - 3)(x + 1) = 1$

33. A rule for estimating the number of board feet of lumber that can be cut from a log depends on the diameter of the log. The diameter d required to get 9 board feet is found from the equation

$$\left(\frac{d - 4}{4}\right)^2 = 9.$$

Solve this equation for d. Are both answers reasonable?

34. A rancher has determined that the number of cattle in his herd has increased over a two-year period at a rate r given by the equation

$$5r^2 + 10r = 1.$$

Find r. Do both answers make sense?

35. Two painters are painting a house in a development of new homes. One of the painters takes 2 hours longer to paint a house working alone than the other painter. When they do the job together, they can complete it in 4.8 hours. How long would it take the faster painter alone to paint the house?

36. Two cars travel at right angles to each other from an intersection until they are 17 miles apart. At that point one car has gone 7 miles farther than the other. How far did the slower car travel?

Review Exercises *Write each quotient in lowest terms. Simplify the radicals if necessary. See Section 8.5.*

37. $\dfrac{2 + 2\sqrt{3}}{2}$ **38.** $\dfrac{3 + 6\sqrt{5}}{3}$ **39.** $\dfrac{4 + 2\sqrt{7}}{8}$ **40.** $\dfrac{5 + 5\sqrt{2}}{10}$

41. $\dfrac{8 + 6\sqrt{3}}{4}$ **42.** $\dfrac{4 + \sqrt{28}}{6}$ **43.** $\dfrac{6 + \sqrt{45}}{12}$ **44.** $\dfrac{8 + \sqrt{32}}{8}$

9.3 Solving Quadratic Equations by the Quadratic Formula

Objectives

1 Identify the values of *a, b,* and *c* in a quadratic equation.

2 Use the quadratic formula to solve quadratic equations.

3 Solve quadratic equations with fractions.

Any quadratic equation can be solved by completing the square, but the method is not very handy. This section introduces a general formula, the quadratic formula, that gives the solution for any quadratic equation.

Get the quadratic formula by starting with the general form of a quadratic equation,

$$ax^2 + bx + c = 0, \qquad a \neq 0.$$

The restriction $a \neq 0$ is important in order to make sure that the equation is quadratic. If $a = 0$, then the equation becomes $0x^2 + bx + c = 0$, or $bx + c = 0$, which is a linear, and not a quadratic, equation.

1 The first step in solving a quadratic equation by this new method is to identify the values of *a, b,* and *c* in the general form of the quadratic equation.

Example 1 Match the coefficients of each of the following quadratic equations with the letters *a, b,* and *c* of the general quadratic equation

$$ax^2 + bx + c = 0.$$

(a) $2x^2 + 3x - 5 = 0$

In this example $a = 2$, $b = 3$, and $c = -5$.

(b) $-x^2 + 2 = 6x$

First rewrite the equation with 0 on one side to match the general form of $ax^2 + bx + c = 0$.

$$-x^2 + 2 = 6x$$
$$-x^2 - 6x + 2 = 0$$

Now identify $a = -1$, $b = -6$, and $c = 2$.

(c) $2(x + 3)(x - 1) = 5$

First multiply on the left.

$$2(x + 3)(x - 1) = 2(x^2 + 2x - 3) = 2x^2 + 4x - 6$$

The equation becomes

$$2x^2 + 4x - 6 = 5, \qquad \text{or} \qquad 2x^2 + 4x - 11 = 0,$$

so that $a = 2$, $b = 4$, and $c = -11$. ✦

The solutions of $2x^2 - 7x - 9 = 0$ are 9/2 and -1. Check by substituting in the original equation. ✛

Example 3   Solve $x^2 = 2x + 1$.

One side of the equation must be 0 before a, b, and c can be found. Subtract $2x$ and 1 from both sides of the equation to get

$$x^2 - 2x - 1 = 0.$$

Then $a = 1$, $b = -2$, and $c = -1$. The solution is found by substituting these values into the quadratic formula.

$$x = \frac{-b \pm \sqrt{b^2 - 4ac}}{2a}$$

$$= \frac{-(-2) \pm \sqrt{(-2)^2 - 4(1)(-1)}}{2(1)} \qquad \text{Let } a = 1, b = -2, \\ c = -1$$

$$= \frac{2 \pm \sqrt{4 + 4}}{2} = \frac{2 \pm \sqrt{8}}{2}$$

Since $\sqrt{8} = \sqrt{4 \cdot 2} = \sqrt{4} \cdot \sqrt{2} = 2\sqrt{2}$,

$$x = \frac{2 \pm 2\sqrt{2}}{2}.$$

Write the solutions in lowest terms by factoring $2 \pm 2\sqrt{2}$ as $2(1 \pm \sqrt{2})$ to get

$$x = \frac{2(1 \pm \sqrt{2})}{2} = 1 \pm \sqrt{2}.$$

The two solutions of the given equation are

$$1 + \sqrt{2} \qquad \text{and} \qquad 1 - \sqrt{2}. \quad ✛$$

When the quantity under the radical, $b^2 - 4ac$, equals zero, the equation has just one rational number solution. In this case, the trinomial $ax^2 + bx + c$ is a perfect square.

Example 4   Solve $4x^2 + 25 = 20x$.

Write the equation as

$$4x^2 - 20x + 25 = 0.$$

Here, $a = 4$, $b = -20$, and $c = 25$. By the quadratic formula,

$$x = \frac{-(-20) \pm \sqrt{(-20)^2 - 400}}{8} = \frac{20 \pm 0}{8} = \frac{5}{2}.$$

Since there is just one solution, 5/2, the trinomial $4x^2 - 20x + 25$ is a perfect square. ✛

3   The final example shows how to solve quadratic equations with fractions.

Example 5   Solve the equation $\dfrac{1}{10}t^2 = \dfrac{2}{5}t - \dfrac{1}{2}$.

Eliminate the denominators by multiplying both sides of the equation by the common denominator, 10.

$$10\left(\frac{1}{10}t^2\right) = 10\left(\frac{2}{5}t - \frac{1}{2}\right)$$
$$t^2 = 4t - 5$$

Subtract $4t$ and add 5 on both sides of the equation to get

$$t^2 - 4t + 5 = 0.$$

From this form identify $a = 1$, $b = -4$, and $c = 5$. Use the quadratic formula to complete the solution.

$$t = \frac{-(-4) \pm \sqrt{(-4)^2 - 4(1)(5)}}{2(1)} = \frac{4 \pm \sqrt{16 - 20}}{2} = \frac{4 \pm \sqrt{-4}}{2}$$

The radical $\sqrt{-4}$ is not a real number, so the equation has no real number solution.   ✚

9.3   Exercises

For each equation, identify the letters a, b, and c of the general quadratic equation $ax^2 + bx + c = 0$. Do not solve. See Example 1.

1. $3x^2 + 4x - 8 = 0$ **2.** $9x^2 + 2x - 3 = 0$ **3.** $-8x^2 - 2x - 3 = 0$

4. $-2x^2 + 3x - 8 = 0$ **5.** $2x^2 = 3x - 2$ **6.** $9x^2 - 2 = 4x$

7. $x^2 = 2$ **8.** $x^2 - 3 = 0$ **9.** $3x^2 - 8x = 0$

10. $5x^2 = 2x$ **11.** $(x - 3)(x + 4) = 0$ **12.** $(x + 6)^2 = 3$

13. $9(x - 1)(x + 2) = 8$ **14.** $(3x - 1)(2x + 5) = x(x - 1)$

Use the quadratic formula to solve each equation. Write all radicals in simplified form. Reduce answers to lowest terms. Round answers to the nearest thousandth in Exercises 45–47. See Examples 2–4.

15. $z^2 + 6z + 9 = 0$ **16.** $6k^2 + 6k + 1 = 0$ **17.** $y^2 + 4y + 4 = 0$

18. $3r^2 - 5r + 1 = 0$ **19.** $z^2 = 13 - 12z$ **20.** $x^2 = 8x + 9$

21. $4p^2 - 12p + 9 = 0$ **22.** $k^2 = 20k - 19$ **23.** $5x^2 + 4x - 1 = 0$

24. $5n^2 + n - 1 = 0$ **25.** $2z^2 = 3z + 5$ **26.** $7r - 2r^2 + 30 = 0$

27. $p^2 + 2p - 2 = 0$ **28.** $x^2 - 2x + 1 = 0$ **29.** $2w^2 + 12w + 5 = 0$

30. $9r^2 + 6r + 1 = 0$ **31.** $5m^2 + 5m = 0$ **32.** $4y^2 - 8y = 0$

33. $6p^2 = 10p$ **34.** $3r^2 = 16r$ **35.** $m^2 - 20 = 0$

36. $k^2 - 5 = 0$ **37.** $9r^2 - 16 = 0$ **38.** $4y^2 - 25 = 0$

39. $2x^2 + 2x + 4 = 4 - 2x$ **40.** $3x^2 - 4x + 3 = 8x - 1$ **41.** $2x^2 + x + 7 = 0$

42. $x^2 + x + 1 = 0$ **43.** $2x^2 = 3x - 2$ **44.** $x^2 = 5x - 20$

45. $x^2 = 1 + x$ **46.** $2x^2 + 2x = 5$ **47.** $5x^2 = 3 - x$

In Exercises 48–59, use the quadratic formula to solve each equation. See Example 5.

48. $\dfrac{3}{2}r^2 - r = \dfrac{4}{3}$ **49.** $\dfrac{1}{2}x^2 = 1 - \dfrac{1}{6}x$ **50.** $\dfrac{3}{5}x - \dfrac{2}{5}x^2 = -1$

51. $\dfrac{2}{3}m^2 - \dfrac{4}{9}m - \dfrac{1}{3} = 0$ **52.** $\dfrac{m^2}{4} + \dfrac{3m}{2} + 1 = 0$ **53.** $\dfrac{r^2}{2} = r + \dfrac{1}{2}$

54. $\dfrac{2y^2}{7} + \dfrac{10}{7}y + 1 = 0$ **55.** $k^2 = \dfrac{2k}{3} + \dfrac{2}{9}$ **56.** $9 - \dfrac{24}{y^2} = -\dfrac{17}{y}$

57. $\dfrac{m^2}{2} = \dfrac{m}{2} - 1$ **58.** $3 + \dfrac{1}{p^2} = \dfrac{6}{p}$ **59.** $2 = \dfrac{4}{x} - \dfrac{3}{x^2}$

60. The time t in seconds under certain conditions for a car to skid 48 feet is given (approximately) by

$$48 = 64t - 16t^2.$$

Solve this equation for t. Are both answers reasonable?

61. A certain projectile is located $d = 2t^2 - 5t + 2$ feet from its starting point after t seconds have elapsed. How many seconds will it take the projectile to travel 14 feet?

62. In a bicycle race over a 12-mile route, Russ finished 8 minutes ahead of Kevin. Russ pedaled 3 miles per hour faster than Kevin. What was Russ's speed?

63. Karen and Jessie work at a fast-food restaurant after school. Working alone, Jessie can close up in 1 hour less time than Karen. Together they can close up in 2/3 of an hour. How long does it take Jessie to close up alone?

Review Exercises   *Use the product rule to simplify each expression. See Section 8.2.*

64. $\sqrt{24}$ **65.** $\sqrt{48}$ **66.** $\sqrt{72}$ **67.** $\sqrt{44}$

68. $\sqrt{800}$ **69.** $\sqrt{288}$ **70.** $\sqrt{150}$ **71.** $\sqrt{270}$

Supplementary Quadratic Equations

Four methods have now been introduced for solving quadratic equations written in the form $ax^2 + bx + c = 0$. The chart on the next page gives some advantages and disadvantages of each method.

Methods for Solving Quadratic Equations	*Method*	*Advantages*	*Disadvantages*
	1. Factoring	Usually the fastest method	Not all equations are factorable
			Some factorable equations are hard to factor
	2. Completing the square	None for solving equations (the procedure is useful in other areas of mathematics)	Requires more steps than other methods
	3. Quadratic formula	Can always be used	More difficult than factoring because of the square root
	4. Square root method	Simplest method for solving equations of the form $(x + a)^2 = b$	Few equations are given in this form

Solve each quadratic equation.

1. $y^2 + 3y + 1 = 0$

2. $p^2 + 3p + 2 = 0$

3. $2x^2 - x = 1$

4. $2a^2 + 1 = a$

5. $8m^2 = 2m + 15$

6. $8x^2 + 2x = 15$

7. $(2p - 1)^2 = 10$

8. $2q^2 + 3q = 1$

9. $5k^2 + 8 = 22k$

10. $5k^2 + 2k = 1$

11. $3z^2 = 4z + 1$

12. $2c^2 + 11c + 12 = 0$

13. $(2q + 9)^2 = 48$

14. $4x^2 + 5x = 1$

15. $15t^2 + 58t + 48 = 0$

16. $12d^2 + 19d = 21$

17. $p^2 + 5p + 5 = 0$

18. $(7m - 1)^2 = 32$

19. $3c^2 - 4c = 4$

20. $5k^2 + 17k = 12$

21. $2x^2 - 5x + 1 = 0$

22. $(5r - 7)^2 = -1$

23. $4m^2 - 11m + 10 = 0$

24. $2a^2 - 7a + 4 = 0$

25. $(3p - 1)(p + 2) = -3$

26. $(3r + 2)^2 = 5$

27. $(a + 6)^2 = 121$

28. $(3x - 1)(2x + 5) = x$

29. $(2x + 1)(x - 1) = 5$

30. $(z - 1)(z - 3) = -3z$

9.4 Complex Solutions of Quadratic Equations

Objectives

1. Write complex numbers like $\sqrt{-5}$ as multiples of i.
2. Add and subtract complex numbers.
3. Multiply complex numbers.
4. Write complex number quotients in standard form.
5. Solve quadratic equations with complex number solutions.

As shown earlier in this chapter, some quadratic equations have no real number solutions. For example, the solution

$$x = \frac{-3 + \sqrt{-5}}{2}$$

is not a real number because of $\sqrt{-5}$. For every quadratic equation to have a solution, a new set of numbers, which includes the real numbers, is needed.

1 This new set of numbers is defined using a new number i such that

i

$$i = \sqrt{-1} \quad \text{and} \quad i^2 = -1.$$

Numbers like $\sqrt{-5}$, $\sqrt{-4}$, and $\sqrt{-8}$ can now be written as multiples of i, using a generalization of the product rule for radicals, as in the next example.

Example 1 Write each number as a multiple of i.

(a) $\sqrt{-5} = \sqrt{-1 \cdot 5} = \sqrt{-1} \cdot \sqrt{5} = i\sqrt{5}$

Write $i\sqrt{5}$ rather than $\sqrt{5}\,i$ to avoid confusion with $\sqrt{5i}$.

(b) $\sqrt{-4} = \sqrt{-1} \cdot \sqrt{4} = i\sqrt{4} = i \cdot 2 = 2i$

(c) $\sqrt{-8} = i\sqrt{8} = i \cdot 2 \cdot \sqrt{2} = 2i\sqrt{2}$ ✦

Numbers that are multiples of i are *imaginary numbers*. The *complex numbers* include all real numbers and all imaginary numbers.

Complex
Number

A **complex number** is a number of the form $a + bi$, where a and b are real numbers. If $b \neq 0$, $a + bi$ also is an **imaginary number.**

For example, the real number 2 is a complex number since it can be written as $2 + 0i$. Also, the imaginary number $3i = 0 + 3i$ is a complex number. Other complex numbers are

$$3 - 2i, \quad 1 + i\sqrt{2}, \quad \text{and} \quad -5 + 4i.$$

A complex number written in the form $a + bi$ (or $a + ib$) is in **standard form.** Figure 9.1 shows the relationships among the various types of numbers discussed in this book.

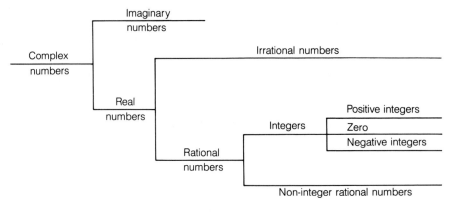

Figure 9.1

[2] All the properties of real numbers also apply to complex numbers. For example, complex numbers are added or subtracted by combining like terms.

Example 2 Add or subtract.

(a) $(2 - 6i) + (7 + 4i) = (2 + 7) + (-6 + 4)i = 9 - 2i$
(b) $3i + (-2 - i) = -2 + (3 - 1)i = -2 + 2i$
(c) $(2 + 6i) - (-4 + i) = [2 - (-4)] + (6 - 1)i = 6 + 5i$
(d) $(-1 + 2i) - 4 = (-1 - 4) + 2i = -5 + 2i$ ✤

[3] Multiplication of complex numbers is performed in the same way as multiplication of polynomials.

Example 3 Find the following products.

(a) $3i(2 - 5i) = 6i - 15i^2$
$= 6i - 15(-1)$ Recall: $i^2 = -1$
$= 6i + 15$
$= 15 + 6i$

The last step gives the result in standard form.

(b) $(4 - 3i)(2 + 5i)$

Use FOIL.
$(4 - 3i)(2 + 5i) = 4(2) + 4(5i) + (-3i)(2) + (-3i)(5i)$
$= 8 + 20i - 6i - 15i^2$
$= 8 + 14i - 15(-1)$
$= 8 + 14i + 15$
$= 23 + 14i$

(c) $(5 - 2i)(5 + 2i) = 25 + 10i - 10i - 4i^2$
$$= 25 - 4(-1)$$
$$= 25 + 4 = 29 \quad \clubsuit$$

4   The quotient of two complex numbers is expressed in standard form by changing the denominator into a real number. For example, to write

$$\frac{1 + 3i}{5 - 2i}$$

in standard form, the denominator must be a real number. In Example 3(c), the product $(5 - 2i)(5 + 2i)$ was 29, a real number. This suggests multiplying the numerator and denominator of the given quotient by $5 + 2i$, as follows.

$$\frac{1 + 3i}{5 - 2i} = \frac{1 + 3i}{5 - 2i} \cdot \frac{5 + 2i}{5 + 2i}$$

$$= \frac{5 + 2i + 15i + 6i^2}{25 - 4i^2}$$

$$= \frac{5 + 17i + 6(-1)}{25 - 4(-1)} \qquad i^2 = -1$$

$$= \frac{5 + 17i - 6}{25 + 4}$$

$$= \frac{-1 + 17i}{29}$$

$$= -\frac{1}{29} + \frac{17}{29}i$$

The last step gives the result in standard form. Recall that this is the method used to rationalize certain expressions in Chapter 8. The complex numbers $5 - 2i$ and $5 + 2i$ are *conjugates*. That is, the **conjugate** of the complex number $a + bi$ is $a - bi$.

Example 4   Write the following quotients in standard form.

(a) $\dfrac{-4 + i}{2 - i}$

Multiply numerator and denominator by the conjugate of the denominator, $2 + i$.

$$\frac{-4 + i}{2 - i} \cdot \frac{2 + i}{2 + i} = \frac{-8 - 4i + 2i + i^2}{4 - i^2}$$

$$= \frac{-8 - 2i - 1}{4 - (-1)}$$

$$= \frac{-9 - 2i}{5} = -\frac{9}{5} - \frac{2}{5}i$$

(b) $\dfrac{3 + i}{-i}$

Here, the conjugate of $0 - i$ is $0 + i$, or i.

$$\frac{3 + i}{-i} \cdot \frac{i}{i} = \frac{3i + i^2}{-i^2} = \frac{-1 + 3i}{-(-1)} = -1 + 3i \quad \clubsuit$$

5   Now quadratic equations that have no real number solutions can be solved, as shown in the next examples.

Example 5   Solve $(x + 3)^2 = -25$.
   Use the square root property.

$$(x + 3)^2 = -25$$

$$x + 3 = \sqrt{-25} \qquad \text{or} \qquad x + 3 = -\sqrt{-25}$$

Since $\sqrt{-25} = 5i$,

$$x + 3 = 5i \qquad \text{or} \qquad x + 3 = -5i$$
$$x = -3 + 5i \qquad \text{or} \qquad x = -3 - 5i.$$

In standard form, the two complex solutions are $-3 + 5i$ and $-3 - 5i$. $\quad \clubsuit$

Example 6   Solve $t^2 - 4t + 5 = 0$.
   The polynomial on the left cannot be factored, so use the quadratic formula, with $a = 1$, $b = -4$, and $c = 5$. The solutions are

$$t = \frac{-(-4) \pm \sqrt{(-4)^2 - 4(1)(5)}}{2(1)}$$

$$= \frac{4 \pm \sqrt{16 - 20}}{2} = \frac{4 \pm \sqrt{-4}}{2} = \frac{4 \pm 2i}{2}.$$

Factor out the common factor of 2 in the numerator and reduce to get

$$t = \frac{2(2 \pm i)}{2} = 2 \pm i.$$

The two complex solutions are $2 + i$ and $2 - i$. $\quad \clubsuit$

Example 7   Solve $2p^2 = 4p - 5$.
   Write the equation as $2p^2 - 4p + 5 = 0$. Then $a = 2$, $b = -4$, and $c = 5$. The solutions are

$$p = \frac{-(-4) \pm \sqrt{(-4)^2 - 4(2)(5)}}{2(2)}$$

$$= \frac{4 \pm \sqrt{16 - 40}}{4} = \frac{4 \pm \sqrt{-24}}{4}.$$

Since $\sqrt{-24} = i\sqrt{24} = i \cdot \sqrt{4} \cdot \sqrt{6} = i \cdot 2 \cdot \sqrt{6} = 2i\sqrt{6}$,

$$p = \frac{4 \pm 2i\sqrt{6}}{4}$$

$$= \frac{2(2 \pm i\sqrt{6})}{4} = \frac{2 \pm i\sqrt{6}}{2}.$$

In standard form, the solutions are the complex numbers

$$1 + \frac{\sqrt{6}}{2}i \quad \text{and} \quad 1 - \frac{\sqrt{6}}{2}i. \quad \clubsuit$$

9.4   Exercises

Write the following numbers as multiples of i. See Example 1.

1. $\sqrt{-9}$ **2.** $\sqrt{-18}$ **3.** $\sqrt{-20}$ **4.** $\sqrt{-27}$

5. $\sqrt{-36}$ **6.** $\sqrt{-50}$ **7.** $\sqrt{-125}$ **8.** $\sqrt{-98}$

Add or subtract as indicated. See Example 2.

9. $(2 + 8i) + (3 - 5i)$ **10.** $(4 - 5i) + (7 - 2i)$

11. $(8 - 3i) - (2 + 6i)$ **12.** $(1 + i) - (3 - 2i)$

13. $(3 - 4i) + (6 - i) - (3 + 2i)$ **14.** $(5 + 8i) - (4 + 2i) + (3 - i)$

Find each product. See Example 3.

15. $(3 + 2i)(4 - i)$ **16.** $(9 - 2i)(3 + i)$ **17.** $(5 - 4i)(3 - 2i)$

18. $(10 + 6i)(8 - 4i)$ **19.** $(3 + 6i)(3 - 6i)$ **20.** $(11 - 2i)(11 + 2i)$

Write each quotient in standard form. See Example 4.

21. $\dfrac{-2 + i}{1 - i}$ **22.** $\dfrac{6 + i}{2 + i}$ **23.** $\dfrac{3 - 4i}{2 + 2i}$

24. $\dfrac{7 - i}{3 - 2i}$ **25.** $\dfrac{4 - 3i}{i}$ **26.** $\dfrac{-i}{1 + 2i}$

Solve each quadratic equation by the square root property. Write solutions in standard form. See Example 5.

27. $(a + 1)^2 = -4$ **28.** $(p - 5)^2 = -36$ **29.** $(k - 3)^2 = -5$

30. $(y + 6)^2 = -12$ **31.** $(3x + 2)^2 = -18$ **32.** $(4z - 1)^2 = -20$

Solve each quadratic equation by the quadratic formula. Write solutions in standard form. See Examples 6 and 7.

33. $m^2 - 2m + 2 = 0$ **34.** $b^2 + b + 3 = 0$ **35.** $2r^2 + 3r + 5 = 0$

36. $3q^2 = 2q - 3$ **37.** $p^2 - 3p + 4 = 0$ **38.** $2a^2 = -a - 3$

39. $5x^2 + 3 = 2x$ **40.** $6y^2 + 2y + 1 = 0$ **41.** $2m^2 + 7 = -2m$

42. $4z^2 + 2z + 3 = 0$ **43.** $2r^2 + 5 = 4r$ **44.** $4q^2 - 2q + 3 = 0$

Review Exercises   *Complete the ordered pairs using the given equation. Then plot the*
ordered pairs. See Section 6.1.

Equation	*Ordered Pairs*
45. $2x - 3y = 6$	$(0,\ \ )\ \ (\ \ , 0)\ \ (\ \ , 2)\ \ (-3,\ \ )$
46. $4x - y = 8$	$(0,\ \ )\ \ (\ \ , 0)\ \ (4,\ \ )\ \ (\ \ , -4)$
47. $3x + 5y = 15$	$(0,\ \ )\ \ (\ \ , 0)\ \ (-5,\ \ )\ \ (\ \ , -3)$
48. $6x + 4y = 12$	$(0,\ \ )\ \ (\ \ , 0)\ \ (-2,\ \ )\ \ (\ \ , -3)$
49. $y - 2x = 0$	$(0,\ \ )\ \ (3,\ \ )\ \ (\ \ , 5)\ \ (\ \ , 2)$
50. $x - 3y = 0$	$(0,\ \ )\ \ (3,\ \ )\ \ (\ \ , 4)\ \ (4,\ \ )$

9.5 Graphing Quadratic Equations

Objectives

1 Graph quadratic equations.

2 Find the vertex of a parabola.

1   Chapter 6 showed that the graph of a linear equation in two variables is a straight line that represents all the solutions of the linear equation. Quadratic equations in two variables of the form $y = ax^2 + bx + c$ are graphed in this section. Perhaps the simplest quadratic equation is

$$y = x^2$$

(or $y = 1x^2 + 0x + 0$). The graph of this equation cannot be a straight line since only linear equations of the form

$$Ax + By = C$$

have graphs that are straight lines. However, $y = x^2$ can be graphed in much the same way as straight lines, by finding ordered pairs that satsify the equation $y = x^2$.

Example 1   Graph $y = x^2$.
    Select several values for x; then find the corresponding y-values. For example, selecting $x = 2$ gives

$$y = 2^2 = 4,$$

and the point (2, 4) belongs to the graph of $y = x^2$. (Recall that in an ordered pair such as (2, 4), the x-value comes first and the y-value second.) In the same way find the ordered pairs $(-3, 9), (-2, 4), (-1, 1),$ (0, 0), (1, 1), and (3, 9). These ordered pairs were used to get the table of values in Figure 9.2. If these ordered pairs are plotted on a coordinate system and a smooth curve drawn through them, the graph is as shown in Figure 9.2.   ✛

x	y
−3	9
−2	4
−1	1
0	0
1	1
2	4
3	9

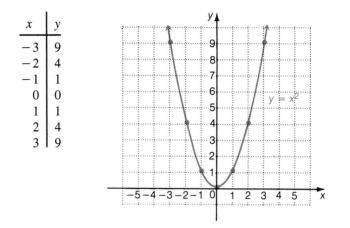

Figure 9.2

The curve in Figure 9.2 is called a **parabola.** The point (0, 0), the lowest point on this graph, is the **vertex** of the parabola.

Because of its many useful properties, the parabola occurs frequently in real-life applications. For example, if an object is thrown into the air, the path that the object follows is a parabola (discounting wind resistance). The cross sections of radar, spotlight, and telescope reflectors also form parabolas.

Every equation of the form

$$y = ax^2 + bx + c,$$

$a \neq 0$, has a graph that is a parabola. By the vertical line test, the parabola shown in Figure 9.2 is a function. A function defined by an equation of the form $y = ax^2 + bx + c$ is called a **quadratic function** since $ax^2 + bx + c = 0$ is a *quadratic equation*. The domain of a quadratic function includes all real numbers. The range of the quadratic function $y = x^2$, as suggested by the graph in Figure 9.2, is $y \geq 0$.

Example 2   Graph the parabola $y = -x^2$.

By selecting values for x and then finding the corresponding y-values, the parabola could be graphed by plotting those points. But for a given x-value, the y-value here will be the negative of the corresponding y-value of the parabola $y = x^2$ discussed above. Those ordered pairs can be used to sketch the graph by making each y-value negative as in the table of values given with Figure 9.3. Because of this, the new parabola has the same shape as the one in the preceding figure but is turned in the opposite direction, so it opens downward, as shown in Figure 9.3. Here the vertex, $(0, 0)$, is the *highest* point on the graph. ✚

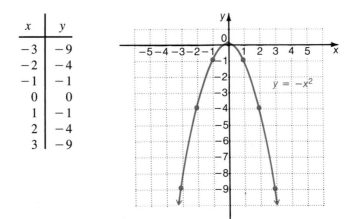

x	y
-3	-9
-2	-4
-1	-1
0	0
1	-1
2	-4
3	-9

Figure 9.3

Example 3   Graph the parabola $y = \dfrac{1}{2}x^2$.

For any value of x that might be chosen, the value of y will be 1/2 the corresponding value of y from $y = x^2$. For this reason, the graph of $y = (1/2)x^2$ will be *broader* than the graph of $y = x^2$. As shown in Figure 9.4, the graph opens upward and has $(0, 0)$ as vertex. ✚

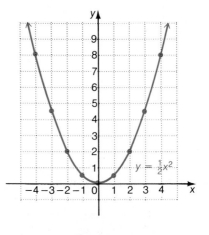

Figure 9.4

2   The next examples show how to find the vertex of a parabola from its equation.

Example 4   Graph the parabola $y = (x - 2)^2$.

     The vertex of a parabola is the point where y has its largest or smallest value. Since $(x - 2)^2$ is always nonnegative, y will be nonnegative. The smallest y-value, the y-value of the vertex, will be 0. Find x at the vertex by letting $y = 0$ in the equation and solving for x.

$$y = (x - 2)^2$$
$$0 = (x - 2)^2$$
$$0 = x - 2$$
$$x = 2$$

The vertex is (2, 0).

     Find some other points on the graph by selecting values for x and finding the corresponding y-values. For example, if $x = -1$, then

$$y = (-1 - 2)^2 = (-3)^2 = 9.$$

Continuing in this way gives the ordered pairs $(-1, 9)$ $(3, 1)$, $(0, 4)$, $(4, 4)$, $(1, 1)$, $(5, 9)$, and $(2, 0)$. Plotting these points and joining them gives the graph shown in Figure 9.5. This parabola has the same shape as the graph of $y = x^2$, but is shifted two units to the right, with vertex $(2, 0)$. ✢

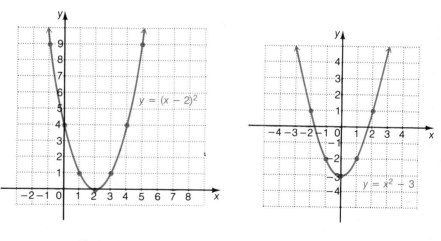

Figure 9.5   Figure 9.6

Example 5   Graph the parabola $y = x^2 - 3$.

     The smallest value of y will occur when $x = 0$ so that $y = 0 - 3$, or $y = -3$, making the vertex of this parabola $(0, -3)$. The graph is shown in Figure 9.6. This time the graph is shifted three units downward, compared with the graph of $y = x^2$. ✢

Example 6   Graph $y = (x - 2)^2 - 3$.

As shown in Figure 9.7, the vertex is $(2, -3)$. The graph is the same shape as that of $y = x^2$, but is shifted two units to the right and three units downward. See Examples 4 and 5 to see why. ✣

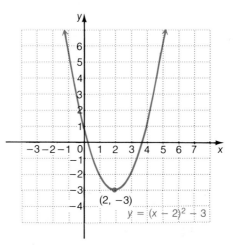

Figure 9.7

These results on graphing parabolas are summarized below.

Graphing Parabolas

The graph of

$$y = a(x - h)^2 + k$$

is a parabola with vertex at (h, k).
The parabola opens upward if $a > 0$ and downward if $a < 0$.
The parabola is broader than $y = x^2$ if $0 < |a| < 1$, and it is narrower than $y = x^2$ if $|a| > 1$.

If a quadratic function is defined by an equation such as

$$y = x^2 + 6x + 5,$$

the equation must be put in the form $y = (x - h)^2 + k$ to identify the vertex. This is done by completing the square (as in Section 9.2).

Example 7   Complete the square of $y = x^2 + 6x + 5$ and write the equation in the form $y = (x - h)^2 + k$.

The left side of the equation must be left alone; only the right side is to be rearranged. Begin by grouping the two terms with x.

$$y = (x^2 + 6x) + 5$$

Take half the coefficient of x and square it:

$$\frac{1}{2}(6) = 3 \qquad \text{and} \qquad 3^2 = 9.$$

Add 9 and subtract 9 on the right side. (Adding $9 + (-9) = 0$; the value of the equation remains unchanged.)

$$y = (x^2 + 6x + 9) + 5 - 9$$

Write the first three terms as $(x + 3)^2$ and combine the last two terms.

$$y = (x + 3)^2 - 4$$

Writing the equation in this form shows that the vertex is $(-3, -4)$. The graph is given in Figure 9.8. ✚

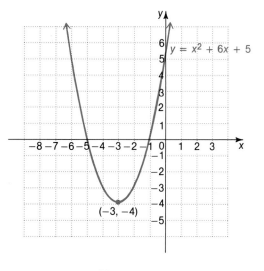

Figure 9.8

In the last example we graphed the equation $y = x^2 + 6x + 5$. Recall that setting y equal to 0 gives the x-intercepts. If $y = 0$, the equation becomes

$$0 = x^2 + 6x + 5.$$

The solutions of this equation can be found by factoring.

$$0 = (x + 5)(x + 1)$$

$$x + 5 = 0 \qquad \text{or} \qquad x + 1 = 0$$

$$x = -5 \qquad \text{or} \qquad x = -1$$

Now look at Figure 9.8, which gives the graph of $y = x^2 + 6x + 5$. As expected, the x-intercepts (the points where $y = 0$) are -5 and -1. The solutions of a quadratic equation always give the intercepts of the graph of the corresponding function.

Intercepts of a Quadratic Function	The real number solutions of a quadratic equation $ax^2 + bx + c = 0$ are the x-intercepts of the graph of the corresponding quadratic function $y = ax^2 + bx + c$.

Since a parabola can cross the x-axis in two, one, or no places, this result shows why some quadratic equations have two solutions, some have one, and some have no real solutions.

Example 8   Decide from the indicated graph how many real number solutions the corresponding quadratic equation with $y = 0$ will have.

(a) Figure 9.6 shows the graph of an equation with two solutions, $x = \sqrt{3}$ and $x = -\sqrt{3}$.

(b) Figure 9.5 shows the graph of an equation with one solution, $x = 2$.

(c) Figure 9.9 gives the graph of $y = x^2 + 2$. Since there are no x-intercepts, the equation $x^2 + 2 = 0$ has no *real* number solutions. The equation *does* have two complex solutions, $i\sqrt{2}$ and $-i\sqrt{2}$.   ❖

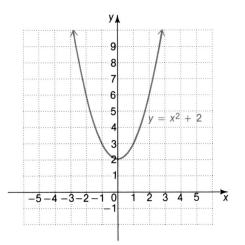

Figure 9.9

9.5 Exercises

Sketch the graph of each quadratic function. Identify each vertex. See Examples 1–6.

1. $y = 2x^2$

2. $y = -2x^2$

3. $y = (x + 1)^2$

4. $y = (x - 2)^2$

5. $y = -(x + 1)^2$

6. $y = -(x - 2)^2$

7. $y = x^2 + 1$

8. $y = -x^2 - 2$

9. $y = 2 - x^2$

10. $y = -4 + x^2$

11. $y = \dfrac{1}{2}x^2 + 2$

12. $y = -\dfrac{1}{3}x^2 + 1$

13. $y = (x + 1)^2 + 2$

14. $y = (x - 2)^2 - 1$

15. $y = (x - 4)^2 - 1$

16. $y = (x + 3)^2 + 3$

Sketch the graph of each function. Identify each vertex. See Example 7.

17. $y = x^2 - 6x + 5$

18. $y = x^2 - 4x + 2$

19. $y = x^2 - 8x + 9$

20. $y = x^2 - 2x - 1$

21. $y = x^2 + 3x + 2$

22. $y = x^2 + 5x - 2$

Decide from the graph how many real number solutions the corresponding equation has. Find any real solutions from the graph. See Example 8.

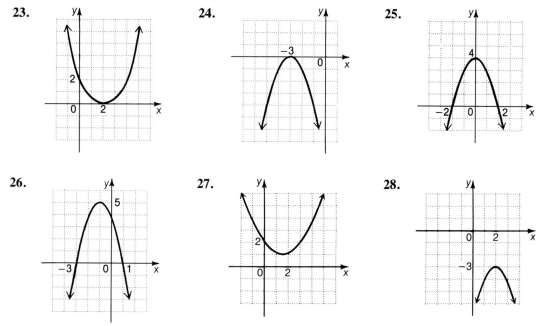

23.   **24.**   **25.**

26.   **27.**   **28.**

Find the domain and range of each of the functions graphed in the indicated exercises.

29. Exercise 23

30. Exercise 24

31. Exercise 25

32. Exercise 26

33. Exercise 27

34. Exercise 28

Given $f(x) = 2x^2 - 5x + 3$, find each of the following.

35. $f(0)$ **36.** $f(1)$ **37.** $f(-2)$ **38.** $f(-1)$

Chapter 9 *Summary*

Square Root Property	If b is a positive number and if $a^2 = b$, then $a = \sqrt{b}$ or $a = -\sqrt{b}$.

Solving a Quadratic Equation by Completing the Square	*Step 1* If the coefficient of the squared term is 1, proceed to Step 2. If the coefficient of the squared term is not 1 but some other nonzero number a, divide both sides of the equation by a. This gives an equation that has 1 as coefficient of the squared term.
	Step 2 Make sure that all terms with variables are on one side of the equals sign and that all numbers are on the other side.
	Step 3 Take half the coefficient of x and square the result. Add the square to both sides of the equation. The side containing the variables now can be written as a perfect square.
	Step 4 Apply the square root property.

Quadratic Formula	The solutions of the quadratic equation $ax^2 + bx + c$, $a \neq 0$, are $$x = \frac{-b + \sqrt{b^2 - 4ac}}{2a} \quad \text{and} \quad x = \frac{-b - \sqrt{b^2 - 4ac}}{2a}.$$

Graphing Parabolas	The graph of $$y = a(x - h)^2 + k$$ is a parabola with vertex at (h, k). The parabola opens upward if $a > 0$ and downward if $a < 0$. The parabola is broader than $y = x^2$ if $0 < |a| < 1$, and it is narrower than $y = x^2$ if $|a| > 1$.

Chapter 9 *Review Exercises*

[9.1] *Solve each equation by using the square root property. Express all radicals in simplest form.*

1. $y^2 = 49$ 2. $x^2 = 15$ 3. $m^2 = 48$ 4. $(k + 2)^2 = 9$

5. $(r - 3)^2 = 7$ 6. $(2p + 1)^2 = 11$ 7. $(3k + 2)^2 = 12$ 8. $\left(x + \frac{1}{2}\right)^2 = \frac{3}{4}$

[9.2] *Solve each equation by completing the square.*

9. $m^2 + 6m + 5 = 0$ 10. $p^2 + 4p = 7$ 11. $-x^2 + 5 = 2x$

12. $2y^2 + 8y = 3$ 13. $5k^2 - 3k - 2 = 0$ 14. $(4a + 1)(a - 1) = -3$

[9.3] *Use the quadratic formula to solve each equation.*

15. $x^2 - 2x - 4 = 0$ **16.** $-m^2 + 3m + 5 = 0$ **17.** $3k^2 + 2k + 3 = 0$

18. $5p^2 = p + 1$ **19.** $2p^2 - 3 = 4p$ **20.** $-4a^2 + 7 = 2a$

21. $\dfrac{c^2}{4} = 2 - \dfrac{3}{4}c$ **22.** $\dfrac{3}{2}r^2 = \dfrac{1}{2} - r$

[9.4] *Perform the indicated operation.*

23. $(3 + 5i) + (2 - 6i)$ **24.** $(-2 - 8i) - (4 - 3i)$ **25.** $(-1 + i) - (2 - i)$

26. $(4 + 3i) + (-2 + 3i)$ **27.** $(6 - 2i)(3 + i)$ **28.** $(2 + 3i)(4 - 2i)$

29. $(5 + 2i)(5 - 2i)$ **30.** $(8 - i)(8 + i)$ **31.** $\dfrac{1 + i}{1 - i}$

32. $\dfrac{3 + 2i}{1 + i}$ **33.** $\dfrac{2 - 4i}{3 + i}$ **34.** $\dfrac{5 + 6i}{2 + 3i}$

Solve each quadratic equation.

35. $(m + 2)^2 = -3$ **36.** $(x - 1)^2 = -2$ **37.** $(3p - 2)^2 = -8$ **38.** $(4p + 1)^2 = -12$

39. $3k^2 = 2k - 1$ **40.** $h^2 + 3h = -8$ **41.** $4q^2 + 2 = 3q$ **42.** $9z^2 + 2z + 1 = 0$

[9.5] *Sketch the graph of each equation. Identify each vertex.*

43. $y = 3x^2$ **44.** $y = -3x^2$ **45.** $y = -x^2 + 3$

46. $y = x^2 - 1$ **47.** $y = (x - 4)^2$ **48.** $y = -(2x + 1)^2$

49. $y = (x + 2)^2 - 3$ **50.** $y = -(x - 1)^2 + 4$ **51.** $y = \dfrac{1}{2}(x - 3)^2 + 1$

52. $y = x^2 + 8x + 14$ **53.** $y = x^2 - 10x + 24$ **54.** $y = x^2 + 2x + 1$

Decide from the graph how many real number solutions the corresponding equation has. Find any real solutions from the graph.

55.   **56.**   **57.**

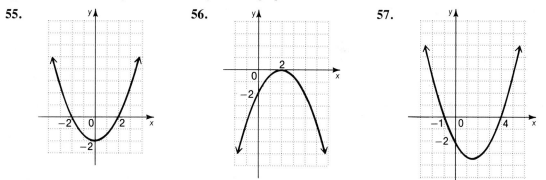

❷ *Solve each equation.*

58. $(2t - 1)(t + 5) = 0$

59. $(2p + 1)^2 = 4$

60. $(k + 2)(k - 1) = 3$

61. $6t^2 + 7t - 3 = 0$

62. $2x^2 + 3x + 2 = x^2 - 2x$

63. $x^2 + 2x + 1 = 3$

64. $(3x + 5)(3x + 5) = 7$

65. $\frac{1}{2}r^2 = \frac{7}{2} - r$

66. $\frac{2}{m} - 3 = \frac{-1}{m^2}$

67. $(2y - 3)^2 = 7$

68. $x^2 + 4x - 1 = 0$

69. $7x^2 - 8 = 5x^2 + 16$

Chapter 9 *Test*

Solve by completing the square.

1. $2z^2 - 5z = 0$

2. $5x^2 = 2 - 9x$

Solve by using the square root property.

3. $x^2 - 5 = 0$

4. $(k - 3)^2 = 49$

5. $(3r - 2)^2 = 35$

Solve by the quadratic formula.

6. $m^2 = 3m + 10$

7. $3z^2 + 2 = 7z$

8. $y^2 - \frac{5}{3}y + \frac{1}{3} = 0$

Solve by the most appropriate method.

9. $m^2 - 2m = 1$

10. $(2x - 1)^2 = 18$

11. $(x - 5)(3x + 2) = 0$

12. $(q - 5)(3q - 2) = 4$

13. $(x - 5)(x - 5) = 8$

14. $y^2 = 6y - 2$

15. $p^2 = 5p - 7$

16. $3m^2 + 1 = 2m$

Perform the indicated operations.

17. $(3 + i) + (5 - 2i) - (1 + i)$

18. $(4 - 3i)(6 + i)$

19. $(2 + 5i)(2 - 5i)$

20. $\dfrac{1 + 2i}{3 - i}$

Sketch the graph and identify the vertex of each parabola in Exercises 21–23.

21. $y = -(x + 3)^2$

22. $y = (x - 4)^2 + 1$

23. $y = x^2 + 6x + 7$

24. From your answer to Exercise 23, decide how many real number solutions the equation $x^2 + 6x + 7 = 0$ has.

Appendices

Objectives

1️⃣ Add and subtract decimals.

2️⃣ Multiply and divide decimals.

3️⃣ Convert percents to decimals and decimals to percents.

4️⃣ Find percentages by multiplication.

1️⃣   A **decimal** is a number written with a decimal point, such as 4.2. The operations on decimals—addition, subtraction, multiplication, and division—are explained in the next examples.

Example 1   Add or subtract as indicated.

(a) 6.92 + 14.8 + 3.217

Place the numbers in a column, with decimal points lined up. Then add. If you want, attach 0's to make all the numbers the same length; this is a good way to avoid errors. For example,

$$
\begin{array}{ccc}
\text{Decimal} & 6.92 & 6.92\mathbf{0} \\
\text{points} & 14.8 \quad \text{or} & 14.8\mathbf{00} \\
\text{lined up} & +\ 3.217 & +\ 3.217 \\
\hline
 & 24.937 & 24.937.
\end{array}
$$

(b) 47.6 − 32.509

Write the numbers in a column, attaching 0's to 47.6.

$$
\begin{array}{ccc}
47.6 & & 47.6\mathbf{00} \\
-\,32.509 & \text{becomes} & -\,32.509 \\
\hline
& & 15.091
\end{array}
$$

(c) 3 − .253

$$
\begin{array}{r}
3.000 \\
-\ .253 \\
\hline
2.747
\end{array}
$$ ✚

405

2 Multiplication and division of decimals is explained next.

Example 2 Multiply.

(a) 29.3×4.52

Multiply as if the numbers were whole numbers. Then, place the decimal points as shown.

$$
\begin{array}{r}
29.3 \\
\times 4.52 \\
\hline
5\ 86 \\
14\ 6\ 5 \\
117\ 2 \\
\hline
132.4\ 36
\end{array}
$$

1 decimal place
2 decimal places

3 decimal places in answer

(b) 7.003×55.8

$$
\begin{array}{r}
7.003 \\
\times\ \ 55.8 \\
\hline
5\ 602\ 4 \\
35\ 015 \\
350\ 15 \\
\hline
390.767\ 4
\end{array}
$$

3 decimal places
1 decimal place

4 decimal places in answer ✦

Example 3 Divide: $279.45 \div 24.3$.

Move the decimal point in 24.3 one place to the right, to get 243. Move the decimal point the same number of places in 279.45. By doing this, 24.3 is converted into the whole number 243.

$$24.3\ \overline{)279.4\,5}$$

Bring the decimal point straight up and divide as with whole numbers.

$$
\begin{array}{r}
11.5 \\
243\overline{)2794.5} \\
243 \\
\hline
364 \\
243 \\
\hline
121\ 5 \\
121\ 5 \\
\hline
0
\end{array}
$$ ✦

3 One of the main uses of decimals comes from percent problems. The word **percent** means "per one hundred." Percent is written with the sign %. One percent means "one per one hundred," or

Percent

$$1\% = .01 \quad \text{or} \quad 1\% = \frac{1}{100}.$$

Example 4   Convert.

(a) 75% to a decimal

Since $1\% = .01$,

$$75\% = 75 \cdot 1\% = 75 \cdot (.01) = .75.$$

The fraction form $1\% = 1/100$ can also be used to convert 75% to a decimal.

$$75\% = 75 \cdot 1\% = 75 \cdot \left(\frac{1}{100}\right) = .75$$

(b) 2.63 to a percent

$$2.63 = 263 \cdot (.01) = 263 \cdot 1\% = 263\% \quad \clubsuit$$

4   A part of a whole is called a **percentage**. For example, since 50% represents $50/100 = 1/2$ of a whole, 50% of 800 is half of 800, or 400. Percentages are found by multiplication, as in the next example.

Example 5   Find the percentages.

(a) 15% of 600

The word *of* indicates multiplication here. For this reason, 15% of 600 is found by multiplying.

$$15\% \cdot 600 = (.15) \cdot 600 = 90$$

(b) 125% of 80

$$125\% \cdot 80 = (1.25) \cdot 80 = 100 \quad \clubsuit$$

Appendix A   Exercises

Perform the indicated operations. See Examples 1–3.

1. $14.23 + 9.81 + 74.63 + 18.715$

2. $89.416 + 21.32 + 478.91 + 298.213$

3. $19.74 - 6.53$

4. $27.96 - 8.39$

5. $219 - 68.51$

6. $283 - 12.42$

7.
$$\begin{array}{r} 48.96 \\ 37.421 \\ + \ 9.72 \end{array}$$

8.
$$\begin{array}{r} 9.71 \\ 4.8 \\ 3.6 \\ 5.2 \\ + 8.17 \end{array}$$

9.
$$\begin{array}{r} 8.6 \\ -3.751 \end{array}$$

10.
$$\begin{array}{r} 27.8 \\ -13.582 \end{array}$$

11. 39.6×4.2 **12.** 18.7×2.3 **13.** 42.1×3.9 **14.** 19.63×4.08

15. $.042 \times 32$ **16.** 571×2.9 **17.** $24.84 \div 6$ **18.** $32.84 \div 4$

19. $7.6266 \div 3.42$ **20.** $14.9202 \div 2.43$ **21.** $2496 \div .52$ **22.** $.56984 \div .034$

Convert the following percents to decimals. See Example 4(a).

23. 53%   **24.** 38%   **25.** 129%   **26.** 174%

27. 96%   **28.** 11%   **29.** .9%   **30.** .1%

Convert the following decimals to percents. See Example 4(b).

31. .80   **32.** .75   **33.** .007   **34.** 1.4

35. .67   **36.** .003   **37.** .125   **38.** .983

Find each of the following. Round your answers to the nearest hundredth. See Example 5.

39. What is 14% of 780?   **40.** Find 12% of 350.

41. Find 22% of 1086.   **42.** What is 20% of 1500?

43. 4 is what percent of 80?   **44.** 1300 is what percent of 2000?

45. What percent of 5820 is 6402?   **46.** What percent of 75 is 90?

47. 121 is what percent of 484?   **48.** What percent of 3200 is 64?

49. Find 118% of 125.8.   **50.** Find 3% of 128.

51. What is 91.72% of 8546.95?   **52.** Find 12.741% of 58.902.

53. What percent of 198.72 is 14.68?   **54.** 586.3 is what percent of 765.4?

Solve the word problems.

55. A retailer has $23,000 invested in her business. She finds that she is earning 12% per year on this investment. How much money is she earning per year?

56. Harley Dabler recently bought a duplex for $144,000. He expects to earn 16% per year on the purchase price. How many dollars per year will he earn?

57. For a recent tour of the eastern United States, a travel agent figured that the trip totaled 2300 miles, with 35% of the trip by air. How many miles of the trip were by air?

58. Capitol Savings Bank pays 8.9% interest per year. What is the annual interest on an account of $3000?

59. An ad for steel-belted radial tires promises 15% better mileage when the tires are used. Alexandria's Escort now goes 420 miles on a tank of gas. If she switched to the new tires, how many extra miles could she drive on a tank of gas?

60. A home worth $77,000 is located in an area where home prices are increasing at a rate of 12% per year. By how much would the value of this home increase in one year?

61. A family of four with a monthly income of $2000 spends 90% of its earnings and saves the rest. Find the *annual* savings of this family.

62. Beth's Bargain Basement is having a sale this week. A purchase of $250 was discounted by $37.50. What percent was the discount?

63. When installing carpet, Delta Carpet Layers wasted 249 yards of a total of 4150 yards laid in April. What percent was wasted?

64. Scott must pay 6.5% sales tax on a new car. The cost of the car is $8600. Find the amount of the tax.

65. A small business takes in $274,600 per year and spends $30,755.20 for advertising. What percent of the income is spent on advertising?

66. A piece of property contains 126,000 square feet. Before the county will give a building permit, the owner of the property must donate 9.4% of the land for a park. How much land must be donated?

67. Self-employed people now must pay a Social Security tax of 11.7%. Find the tax due on earnings of $1756.

68. The Social Security tax on people who work for others is 7.15%. Find the tax on earnings of $2109.

Appendix B   Sets

Objectives

1 List the elements of a set.

2 Learn the vocabulary and symbols used to discuss sets.

3 Decide whether a given set is a subset of another set.

4 Find the union and the intersection of two sets.

1 A **set** is a collection of things. The objects in a set are called the **elements** of the set. A set is represented by listing its elements between **set braces, { }.** The order in which the elements of a set are listed is unimportant.

Example 1 Represent the following sets by listing the elements.

(a) The set of states in the United States that border on the Pacific Ocean = {California, Oregon, Washington, Hawaii, Alaska}.

(b) The set of all counting numbers less than 6 = {1, 2, 3, 4, 5}. ✤

2 Capital letters are used to name sets. To state that 5 is an element of

$$S = \{1, 2, 3, 4, 5\},$$

we write $5 \in S$. The statement $6 \notin S$ means that 6 is not an element of S.

A set with no elements is called the **empty set,** or the **null set.** The symbols $\emptyset$ or { } are used for the empty set. Let A be the set of all cats that fly, then A is the empty set.

$$A = \emptyset \qquad \text{or} \qquad A = \{ \ \}$$

The **universal set,** denoted U, includes all the elements under consideration. For example, if a discussion is about presidents of the United States, then the set of all presidents of the United States is the universal set.

Two sets are **equal** if they have exactly the same elements. Thus, the set of natural numbers and the set of positive integers are equal sets. Also, the sets

$$\{1, 2, 4, 7\} \qquad \text{and} \qquad \{4, 2, 7, 1\}$$

are equal. The order of the elements does not make a difference.

3   If all the elements of a set A are also elements of some set B, then A is a **subset** of B, written $A \subset B$. The symbol $A \not\subset B$ is used to indicate that A is not a subset of B.

Example 2   Let $A = \{1, 2, 3, 4\}$, $B = \{1, 4\}$, and $C = \{1\}$. Then $B \subset A$, $C \subset A$, and $C \subset B$, but $A \not\subset B$, $A \not\subset C$, and $B \not\subset C$. ❖

4   The **union** of two sets A and B, written $A \cup B$, is the set of all elements of A together with all elements of B. For example, if set $A = \{a, b, c\}$ and set $B = \{a, d, f, g\}$,

$$A \cup B = \{a, b, c, d, f, g\}.$$

Example 3   If $M = \{2, 5, 7\}$ and $N = \{1, 2, 3, 4, 5\}$, then $M \cup N = \{1, 2, 3, 4, 5, 7\}$. ❖

The **intersection** of two sets A and B, written $A \cap B$, is the set of all elements that belong to both A and B. For example, if

$$A = \{\text{Jose, Ellen, Marge, Kevin}\}$$

and $$B = \{\text{Jose, Patrick, Ellen, Sue}\},$$

then $$A \cap B = \{\text{Jose, Ellen}\}.$$

Example 4   Suppose that $P = \{3, 9, 27\}$, $Q = \{2, 3, 10, 18, 27, 28\}$, and $R = \{2, 10, 28\}$.

(a) $P \cap Q = \{3, 27\}$

(b) $Q \cap R = \{2, 10, 28\} = R$

(c) $P \cap R = \varnothing$ ❖

A pair of sets like P and R in Example 4 that have no elements in common are called **disjoint sets.**

Appendix B   Exercises

List the elements of each of the following sets. See Example 1.

1. The set of all natural numbers less than 8

2. The set of all integers between 4 and 10

3. The set of seasons

4. The set of months of the year

5. The set of women presidents of the United States

6. The set of all living humans more than 200 years old

Tell whether each of the following is true or false.

7. $5 \in \{1, 2, 5, 8\}$

8. $6 \in \{1, 2, 3, 4, 5\}$

9. $7 \notin \{2, 4, 6, 8\}$

10. $7 \notin \{1, 3, 5, 7\}$

11. $\{2, 4, 9, 12, 13\} = \{13, 12, 9, 4, 2\}$

12. $\{7, 11, 4\} = \{7, 11, 4, 0\}$

Let $A = \{1, 3, 4, 5, 7, 8\}$, $B = \{2, 4, 6, 8\}$, $C = \{1, 3, 5, 7\}$, $D = \{1, 2, 3\}$, $E = \{3, 7\}$, and $U = \{1, 2, 3, 4, 5, 6, 7, 8, 9, 10\}$. Tell whether each of the following is true or false. See Examples 2–4.

13. $D \subset B$

14. $E \subset C$

15. $D \not\subset E$

16. $E \not\subset A$

17. $\{3, 1, 0\} \cap \{0, 2, 4\} = \{0\}$

18. $\{3, 9, 12\} \cup \emptyset = \emptyset$

19. $\{1, 2, 3\} \cup \{1, 2, 3\} = \{1, 2, 3\}$

20. $\{3, 5, 7, 9\} \cup \{4, 6, 8\} = \emptyset$

Let $U = \{a, b, c, d, e, f, g, h\}$, $A = \{a, b, c, d, e, f\}$, $B = \{a, c, e\}$, $C = \{a, f\}$, $D = \{d\}$. List the elements in the following sets. See Examples 3 and 4.

21. $A \cap B$

22. $B \cap D$

23. $B \cap C$

24. $A \cup B$

25. $B \cup D$

26. $B \cup C$

27. $A \cap \emptyset$

28. $B \cup \emptyset$

Appendix C   The Metric System

The basic units used in the metric system are listed in the following table.

Basic Unit	Used For
meter	length
gram	weight
liter	volume

Prefixes are used with the basic units to indicate larger or smaller measures. The most common prefixes are listed below.

Prefix	Definition	Example
milli-	1/1000	1 millimeter = 1/1000 meter
centi-	1/100	1 centiliter = 1/100 liter
deci-	1/10	1 decigram = 1/10 gram
kilo-	1000	1 kilogram = 1000 grams

Eventually, people will think in the metric system as easily as they now think in the English system. To help you "think metric," get in the habit of estimating in the metric system. As an aid, use the *approximate* conversion table shown on the next page.

METRIC TO ENGLISH			ENGLISH TO METRIC		
From	to	Multiply by	From	to	Multiply by
meters	yards	1.094	yards	meters	.9144
meters	feet	3.281	feet	meters	.3048
meters	inches	39.37	inches	meters	.0254
kilometers	miles	.6214	miles	kilometers	1.6093
grams	pounds	.00220	pounds	grams	454
kilograms	pounds	2.20	pounds	kilograms	.454
liters	quarts	1.057	quarts	liters	.946
liters	gallons	.264	gallons	liters	3.785

Answers to Selected Exercises

If you need further help with algebra, you may want to get copies of both the *Study Guide* and the *Student's Solutions Manual* that accompany this textbook. The additional worked-out examples and practice problems that these books provide can help you study and understand the course material. Your college bookstore either has these books or can order them for you.

CHAPTER 1

Section 1.1 (page 6)
1. $2 \cdot 3 \cdot 5$ **3.** $2 \cdot 5 \cdot 5$ **5.** $5 \cdot 13$ **7.** $2 \cdot 2 \cdot 5 \cdot 5$ **9.** 17 **11.** $2 \cdot 2 \cdot 31$
13. 1/2 **15.** 5/6 **17.** 8/9 **19.** 2/3 **21.** 2/3 **23.** 5/6 **25.** 9/20 **27.** 3/25
29. 6/5 or 1 1/5 **31.** 3/10 **33.** 1/9 **35.** 3/20 **37.** 7/18 **39.** 13/6 or 2 1/6
41. 8 **43.** 2/3 **45.** 1/2 **47.** 4/5 **49.** 10/9 **51.** 19/22 **53.** 1/15 **55.** 8/15
57. 27/8 **59.** 49/6 **61.** 17/12 **63.** 49/30 **65.** 26/63 **67.** 7/24 **69.** 17/24 of
the debt **71.** 14 7/16 tons **73.** 8 23/24 hours **75.** 36 yards **77.** 10 chairs

Section 1.2 (page 9)
1. $<$ **3.** $<$ **5.** $>$ **7.** $<$ **9.** $<$ **11.** $>$ **13.** $\le$ **15.** $\ge$ **17.** $\le$
19. $\le$ **21.** $<, \le$ **23.** $\ge, >$ **25.** $\le, \ge$ **27.** $\ge, >$ **29.** $>, \ge$ **31.** $<, \le$
33. $<, \le$ **35.** $<, \le$ **37.** $7 = 5 + 2$ **39.** $3 < 50/5$ **41.** $12 \ne 5$ **43.** $0 \ge 0$
45. True **47.** True **49.** True **51.** True **53.** False **55.** False **57.** False
59. False **61.** False **63.** True **65.** $14 > 6$ **67.** $3 \le 15$ **69.** $8 < 9$
71. $6 \ge 0$ **73.** $15/7 \le 18/5$ **75.** $.439 \le .481$

Section 1.3 (page 14)
1. 36 **3.** 64 **5.** 289 **7.** 125 **9.** 1296 **11.** 32 **13.** 729 **15.** 1/4
17. 8/125 **19.** 64/125 **21.** .475 **23.** 3.112 **25.** 16 **27.** 2 **29.** 0
31. 60 **33.** 65 **35.** 15 **37.** 10/3 **39.** 66 **41.** $16 \le 16$; true **43.** $61 \le 60$;
false **45.** $0 \ge 0$; true **47.** $45 \ge 46$; false **49.** $66 > 72$; false **51.** $2 \ge 3$; false
53. $3 \ge 3$; true **55.** $21.912 \ge 21.92$; false **57.** $10 - (7 - 3) = 6$ **59.** No parentheses
needed, or $(3 \cdot 5) + 7 = 22$ **61.** $3 \cdot (5 - 4) = 3$ **63.** No parentheses needed, or
$(3 \cdot 5) + (2 \cdot 4) = 23$ **65.** $(3 \cdot 5 + 2) \cdot 4 = 68$ **67.** No parentheses needed, or
$(360 \div 18) \div 4 = 5$ **69.** No parentheses needed, or $(4096 \div 256) \div 4 = 4$
71. $(6 + 5) \cdot 3^2 = 99$ **73.** $(8 - 2^2) \cdot 2 = 8$ **75.** $(3 \cdot 600) - 150 = 1650$ dollars
77. $(63 - 23) + 17 = 57$ passengers

Section 1.4 (page 19)
1. (a) 12 (b) 24 **3.** (a) 15 (b) 75 **5.** (a) 14 (b) 38 **7.** (a) 4/3 (b) 16/3 **9.** (a) 2/3
(b) 4/3 **11.** (a) 30 (b) 690 **13.** (a) 19.377 (b) 96.885 **15.** (a) .8195 (b) 16.9115
17. (a) 24 (b) 33 **19.** (a) 6 (b) 9/5 **21.** (a) 23/6 (b) 4/3 **23.** (a) 150 (b) 195
25. (a) 2 (b) 17/7 **27.** (a) 5/6 (b) 16/27 **29.** (a) 10/3 (b) 13/3 **31.** (a) 14.736

(b) 8.841　　**33.** $8x$　　**35.** $\dfrac{5}{x}$　　**37.** $8 - x$　　**39.** $3x + 8$ or $8 + 3x$　　**41.** $8x + 52$ or $52 + 8x$　　**43.** Yes　　**45.** No　　**47.** Yes　　**49.** Yes　　**51.** Yes　　**53.** Yes　　**55.** No　　**57.** $x + 8 = 12$; 4　　**59.** $2x + 2 = 10$; 4　　**61.** $5 + 2x = 13$; 4　　**63.** $3x = 2 + 2\times$; 2　　**65.** $\dfrac{20}{5x} = 2$; 2　　**67.** 110/3　　**69.** 24/5

Section 1.5 (page 27)

1. -8　　**3.** 9　　**5.** 2　　**7.** -15　　**9.** -8　　**11.** -12　　**13.** -12　　**15.** -8　　**17.** 3　　**19.** $|-3|$ or 3　　**21.** $-|-6|$ or -6　　**23.** $|5 - 3|$ or 2　　**25.** True　　**27.** True　　**29.** False　　**31.** True　　**33.** True　　**35.** True　　**37.** True　　**39.** True　　**41.** False　　**43.** True　　**45.** False　　**47.** (a) 0, 3, 7　(b) -9, 0, 3, 7　(c) -9, -1 1/4, $-3/5$, 0, 3, 5, 9, 7　(d) $-\sqrt{7}$, $\sqrt{5}$　(e) All the numbers are real numbers.

49.

51.

53.

55.

Many answers are possible in Exercises 57–61. We give three possible answers for each.

57. 1/2, 5/8, 1 3/4　　**59.** -3 1/2, $-2/3$, 3/7　　**61.** $\sqrt{5}$, π, $-\sqrt{3}$　　**63.** True　　**65.** False　　**67.** False (0 is a whole number, but not positive)　　**69.** True　　**71.** False　　**73.** True: $a = 0$ or $b = 0$ or both $a = 0$ and $b = 0$; false: choose any values for a and b so that not both are zero　　**75.** True: a is equal to the opposite of b $(a = -b)$; false: a is not equal to the opposite of b $(a \neq -b)$

Section 1.6 (page 31)

1. 2　　**3.** -2　　**5.** -8　　**7.** -12　　**9.** 4　　**11.** 12　　**13.** 5　　**15.** 2　　**17.** -9　　**19.** -11　　**21.** 1/2　　**23.** $-19/24$　　**25.** $-3/4$　　**27.** -7.7　　**29.** -8　　**31.** 0　　**33.** -20　　**35.** 1.7125　　**37.** False　　**39.** True　　**41.** False　　**43.** True　　**45.** True　　**47.** False　　**49.** True　　**51.** True　　**53.** -3　　**55.** -2　　**57.** -2　　**59.** 2　　**61.** $-9 + 2 + 6 = -1$　　**63.** $12 + [-17 + (-6)] = -11$　　**65.** $[-11 + (-4)] + (-5) = -20$　　**67.** \$9　　**69.** -135 feet　　**71.** 90　　**73.** 13°　　**75.** \$943.72　　**77.** 17　　**79.** -16

Section 1.7 (page 35)

1. -3　　**3.** -4　　**5.** -8　　**7.** -14　　**9.** 9　　**11.** -4　　**13.** 4　　**15.** 3/4　　**17.** $-11/8$　　**19.** 15/8　　**21.** 11.6　　**23.** -9.9　　**25.** -1.72448　　**27.** 3.6863　　**29.** 10　　**31.** -5　　**33.** 11　　**35.** -10　　**37.** -22　　**39.** 2　　**41.** -6　　**43.** -12　　**45.** -5.90617　　**47.** -41.4616　　**49.** $12 - (-6) = 18$　　**51.** $-25 - (-4) = -21$　　**53.** $-24 - (-27) = 3$　　**55.** $-7.3 - (-8.4) = 1.1$　　**57.** $8 - (-2) = 10$　　**59.** $-11 + (-4 - 2) = -17$　　**61.** $[-12 + (-3)] - 4 = -19$　　**63.** $-15°$　　**65.** 14,776 feet　　**67.** $-172.3°$　　**69.** \$105,000　　**71.** -9　　**73.** -17

Section 1.8 (page 40)

1. 12　　**3.** -12　　**5.** 120　　**7.** 0　　**9.** -165　　**11.** 5/12　　**13.** $-1/6$　　**15.** 6　　**17.** $-.102$　　**19.** -12.96　　**21.** -11.9　　**23.** -52.752 (rounded)　　**25.** 42.100 (rounded)　　**27.** -14　　**29.** -3　　**31.** -36　　**33.** 12　　**35.** 5　　**37.** 12　　**39.** 18　　**41.** -14　　**43.** -10　　**45.** -28　　**47.** 12　　**49.** -360　　**51.** 0　　**53.** -24　　**55.** 23

57. -2 **59.** 0   **61.** -3 **63.** -2 **65.** -1 **67.** -37 **69.** 28   **71.** -9
73. -34 **75.** 42   **77.** 34   **79.** True   **81.** True

Section 1.9 (page 45)

1. 1/9   **3.** $-1/4$ **5.** None exists   **7.** $-10/9$ **9.** 1.150   **11.** -2 **13.** -6
15. Not defined   **17.** 0   **19.** 2/3   **21.** 2.1   **23.** -5 **25.** -4 **27.** -10
29. $-5/2$ **31.** -60 **33.** 2   **35.** 5   **37.** 4   **39.** -23 **41.** 52/75   **43.** 7/3
45. $-32, -16, -8, -4, -2, -1, 1, 2, 4, 8, 16, 32$ **47.** $-40, -20, -10, -8, -5, -4, -2, -1,$
$1, 2, 4, 5, 8, 10, 20, 40$ **49.** $-29, -1, 1, 29$ **51.** -2 **53.** 4   **55.** 0   **57.** -2
59. $4x = -32; -8$ **61.** $\dfrac{x}{-3} = -4; 12$ **63.** $\dfrac{x}{5} = -2; -10$ **65.** $\dfrac{x^2}{3} = 12; 6$ or -6
67. $4 + \dfrac{(-28)}{4} = -3$ **69.** $\dfrac{-16}{4} - 3 = -7$ **71.** 10   **73.** $-7/10$ **75.** 32.212
77. -1.160

Section 1.10 (page 51)

1. Associative   **3.** Commutative   **5.** Commutative   **7.** Commutative   **9.** Inverse
11. Identity   **13.** Inverse   **15.** Identity   **17.** Distributive   **19.** $k \cdot 9$ **21.** m
23. $3r + 3m$ **25.** 1   **27.** 0   **29.** $-5 + 5 = 0$ **31.** $-3r - 6$ **33.** 9
35. $5[k(-6)]$ **37.** $5m + 10$ **39.** $-4r - 8$ **41.** $-8k + 16$ **43.** $-9a - 27$
45. $4r + 32$ **47.** $-16 + 2k$ **49.** $10r + 12m$ **51.** $-12x + 16y$
53. $5(8 + 9) = 5(17) = 85$ **55.** $7(2 + 8) = 7(10) = 70$ **57.** $9(p + q)$ **59.** $5(7z + 8w)$
61. $-3k - 5$ **63.** $-4y + 8$ **45.** $4 - p$ **67.** $1 + 15r$ **69.** No   **71.** Yes
73. It is not.   **75.** True   **77.** False; let $a = 3, b = 2, c = 5$. Then,
$a + (b \cdot c) = 3 + (2 \cdot 5) = 13$ and $(a + b)(a + c) = (3 + 2)(3 + 5) = (5)(8) = 40$; $13 \neq 40$, so the
statement is false.

Chapter 1 Review Exercises (page 54)

1. 1/3   **3.** 1/5   **5.** 7/50   **7.** 7/8   **9.** 7/12   **11.** $>, \geq$ **13.** $<, \leq$ **15.** 64
17. $70 < -68$; false   **19.** $8/9 < 1$; true   **21.** $41 \leq 40$; false   **23.** 2/3   **25.** 137/3
27. $4 - x$ **29.** $|-7|$ or 7   **31.** $-|-7|$ or -7 **33.** False   **35.** True   **37.** False
39. False   **41.** $-2\frac{4}{5}$ $-1\frac{1}{8}$ $\frac{2}{3}$ $3\frac{1}{4}$ **43.** -13 **45.** 23/40   **47.** 5.6   **49.** $-\$2$

$$\begin{array}{c}\vdash\!\!+\!\!+\!\!+\!\!+\!\!+\!\!+\!\!+\!\!+\!\!+\!\!\dashv\!\!\rightarrow\\[-4pt] -4 \quad -2 \quad\ 0 \quad\ 2 \quad\ 4\end{array}$$

51. 18   **53.** 17/12   **55.** 0   **57.** 8/7   **59.** -28.25 **61.** -60 **63.** -2
65. -73 **67.** 7/4   **69.** Identity   **71.** Commutative   **73.** Inverse   **75.** $18k + 30r$
77. -15 **79.** -38 **81.** -3 **83.** 2   **85.** 65   **87.** 9   **89.** -6 **91.** 5/14
93. $\$28,000 - \$7000 = \$21,000$ **95.** $\dfrac{x + 4}{8}$

Chapter 1 Test (page 56)

1. 7/11   **2.** 203/120   **3.** 18/19   **4.** False   **5.** True   **6.** True   **7.** True
8. False   **9.** 46   **10.** 124/5   **11.** $-|-8|$ or -8 **12.** .705   **13.** $11 - 2x$
14. $\dfrac{9}{x - 8}$ **15.** Yes   **16.** No   **17.** -7 **18.** 61/20   **19.** 0   **20.** -1
21. -14 **22.** 3   **23.** 4   **24.** Not defined   **25.** -9 **26.** -4 **27.** -3
28. A, E   **29.** H   **30.** C, I   **31.** B, G   **32.** D, F   **33.** $-3 + 4m$

CHAPTER 2

Section 2.1 (page 61)

1. 15   **3.** -22 **5.** 35   **7.** -9 **9.** 1   **11.** -1 **13.** Like terms   **15.** Unlike terms   **17.** Like terms   **19.** Like terms   **21.** $17y$ **23.** $-6a$ **25.** $13b$
27. $7k + 15$ **29.** $-4y$ **31.** $2x + 6$ **33.** $14 - 7m$ **35.** $-17 + x$ **37.** $23x$
39. $9y^2$ **41.** $-14p^3 + 5p^2$ **43.** $-2.844q^2 - 5.32q - 18.7$ **45.** $30t + 66$
47. $-3n - 15$ **49.** $4r + 15$ **51.** $12k - 5$ **53.** $-2k - 3$ **55.** $-.148y - 20.7872$
57. $2 - x$ **59.** $10x$ **61.** $-32 - 47x$ **63.** $-2x + 6y - 2$ **65.** $11p^2 + 5p + 1$
67. $10r - 34$ **69.** $-z^2 - 3z + 7$ **71.** -6 **73.** 4   **75.** Add 7   **77.** Add -6

Section 2.2 (page 66)

1. 10   **3.** -2 **5.** 10   **7.** -8 **9.** -5 **11.** -2 **13.** 4   **15.** -5
17. -11 **19.** -6 **21.** 1/2   **23.** 19   **25.** -1 **27.** -8 **29.** 7   **31.** 0
33. 2   **35.** 4   **37.** 2   **39.** 34   **41.** -14 **43.** 67   **45.** -65 **47.** -5
49. -2 **51.** 7/6   **53.** -5 **55.** $3x = 17 + 2x$; 17   **57.** $5x + 3x = 7x + 9$; 9
59. $6(2x + 5) = 13x - 8$; 38   **61.** m **63.** y **65.** q **67.** r

Section 2.3 (page 71)

1. 5   **3.** -8 **5.** -7 **7.** $-8/3$ **9.** -6 **11.** 0   **13.** -6 **15.** 4
17. 4   **19.** 4   **21.** 8   **23.** 20   **25.** -3 **27.** -3 **29.** 7   **31.** 15
33. 9   **35.** 8/3   **37.** -80 **39.** 15/2   **41.** 49/2   **43.** 3   **45.** -6.1 **47.** .7
49. -1.1 **51.** $-5/3$ **53.** $-2/3$ **55.** No solution   **57.** $4x = 6$; 3/2
59. $\dfrac{x}{4} = 62$; \$248   **61.** $\dfrac{2x}{1.74} = -8.38$; -7.2906 **63.** $18q + 63$ **65.** $-20p + 10$
67. $-2 + 11r$ **69.** $-3 - 12a$

Section 2.4 (page 77)

1. 2   **3.** 24   **5.** 9   **7.** -2 **9.** -1 **11.** -4 **13.** -12.5 **15.** 2
17. -1.5 **19.** -3 **21.** -3 **23.** 8   **25.** -2 **27.** 6   **29.** -5 **31.** 0
33. 4   **35.** 0   **37.** $-1/5$ **39.** $-5/7$ **41.** 1.4   **43.** -5.2 **45.** 1.6
47. $8 + x$ **49.** $-1 + x$ **51.** $x + (-18)$ **53.** $x - 5$ **55.** $x - 9$ **57.** $9x$
59. $3x$ **61.** $\dfrac{1}{x} - x$ **63.** $\dfrac{x + 6}{-4}$ **65.** $8(x + 3)$ **67.** $7 - \dfrac{x}{6}$ **69.** $8(x - 8)$

71. $11 - q$ **73.** $x + 7$ **75.** $a + 12$ **77.** $\dfrac{t}{5}$ **79.** $n + 2$

Section 2.5 (page 82)

1. 9   **3.** -4 **5.** 0   **7.** 48 points   **9.** 36 prescriptions   **11.** 613 gallons   **13.** 10
15. 55, 57   **17.** 11, 13, 15   **19.** Rick, 30 miles; Larry, 60 miles   **21.** 45, 47   **23.** 15
25. 4   **27.** 19 inches   **29.** Bob, 7 years old; Kevin, 21 years old   **31.** width, 3 feet;
length, 9 feet   **33.** \$650   **35.** 42   **37.** 640   **39.** 60

Section 2.6 (page 89)

1. 60   **3.** 4   **5.** 8   **7.** 14   **9.** 10   **11.** 21   **13.** 1.5   **15.** 254.34
17. 113.04   **19.** 2   **21.** 3   **23.** 7   **25.** 1   **27.** $L = \dfrac{A}{W}$ **29.** $H = \dfrac{V}{LW}$

31. $r = \dfrac{C}{2\pi}$ **33.** $W = \dfrac{1}{2}(P - 2L)$ or $W = \dfrac{P - 2L}{2}$ **35.** $b = \dfrac{2A}{h} - B$ **37.** $h = \dfrac{S - 2\pi r^2}{2\pi r}$ or

$h = \dfrac{S}{2\pi r} - r$ **39.** $r^2 = \dfrac{A}{4\pi}$ **41.** $r^2 = \dfrac{3V}{\pi h}$ **43.** 20 centimeters   **45.** 10 inches

47. 20 centimeters   **49.** $x = \dfrac{6 - y}{5}$ **51.** $x = \dfrac{9 - 3y}{2}$ **53.** $x = \dfrac{5y + 15}{3}$ or $x = \dfrac{5y}{3} + 5$

55. $x = \dfrac{48 - 2y}{3}$ or $x = 16 - \dfrac{2y}{3}$ **57.** $x = \dfrac{y - b}{a}$ **59.** $x = \dfrac{c - by}{a}$ **61.** 2   **63.** 8/3

65. $-15/7$ **67.** -20

Section 2.7 (page 95)

1. 3/2   **3.** 36/55   **5.** 2/5   **7.** 5/8   **9.** 1/10   **11.** 8/5   **13.** 1/6   **15.** 4/15
17. True   **19.** True   **21.** False   **23.** False   **25.** True   **27.** False   **29.** 175
31. 4   **33.** 16   **35.** 14/3   **37.** 25/3   **39.** 35/8   **41.** 15/2   **43.** 7/15
45. 2 1/2 bars   **47.** 9 pounds   **49.** 9 inches   **51.** $67.50   **53.** 420 miles   **55.** $144
57. -1 **59.** 10   **61.** 19.2 yards   **63.** $90 for 750 miles, $144 for 1200 miles
65. 40 miles per hour   **67.** 2 1/2 years

Section 2.8 (page 101)

1. 4   **3.** 3   **5.** 2.5 centimeters   **7.** 25 fives   **9.** 18 of the 16¢ stamps   **11.** 84 fives, 42
tens   **13.** $12,000 at 16%   **15.** $21,000 at 10%, $15,000 at 14%   **17.** 2 hours   **19.** 2 1/2
hours   **21.** 160 gallons   **23.** 160 kilograms   **25.** 3 days   **27.** 25 miles per hour
29. 16 meters   **31.** 20 pounds   **33.** 10 inches   **35.** 13 1/3 liters   **37.** $<$ **39.** $<$
41. $>$ **43.** $>$

Section 2.9 (page 112)

1.   **3.**   **5.**

7.   **9.** $a < 2$ **11.** $z \geq 1$ **13.** $k \geq 5$ **15.** $x < 9$

17. $k \geq -6$ **19.** $y > 9$

21. $n \leq -11$ **23.** $z > 5$ **25.** $k \geq -5$

27. $r < 30$ **29.** $q > -\dfrac{1}{2}$ **31.** $p > -5$

33. $k \leq 0$ **35.** $r \geq -1$ **37.** $-1 \leq x \leq 6$

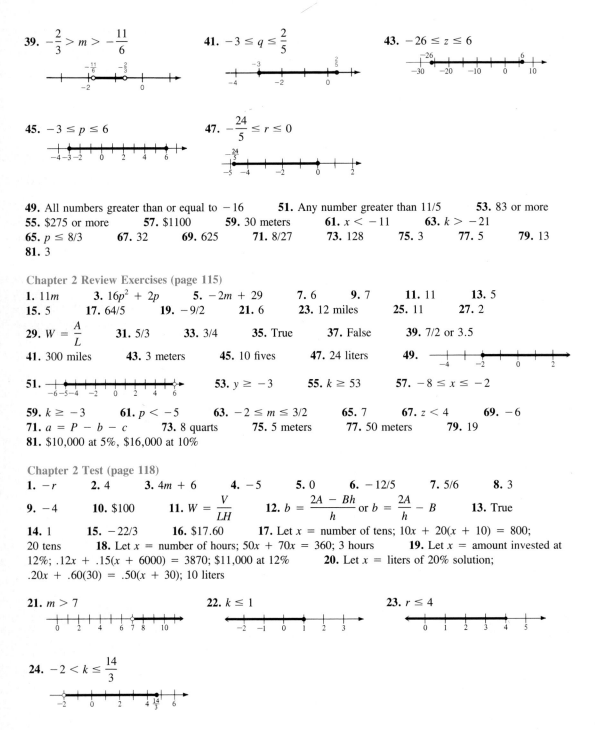

39. $-\dfrac{2}{3} > m > -\dfrac{11}{6}$

41. $-3 \le q \le \dfrac{2}{5}$

43. $-26 \le z \le 6$

45. $-3 \le p \le 6$

47. $-\dfrac{24}{5} \le r \le 0$

49. All numbers greater than or equal to -16 **51.** Any number greater than 11/5   **53.** 83 or more
55. \$275 or more   **57.** \$1100   **59.** 30 meters   **61.** $x < -11$ **63.** $k > -21$
65. $p \le 8/3$ **67.** 32   **69.** 625   **71.** 8/27   **73.** 128   **75.** 3   **77.** 5   **79.** 13
81. 3

Chapter 2 Review Exercises (page 115)
1. $11m$ **3.** $16p^2 + 2p$ **5.** $-2m + 29$ **7.** 6   **9.** 7   **11.** 11   **13.** 5
15. 5   **17.** 64/5   **19.** $-9/2$ **21.** 6   **23.** 12 miles   **25.** 11   **27.** 2
29. $W = \dfrac{A}{L}$ **31.** 5/3   **33.** 3/4   **35.** True   **37.** False   **39.** 7/2 or 3.5
41. 300 miles   **43.** 3 meters   **45.** 10 fives   **47.** 24 liters   **49.**
51.   **53.** $y \ge -3$ **55.** $k \ge 53$ **57.** $-8 \le x \le -2$
59. $k \ge -3$ **61.** $p < -5$ **63.** $-2 \le m \le 3/2$ **65.** 7   **67.** $z < 4$ **69.** -6
71. $a = P - b - c$ **73.** 8 quarts   **75.** 5 meters   **77.** 50 meters   **79.** 19
81. \$10,000 at 5%, \$16,000 at 10%

Chapter 2 Test (page 118)
1. $-r$ **2.** 4   **3.** $4m + 6$ **4.** -5 **5.** 0   **6.** $-12/5$ **7.** 5/6   **8.** 3
9. -4 **10.** \$100   **11.** $W = \dfrac{V}{LH}$ **12.** $b = \dfrac{2A - Bh}{h}$ or $b = \dfrac{2A}{h} - B$ **13.** True
14. 1   **15.** $-22/3$ **16.** \$17.60   **17.** Let $x =$ number of tens; $10x + 20(x + 10) = 800$;
20 tens   **18.** Let $x =$ number of hours; $50x + 70x = 360$; 3 hours   **19.** Let $x =$ amount invested at
12%; $.12x + .15(x + 6000) = 3870$; \$11,000 at 12%   **20.** Let $x =$ liters of 20% solution;
$.20x + .60(30) = .50(x + 30)$; 10 liters

21. $m > 7$ **22.** $k \le 1$ **23.** $r \le 4$

24. $-2 < k \le \dfrac{14}{3}$

CHAPTER 3

Section 3.1 (page 124)

1. Base 5, exponent 12   **3.** Base 2, exponent 4   **5.** Base -24, exponent 2   **7.** Base r, exponent 5   **9.** 3^5 **11.** $(-2)^5$ **13.** $1/(-2)^3$ **15.** p^5 **17.** $1/y^4$ **19.** $(-2z)^3$ **21.** 90   **23.** 80   **25.** 36   **27.** 12   **29.** 4^5 **31.** 3^{11} **33.** 4^{18} **35.** $(-3)^5$ **37.** y^{14} **39.** r^{17} **41.** $-63r^9$ **43.** $10p^{13}$ **45.** $13m^3$; $36m^6$ **47.** $-p$; $-132p^2$ **49.** $15r$; $105r^3$ **51.** $-2a^2$; $-15a^4$ **53.** 6^6 **55.** 9^6 **57.** 5^8 **59.** -4^6 **61.** 5^3m^3 or $125\ m^3$ **63.** $(-2)^4p^4q^4$ or $16p^4q^4$ **65.** $\dfrac{(-3)^2x^{10}}{4^2}$ or $\dfrac{9x^{10}}{16}$ **67.** $\dfrac{5^3a^6b^3}{c^{12}}$ or $\dfrac{125a^6b^3}{c^{12}}$ **69.** $\dfrac{4^8}{3^5}$ **71.** $\dfrac{3^5}{4^5}x^5$ or $\dfrac{3^5x^5}{4^5}$ **73.** 3^7m^7 **75.** 8^9z^9 **77.** $2^3m^7n^5$ or $8m^7n^5$ **79.** $5^7a^{11}b^{12}$ **81.** 5^{8r} **83.** 2^{8q} **85.** 2^pm^p **87.** $4^yr^{2y}5^{xy}$ **89.** $\dfrac{4^{2r}}{3^{3r}}$ **91.** $\dfrac{2^np^{mn}}{q^{rn}}$ **93.** (a) -3 (b) -28 **95.** (a) -8 (b) 32

Section 3.2 (page 130)

1. 2   **3.** 2   **5.** 1/27   **7.** $-1/12$ **9.** 32   **11.** 2   **13.** 5/6   **15.** 9/20   **17.** 1.041 (rounded)   **19.** .017 (rounded)   **21.** 4^5 **23.** $1/8^6$ **25.** $1/6^6$ **27.** 14^3 **29.** x^{15} **31.** z^2 **33.** 2^5 **35.** $2k^2$ **37.** $\dfrac{1}{4^9}$ **39.** 5^9 **41.** 2^6 **43.** 3^5 **45.** $\dfrac{5^2}{m^2}$ or $\dfrac{25}{m^2}$ **47.** $\dfrac{1}{x^3}$ **49.** $\dfrac{a^2}{3^3}$ **51.** $4^2k^4m^8$ **53.** 5^{6r} **55.** $\dfrac{1}{x^{11a}}$ **57.** a^{4y} **59.** q^{5k} **61.** $\dfrac{p^{3y}}{6^y}$ **63.** $\dfrac{m^{2q}}{3^qp^q}$ **65.** 8/3   **67.** $-5/8$ **69.** $\dfrac{a^{11}}{2b^5}$ **71.** $\dfrac{108}{y^5z^3}$ **73.** $\dfrac{9z^2}{400x^3}$ **75.** 64,270   **77.** 1230   **79.** 3.4   **81.** .237

Section 3.3 (page 134)

1. 6.835×10^9 **3.** 8.36×10^{12} **5.** 2.15×10^2 **7.** 2.5×10^4 **9.** 3.5×10^{-2} **11.** 1.01×10^{-2} **13.** 1.2×10^{-5} **15.** 8,100,000,000   **17.** 9,132,000   **19.** 324,000,000   **21.** .00032   **23.** .041   **25.** 800,000   **27.** .000004   **29.** 420   **31.** 1440   **33.** 3,000,000   **35.** .2   **37.** 1300   **39.** .000008   **41.** 4×10^{-4}, 8×10^{-4} **43.** 3.68×10^{15} **45.** 35,000   **47.** 1000, .06102   **49.** $15m$ **51.** $4p$ **53.** $7m - 5$ **55.** -13 **57.** -48

Section 3.4 (page 140)

1. $-r^5$ **3.** x^5 **5.** $-p^7$ **7.** $6y^2$ **9.** 0   **11.** $9y^4 + 7y^2$ **13.** $14z^5 - 9z^3 + 8z^2$ **15.** $1.171q^2 + 1.401q - .252$ **17.** Already simplified; degree 4; binomial   **19.** Already simplified; degree 9; trinomial   **21.** Already simplified; degree 8; trinomial   **23.** Simplifies to x^5; degree 5; monomial   **25.** Already simplified; degree 9; binomial   **27.** Sometimes   **29.** Never   **31.** Sometimes   **33.** (a) 38   (b) -1 **35.** (a) 19   (b) -2 **37.** (a) -5 (b) 1   **39.** (a) -20 (b) 7   **41.** $5m^2 + 3m$ **43.** $4x^4 - 4x^2$ **45.** $-n^5 - 12n^3 - 2$ **47.** $12m^3 + m^2 + 12m - 14$ **49.** $5a^4 - 4a^3 + 3a^2 - a + 1$ **51.** $-2r^2 + 7r - 6$ **53.** $-2x^2 - x + 1$ **55.** $4x^3 + 2x^2 + 5x$ **57.** $25y^4 + 8y^2 + 3y$ **59.** $(4 + x^2) + (-9x + 2) > 8$ **61.** $(5 + x^2) + (3 - 2x) \neq 5$ **63.** $5m^2 + 8m - 10$ **65.** $3.386m^2 - 12.273m + 4.953$ **67.** $-6x^2 + 6x - 7$ **69.** $10x^4 - 6x^2 + 10x - 1$ **71.** $2p^2$ **73.** $10x^3$ **75.** $56m^5$ **77.** $30p^9$

Section 3.5 (page 144)

1. $-32x^7$ **3.** $15y^{11}$ **5.** $30a^9$ **7.** $15pq^2$ **9.** $-18m^3n^2$ **11.** $6m^2 + 4m$
13. $-6p^4 + 12p^3$ **15.** $-16z^2 - 24z^3 - 24z^4$ **17.** $6y^3 + 4y^4 + 10y^7$
19. $28r^5 - 32r^4 + 36r^3$ **21.** $6a^4 - 12a^3b + 15a^2b^2$ **23.** $21m^5n^2 + 14m^4n^3 - 7m^3n^5$
25. $m^2 + 12m + 35$ **27.** $x^2 - 25$ **29.** $12x^2 + 10x - 12$ **31.** $9x^2 - 12x + 4$
33. $10a^2 + 37a + 7$ **35.** $12 + 8m - 15m^2$ **37.** $20 - 7x - 3x^2$ **39.** $8a^2 + 2ab - 3b^2$
41. $15x^2 + xy - 6y^2$ **43.** $12x^3 + 26x^2 + 10x + 1$ **45.** $81a^3 + 27a^2 + 11a + 2$
47. $20m^4 - m^3 - 8m^2 - 17m - 15$ **49.** $6x^6 - 3x^5 - 4x^4 + 4x^3 - 5x^2 + 8x - 3$
51. $5x^4 - 13x^3 + 20x^2 + 7x + 5$ **53.** $x^2 + 14x + 49$ **55.** $a^2 - 8a + 16$
57. $4p^2 - 20p + 25$ **59.** $25k^2 + 30kq + 9q^2$ **61.** $m^3 - 15m^2 + 75m - 125$
63. $8a^3 + 12a^2 + 6a + 1$ **65.** $56m^2 - 14m - 21$ **67.** $-9a^3 + 33a^2 + 12a$
69. $81r^4 - 216r^3s + 216r^2s^2 - 96rs^3 + 16s^4$ **71.** $6p^8 + 15p^7 + 12p^6 + 36p^5 + 15p^4$
73. $-24x^8 - 28x^7 + 32x^6 + 20x^5$ **75.** 2; 3 **77.** -7; 3 **79.** 9; -6

Section 3.6 (page 149)

1. $r^2 + 2r - 3$ **3.** $x^2 - 10x + 21$ **5.** $6x^2 + x - 2$ **7.** $6z^2 - 13z - 15$
9. $8r^2 + 2r - 3$ **11.** $20 + 9x - 20x^2$ **13.** $-12 + 5r + 2r^2$ **15.** $15 + a - 2a^2$
17. $p^2 + 4pq + 3q^2$ **19.** $10y^2 - 3yz - z^2$ **21.** $8y^2 + 31yz - 45z^2$
23. $21m^2 - 23mn - 88n^2$ **25.** $28.595m^2 + 6.306m - 4.32$ **27.** $x^2 + 16x + 64$
29. $z^2 - 10z + 25$ **31.** $9m^2 - 6mp + p^2$ **33.** $64a^2 - 48ab + 9b^2$ **35.** $81y^2 + 18yz + z^2$
37. $.4489m^2 - .2278mk + .0289k^2$ **39.** $k^2 - 25$ **41.** $16 - 9t^2$ **43.** $4b^2 - 25$
45. $36a^2 - p^2$ **47.** $.2304q^2 - .1369r^2$ **49.** $(x + 3)^2 = 5$ **51.** $(3 + x)(x - 4) > 7$
53. $6p^2 - \dfrac{17}{5}pq - \dfrac{4}{3}q^2$ **55.** $4z^2 - 10zx + \dfrac{25}{4}x^2$ **57.** $.04x^2 - 1.2xy + 9y^2$

59. $4m^2 - \dfrac{25}{9}$ **61.** $49y^4 - 100z^2$ **63.** $(a + b)^2 = 49$; $a^2 + b^2 = 29$; not equal **65.** $3y^3$

67. $5m^4$ **69.** $\dfrac{-4}{5z^2}$ **71.** $\dfrac{3s^6}{2r^4}$

Section 3.7 (page 152)

1. $2x$ **3.** $2a^2$ **5.** $\dfrac{9k^3}{m}$ **7.** $\dfrac{-3m^2}{p^2}$ **9.** $30m^3 - 10m$ **11.** $5m^4 - 8m + 4m^2$

13. $4m^4 - 2m^2 + 2m$ **15.** $m^4 - 2m + 4$ **17.** $m - 1 + \dfrac{5}{2m}$ **19.** $-5q^2 + 2q - 1$

21. $8q^4 - 7q^2 + \dfrac{4}{q}$ **23.** $-\dfrac{3q^3}{5} - q + \dfrac{6}{5}$ **25.** $-q^2 + \dfrac{3q}{5} - \dfrac{2}{5} + \dfrac{7}{5q}$ **27.** $x^2 + 3x$

29. $4x^2 + 1 - x$ **31.** $\dfrac{12}{x} + 8 + x$ **33.** $\dfrac{x}{3} + 2 - \dfrac{1}{3x}$ **35.** $4k^3 - 6k^2 - k + \dfrac{7}{2} - \dfrac{3}{2k}$

37. $-10p^3 + 5p^2 - 3p + \dfrac{3}{p}$ **39.** $4y^3 - 2 + \dfrac{3}{y}$ **41.** $\dfrac{12}{x} - \dfrac{6}{x^2} + \dfrac{14}{x^3} - \dfrac{10}{x^4}$ **43.** 8; 13; no
45. $12x^5 + 9x^4 - 12x^3 + 6x^2$ **47.** $-63y^4 - 21y^3 - 35y^2 + 14y$ **49.** $A = LW$; $2x^2 + 3x + 8$
51. $V = LWH$; $6x + 7 - \dfrac{3}{x}$ **53.** $2x^3 - 5x^2 + x$ **55.** $-20k^4 + 12k^3 - 8k^2$
57. $6m^{10} - 12m^8 + 3m^7$ **59.** $5x + 2$ **61.** $-2x^2 - 11$ **63.** $7x^2 - 10x + 1$

Section 3.8 (page 157)

1. $x + 2$ **3.** $2y - 5$ **5.** $p - 4 + \dfrac{4}{p + 6}$ **7.** $r - 5$ **9.** $6m - 1$

11. $x - 2 + \dfrac{11}{2x + 4}$ **13.** $4r + 1 + \dfrac{10}{3r - 5}$ **15.** $2k + 5 + \dfrac{10}{7k - 8}$ **17.** $x^2 - x + 2$

19. $4k^3 - k + 2$ **21.** $3y^2 - 2y + 6 + \dfrac{-5}{y + 1}$ **23.** $3k^2 + 2k - 2 + \dfrac{6}{k - 2}$

25. $2p^3 - 6p^2 + 7p - 4 + \dfrac{14}{3p + 1}$ **27.** $x^2 + 1 + \dfrac{-6x + 2}{x^2 - 2}$

29. $m^3 - 2m^2 - 7 + \dfrac{-7m - 36}{4m^2 - m - 3}$ **31.** $y^2 - y + 1$ **33.** $a^2 - 1$ **35.** $y - \dfrac{9}{2} + \dfrac{15}{2y + 4}$

37. $3w + \dfrac{11}{3} + \dfrac{\frac{52}{3}}{3w - 2}$ **39.** $2p^2 + \dfrac{4}{3}p - \dfrac{5}{3} + \dfrac{-\frac{8}{3}p^2 + \frac{19}{3}p + 2}{3p^3 + 2p}$ **41.** 1, 2, 3, 4, 6, 12

43. 1, 2, 4, 5, 10, 20   **45.** 1, 2, 5, 10, 25, 50   **47.** 1, 29   **49.** $2m^3 - 2m^2 + 2m$
51. $-3y^6 - 2y^5 + y^2$

Chapter 3 Review Exercises (page 159)

1. 80   **3.** 81/256   **5.** $a^4 b^8 c^{12}$ **7.** $-1/49$ **9.** 5^8 **11.** x^2 **13.** $r^2/6$
15. $72r^5$ **17.** 4.251×10^{-4} **19.** 7.84×10^{-1} **21.** .0004253   **23.** 800
25. 4,000,000   **27.** $p^3 - p^2 + 4p + 2$; degree 3   **29.** $-8y^5 - 7y^4 + 9y$; degree 5
31. $y^2 - 10y + 9$ **33.** $5m^3 - 6m^2 - 3$ **35.** $-22y^4 + 4y^3 + 18y^2$ **37.** $a^3 - 2a^2 - 7a + 2$
39. $5p^5 - 2p^4 - 3p^3 + 25p^2 + 15p$ **41.** $2a^2 + 5ab - 3b^2$ **43.** $a^2 + 8a + 16$
45. $36m^2 - 25$ **47.** $-\dfrac{5y^2}{3}$ **49.** $-2m^2 + m + \dfrac{6}{5}$ **51.** $2r + 7$

53. $k^2 + k + \dfrac{7}{2} + \dfrac{-\frac{9}{2}k - \frac{23}{2}}{2k^2 + k + 1}$ **55.** 2   **57.** 1/16   **59.** $4m + 3 + \dfrac{5}{3m - 5}$
61. $6k^3 - 21k - 6$ **63.** $4r^2 + 20rs + 25s^2$ **65.** $10r^2 + 21r - 10$ **67.** $10p^2 - 3p - 5$
69. 5^8

Chapter 3 Test (page 161)

1. 64/27   **2.** 1/25   **3.** 5^9 **4.** $1/8^2$ **5.** $1/(6q)$ **6.** $2^5 p^4$ or $32p^4$ **7.** 3.79×10^{-4}
8. 900   **9.** .0004   **10.** $-x^2 + 6x$; degree 2; binomial   **11.** $2m^4 + 11m^3 - 8m^2$; degree 4;
trinomial   **12.** $-13m^3 + 7m^2 - 16m + 8$ **13.** $x^5 - x^2 - 2x + 12$ **14.** $3y^2 - 2y - 2$
15. $6m^5 + 12m^4 - 18m^3 + 42m^2$ **16.** $15m^2 - 37m - 66$ **17.** $14t^2 + 27ts - 20s^2$
18. $6k^3 - 17k^2 + 10k + 16$ **19.** $4k^2 + 20km + 25m^2$ **20.** $4r^2 + 2rt + \dfrac{1}{4}t^2$

21. $36p^4 - 25r^2$ **22.** $-5y^3 - \dfrac{8y^2}{3} + 2y$ **23.** $2 + \dfrac{5}{r} - \dfrac{3}{r^2} + \dfrac{8}{5r^3}$ **24.** $3x^2 + 4x + 2$

25. $5r^2 + 2r - 5 + \dfrac{5}{2r^2 - 3}$

CHAPTER 4

Section 4.1 (page 166)

1. 12 **3.** $10p^2$ **5.** 1 **7.** $6m^2n$ **9.** $4y^3x$ **11.** x **13.** $3m^2$ **15.** $-2z^4$
17. xy^2 **19.** $3ab$ **21.** $-4m^2n$ **23.** $8(k-8)$ **25.** $7a(7a+2)$ **27.** $5y^5(13y^4-7)$
29. $11p^4(11p-3)$ **31.** No common factor other than 1 **33.** $19y^2p^2(y+2p)$
35. $13y^3(y^3+2y^2-3)$ **37.** $9qp^3(5q^3p^2-4p^3+9q)$ **39.** $(a-2)(a+5)$
41. $(7z+3m)(z+2m)$ **43.** $(m+3p)(5m-2p)$ **45.** $(a^2+b^2)(3a+2b)$
47. $ab^3(a^2b^2-ab^4+1)$ **49.** $5z^3a^3(25z^2-12za^2+17a)$ **51.** $11y^3(3y^5-4y^9+7+y)$
53. $4a^4b^2(9b+8a-12a^2b)$ **55.** $m^2+7m+12$ **57.** $r^2-4r-45$ **59.** $y^2-12y+32$
61. a^2-49

Section 4.2 (page 169)

1. $x+3$ **3.** $r+8$ **5.** $t-12$ **7.** $x-8$ **9.** $m+6$ **11.** $x-5y$
13. $(x+5)(x+1)$ **15.** $(a+4)(a+5)$ **17.** $(x-7)(x-1)$ **19.** Prime
21. $(y-4)(y-2)$ **23.** $(s-5)(s+7)$ **25.** Prime **27.** $(b-8)(b-3)$
29. $(k-5)(k-5)$ or $(k-5)^2$ **31.** $(x+3a)(x+a)$ **33.** $(y+5b)(y-6b)$
35. $(x+6y)(x-5y)$ **37.** $(r-s)(r-s)$ or $(r-s)^2$ **39.** $(p-5q)(p+2q)$
41. $3m(m+3)(m+1)$ **43.** $6(a+2)(a-10)$ **45.** $3r(r-4)(r-6)$ **47.** $3x^2(x-6)(x+5)$
49. Because $2x+4$ has a common factor of 2 **51.** y^2-y-42 **53.** $a^3(a+4b)(a-b)$
55. $yz(y+3z)(y-2z)$ **57.** $z^8(z-7y)(z+3y)$ **59.** $(a+b)(x+4)(x-3)$
61. $(2p+q)(r-9)(r-3)$ **63.** $2y^2+y-28$ **65.** $15z^2-4z-4$ **67.** $8p^2-10p-3$

Section 4.3 (page 175)

1. $x-1$ **3.** $b-3$ **5.** $4y-3$ **7.** $5x+4$ **9.** $m+10$ **11.** $3a-4b$
13. $k+3m$ **15.** $2x^2-5x-3$; $x-3$ **17.** $6m^2+7m-20$; $2m+5$ **19.** $(2x+1)(x+3)$
21. $(3a+7)(a+1)$ **23.** $(4r-3)(r+1)$ **25.** $(3m-1)(5m+2)$ **27.** $(2m-3)(4m+1)$
29. $(5a+3)(a-2)$ **31.** $(3r-5)(r+2)$ **33.** $(y+17)(4y+1)$ **35.** $(19x+2)(2x+1)$
37. $(2x+3)(5x-2)$ **39.** $(2w+5)(3w+2)$ **41.** $(2q+3)(3q+7)$
43. $(5m-4)(2m-3)$ **45.** $(4k-5)(2k+3)$ **47.** $(5m-8)(2m+3)$
49. $(4x-1)(2x-3)$ **51.** $(8m-3)(5m+2)$ **53.** $(4p-3q)(3p+4q)$
55. $(5a+2b)(5a+3b)$ **57.** $(3a-5b)(2a+b)$ **59.** $3n^2(5n-3)(n-2)$
61. $4w^4(z-1)(8z+3)$ **63.** $2k^2(2k-3w)(k+w)$ **65.** $m^4n(3m+2n)(2m+n)$
67. $3zy(3z+7y)(2z-5y)$ **69.** $(5q-2)(5q+1)(m+1)^3$ **71.** $25p^2-9q^2$ **73.** x^3-1
75. $27k^3-1$

Section 4.4 (page 181)

1. $(x+4)(x-4)$ **3.** $(a+b)(a-b)$ **3.** $(3m+1)(3m-1)$ **7.** $(5m+4)(5m-4)$
9. $4(3t+2)(3t-2)$ **11.** $(5a+4r)(5a-4r)$ **13.** Prime **15.** $(p^2+6)(p^2-6)$
17. $(a^2+1)(a+1)(a-1)$ **19.** $(m^2+9)(m+3)(m-3)$ **21.** $(a+2)^2$ **23.** $(x-5)^2$
25. $(7+a)^2$ **27.** $(k+11)^2$ **29.** Prime **31.** $(7x+2y)^2$ **33.** $(2c+3d)^2$
35. $(5h-2y)^2$ **37.** $xy(x+3y)^2$ **39.** $(2a+1)(4a^2-2a+1)$
41. $(3x-5)(9x^2+15x+25)$ **43.** $(2p+q)(4p^2-2pq+q^2)$
45. $(3a-4b)(9a^2+12ab+16b^2)$ **47.** $(4x+5y)(16x^2-20xy+25y^2)$
49. $(5m-2p)(25m^2+10mp+4p^2)$ **51.** $(10z+3x)(100z^2-30zx+9x^2)$
53. $(4y^2+1)(16y^4-4y^2+1)$ **55.** $(2k^2-3q)(4k^4+6k^2q+9q^2)$
57. $(10a-7b^3)(100a^2+70ab^3+49b^6)$ **59.** $4mn$ **61.** $-2b(3a^2+b^2)$ **63.** $4(x+1)^2$
65. $(x-1)(x+3)$ **67.** $(m-p+2)(m+p)$ **69.** 10 **71.** 9 **73.** 2 **75.** 2/3
77. $-9/7$ **79.** $-5/8$

Supplementary Factoring Exercises (page 182)

1. $(a - 6)(a + 2)$ **3.** $6(y - 2)(y + 1)$ **5.** $6(a + 2b + 3c)$ **7.** $(p - 11)(p - 6)$
9. $(5z - 6)(2z + 1)$ **11.** $(m + n + 5)(m - n)$ **13.** $8a^3(a - 3)(a + 2)$
15. $(z - 5a)(z + 2a)$ **17.** $(x - 5)(x - 4)$ **19.** $(3n - 2)(2n - 5)$ **21.** $4(4x + 5)$
23. $(3y - 4)(2y + 1)$ **25.** $(6z + 1)(z + 5)$ **27.** $(2k - 3)^2$ **29.** $6(3m + 2z)(3m - 2z)$
31. $(3k - 2)(k + 2)$ **33.** $7k(2k + 5)(k - 2)$ **35.** $(y^2 + 4)(y + 2)(y - 2)$ **37.** $8m(1 - 2m)$
39. $(z - 2)(z^2 + 2z + 4)$ **41.** Prime **43.** $8m^3(4m^6 + 2m^2 + 3)$ **45.** $(4r + 3m)^2$
47. $(5h + 7g)(3h - 2g)$ **49.** $(k - 5)(k - 6)$ **51.** $3k(k - 5)(k + 1)$
53. $(10p + 3)(100p^2 - 30p + 9)$ **55.** $(2 + m)(3 + p)$ **57.** $(4z - 1)^2$ **59.** $3(6m - 1)^2$
61. Prime **63.** $8z(4z - 1)(z + 2)$ **65.** $(4 + m)(5 + 3n)$ **67.** $2(3a - 1)(a + 2)$
69. $(a - b)(a^2 + ab + b^2 + 2)$ **71.** $(8m - 5n)^2$ **73.** $(4k - 3h)(2k + h)$
75. $(m + 2)(m^2 + m + 1)$ **77.** $(5y - 6z)(2y + z)$ **79.** $(8a - b)(a + 3b)$
81. $(3m + 8)(3m - 8)$ **83.** Prime **85.** $(a + 4)^2$

Section 4.5 (page 187)

1. $2, -4$ **3.** $-5/3, 1/2$ **5.** $-1/5, 1/2$ **7.** $-9/2, 1/3$ **9.** $1, -5/3$ **11.** $7/3, -4$
13. $-2, -3$ **15.** $-1, 6$ **17.** $-7, 4$ **19.** $-8, 3$ **21.** $-1, -2$ **23.** $1/3, -2$
25. $5/2, -2$ **27.** $1/3, -5/2$ **29.** $-4, 5/2$ **31.** $5/3, -4$ **33.** $2/5, -1/3$
35. $3/2, -3$ **37.** $5/4, -5/4$ **39.** $2, -2$ **41.** $5, 2$ **43.** $-12, 11/2$ **45.** $3, -1$
47. $1/4, 4$ **49.** $5/2, 1/3, 5$ **51.** $-7/2, 3, -1$ **53.** $0, -5, 5$ **55.** $0, 7/3, -7/3$
57. $0, 4, -2$ **59.** $0, -5, 4$ **61.** $0, 5, -3$ **63.** $-4/3, 1/2, -1$ **65.** $4, -2/3$
67. 2 or -5 **69.** 7 meters **71.** 18

Section 4.6 (page 193)

1. 6 centimeters **3.** 10 meters **5.** 2 **7.** 5 centimeters **9.** 19 feet **11.** 6 feet
13. 12 centimeters **15.** 8 feet **17.** 4 or -1 **19.** 12, 14, 16 or $-2, 0, 2$ **21.** $-9, -5$ or
6, 10 **23.** 4 inches **25.** 8 feet **27.** -2 **29.** 16 square meters **31.** 256 feet
33. 10 seconds **35.** 48 feet **37.** 48 feet **39.** $y^2 - 4xy + 4x^2$ **41.** False **43.** True
45. True

Section 4.7 (page 199)

1. $-2 < m < 5$

3. $t \le -6$ or $t \ge -5$

5. $-3 < a < 3$

7. $a \le -6$ or $a \ge 7$

9. $m < -3$ or $m > -2$

11. $-1 \le z \le 5$

13. $-1 < m < \dfrac{2}{5}$

15. $-\dfrac{1}{2} < r < \dfrac{4}{3}$

17. $1 < q < 6$

19. $m < -\dfrac{1}{2}$ or $m > \dfrac{1}{3}$

21. $-\dfrac{2}{3} < p < -\dfrac{1}{4}$

23. $m < -2$ or $m > 2$

25. $r < -4$ or $r > 4$ **27.** $-2 \le a \le \dfrac{1}{3}$ or $a \ge 4$ **29.** $r < -1$ or $2 < r < 4$

31. 2/3   **33.** 1/3   **35.** 25/36   **37.** 1/6

Chapter 4 Review Exercises (page 201)

1. $6(1 - 3r^5 + 2r^3)$ **3.** $(2p + 3)(3p + 2)$ **5.** $(r - 9)(r + 3)$ **7.** $(z - 11)(z + 4)$
9. $p^5(p - 2q)(p + q)$ **11.** $(2k - 1)(k - 2)$ **13.** $(3r + 2)(2r - 3)$
15. $(7m - 2n)(m + 3n)$ **17.** $(10a + 3)(10a - 3)$ **19.** $36(2p + q)(2p - q)$ **21.** $(4m + 5n)^2$
23. $6x(3x - 2)^2$ **25.** $(7x + 4)(49x^2 - 28x + 16)$ **27.** 4, 1   **29.** 6   **31.** 0, 3, -3
33. 4, 6 or -2, 0   **35.** $q < -5$ or $q > 3$ **37.** $2 \le m \le 3$ **39.** $p \le -4$ or $p \ge 3/2$
41. $(r - 12s)(r + 8s)$ **43.** $5m^2n^4(3m - 4n + 10mn^2)$ **45.** Prime   **47.** $3m(m - 5)(2m + 3)$
49. $(y - 5z)(y - 3z)$ **51.** $(z + 5y)(2z - y)$ **53.** $(3m + 4p)(5m - 4p)$ **55.** -2, -1, 1/3
57. 4 meters   **59.** Shorter leg, 15 meters; longer leg, 36 meters; hypotenuse, 39 meters

Chapter 4 Test (page 202)

1. $6ab(1 - 6b)$ **2.** $5kt(3k + 5t)$ **3.** $8m^2(2 - 3m + 4m^2)$ **4.** $14p(2q + 1 + 4pq^3)$
5. $(m - 9)(m - 3)$ **6.** $(2 - a)(6 + b)$ **7.** $(x - 9)(x + 5)$ **8.** $3p^2q(p + 7)(p - 1)$
9. $(3a - 2)(a + 5)$ **10.** $3z^3(5z + 1)(2z - 5)$ **11.** $p^2(4r + 5)(3r + 1)$
12. $(3t - 2x)(2t + x)$ **13.** $2(5m + 7)(5m - 7)$ **14.** $(12a + 13b)(12a - 13b)$
15. $(a^2 + 25)(a + 5)(a - 5)$ **16.** $(2p + 3)^2$ **17.** $(5z - 1)^2$ **18.** $4y(y + 2)^2$
19. $(2p - 5)(4p^2 + 10p + 25)$ **20.** $(3r + 4t^2)(9r^2 - 12rt^2 + 16t^4)$ **21.** $(m - n - 4)(m + n)$
22. 1/3, -2 **23.** 5/2, -4 **24.** 3, 5/2, $-2/3$ **25.** 0, 4, -4 **26.** 3 inches   **27.** 8

and 17 or -7 and 2   **28.** 8   **29.** $-3 \le p \le \dfrac{1}{2}$ **30.** $m < -6$ or $m > 4$

CHAPTER 5

Section 5.1 (page 208)

1. 0   **3.** -5 **5.** 5, 3   **7.** None; this rational expression is always defined   **9.** (a) 1
(b) 14/9   **11.** (a) 2   (b) $-14/3$ **13.** (a) 256/15   (b) Undefined   **15.** (a) -5 (b) 5/23

17. (a) -5 (b) 15/2   **19.** (a) Undefined   (b) 1/10   **21.** $2k$ **23.** $\dfrac{-4m}{3p}$ **25.** 3/4

27. $\dfrac{x - 1}{x + 1}$ **29.** $4m^2 - 3$ **31.** $\dfrac{8y + 5}{6}$ **33.** Cannot be written in simpler form

35. $m - n$ **37.** $\dfrac{m}{2}$ **39.** $2(2r + s)$ or $4r + 2s$ **41.** $\dfrac{m - 2}{m + 3}$ **43.** $\dfrac{x + 4}{x + 1}$ **45.** -1

47. 1   **49.** $-(x + 1)$ or $-x - 1$ **51.** -1 **53.** $\dfrac{a + 1}{b}$ **55.** $\dfrac{m + n}{2}$

57. $-\dfrac{b^2 + ab + a^2}{a + b}$ **59.** $\dfrac{z + 3}{z}$ **61.** 15/32   **63.** 32/9   **65.** 4   **67.** 81/10

Section 5.2 (page 213)

1. $\dfrac{3m}{4}$ **3.** $\dfrac{3}{32}$ **5.** $2a^4$ **7.** $\dfrac{1}{4}$ **9.** $\dfrac{1}{6}$ **11.** $\dfrac{2}{r^5}$ **13.** $\dfrac{6}{a+b}$ **15.** $\dfrac{2}{9}$

17. $\dfrac{3}{10}$ **19.** $\dfrac{2r}{3}$ **21.** $(y+4)(y-3)$ **23.** $\dfrac{18}{(m-1)(m+2)}$ **25.** $-\dfrac{7}{8}$ **27.** -1

29. $m-4$ **31.** $\dfrac{k+2}{k+3}$ **33.** $\dfrac{z+4}{z-4}$ **35.** $\dfrac{m-3}{2m-3}$ **37.** $\dfrac{m+4p}{m+p}$ **39.** $(x+2)(x+1)$

41. $\dfrac{10}{x+10}$ **43.** $\dfrac{3-a-b}{2a-b}$ or $\dfrac{a+b-3}{b-2a}$ **45.** $-\dfrac{(x+y)^2(x^2-xy+y^2)}{3y(y-x)(x-y)}$ or

$-\dfrac{(x+y)^2(x^2-xy+y^2)}{3y(x-y)^2}$ **47.** $2^2\cdot 3$ **49.** $2^3\cdot 3^2$ **51.** $x(x-4)$ **53.** $(2y-1)(y+3)$

Section 5.3 (page 217)

1. 60 **3.** 120 **5.** 1800 **7.** x^5 **9.** $30p$ **11.** $180y^4$ **13.** $15a^5b^3$
15. $12p(p-2)$ **17.** $32r^2(r-2)$ **19.** $18(r-2)$ **21.** $12p(p+5)$ **23.** $8(y+2)(y+1)$
25. $m-3$ or $3-m$ **27.** $p-q$ or $q-p$ **29.** $a(a+6)(a-3)$ **31.** $(k+7)(k-5)(k+8)$
33. $(2y-1)(y+4)(y-3)$ **35.** $(3r-5)(2r+3)(r-1)$ **37.** $\dfrac{42}{66}$ **39.** $\dfrac{-88}{8m}$ **41.** $\dfrac{24y^2}{70y^3}$

43. $\dfrac{60m^2k^3}{32k^4}$ **45.** $\dfrac{57z}{6z-18}$ **47.** $\dfrac{-4a}{18a-36}$ **49.** $\dfrac{6(k+1)}{k(k-4)(k+1)}$

51. $\dfrac{36r(r+1)}{(r-3)(r+2)(r+1)}$ **53.** $\dfrac{ab(a+2b)}{2a^3b+a^2b^2-ab^3}$ **55.** $\dfrac{(t-r)(4r-t)}{t^3-r^3}$

57. $\dfrac{2y(z-y)(y-z)}{y^4-z^3y}$ or $\dfrac{-2y(y-z)^2}{y^4-z^3y}$ **59.** $\dfrac{7}{3}$ **61.** $\dfrac{9}{10}$ **63.** $\dfrac{2}{3}$ **65.** $\dfrac{19}{72}$

Section 5.4 (page 223)

1. $\dfrac{7}{p}$ **3.** $\dfrac{-3}{k}$ or $-\dfrac{3}{k}$ **5.** 1 **7.** $\dfrac{-2q}{3}$ or $-\dfrac{2q}{3}$ **9.** m **11.** $q-2$ **13.** $\dfrac{2m+3}{6}$

15. $\dfrac{4y-3}{3y}$ **17.** $\dfrac{m}{6}$ **19.** $\dfrac{k+2}{2}$ **21.** $\dfrac{5m+11}{4m}$ **23.** $\dfrac{6-2y}{y^2}$ **25.** $\dfrac{3x-y}{x^2y}$

27. $\dfrac{4(x+6)}{(x-2)(x+2)}$ **29.** $\dfrac{x}{2(x+y)}$ **31.** $\dfrac{m}{3(m+3)(m-3)}$ **33.** $\dfrac{3}{(m+1)(m-1)(m+2)}$

35. $\dfrac{3}{q-2}$ or $\dfrac{-3}{2-q}$ **37.** $\dfrac{-6}{4p-5}$ or $\dfrac{6}{5-4p}$ **39.** $\dfrac{43m+1}{5m(m-2)}$ **41.** $\dfrac{9r+2}{r(r+2)(r-1)}$

43. $\dfrac{11y^2-y-11}{(2y-1)(y+3)(3y+2)}$ **45.** $\dfrac{2x^2+6xy+8y^2}{(x+y)(x+y)(x+3y)}$ or $\dfrac{2x^2+6xy+8y^2}{(x+y)^2(x+3y)}$

47. $\dfrac{15r^2+10ry-y^2}{(3r+2y)(6r-2y)(6r+2y)}$ **49.** $\dfrac{2p^2+2p^2q^2+4q^2-14p+6}{(1+2q^2)(1-2q^2)(p-4)}$ **51.** $\dfrac{2k^2-10k+6}{(k-3)(k-1)^2}$

53. $\dfrac{7k^2+31k+92}{(k-4)(k+4)^2}$ **55.** (a) $\dfrac{7y-2}{6}$ (b) $\dfrac{y^2-y-2}{12}$ **57.** $\dfrac{14}{3}$ **59.** $\dfrac{1}{5}$

Section 5.5 (page 228)

1. $\dfrac{31}{50}$ **3.** $\dfrac{5}{18}$ **5.** $\dfrac{1}{pq}$ **7.** $\dfrac{1}{x}$ **9.** $\dfrac{mp}{40}$ **11.** $\dfrac{x(x+1)}{4(x-3)}$ **13.** $\dfrac{18q}{1-q}$

15. $\dfrac{5(2x+3)}{4(x-1)}$ **17.** $\dfrac{2}{y}$ **19.** $\dfrac{8}{x}$ **21.** $\dfrac{x^2+1}{4+xy}$ **23.** $\dfrac{5p+15}{5+p}$ **25.** $\dfrac{40-12p}{85p}$

27. $\dfrac{5y - 2x}{3 + 4xy}$ **29.** $\dfrac{a - 2}{2a}$ **31.** $\dfrac{z - 5}{4}$ **33.** $\dfrac{-m}{m + 2}$ **35.** $\dfrac{3m(m - 3)}{(m - 1)(m - 8)}$ **37.** $\dfrac{1}{3}$

39. $\dfrac{19r}{15}$ **41.** $12x + 2$ **43.** $44p^2 - 27p$ **45.** -2 **47.** $-13/3$ **49.** $2/3$

Section 5.6 (page 234)

1. $1/2$ **3.** 12   **5.** $21/8$ **7.** $2/5$ **9.** 24   **11.** 2   **13.** -7 **15.** 1   **17.** 2
19. -8 **21.** 12   **23.** 2   **25.** -2 **27.** 5   **29.** 13   **31.** 12   **33.** 5
35. -3 **37.** 1   **39.** No solution   **41.** No solution   **43.** 3, $4/3$ **45.** 3, $-6/7$
47. $R = \dfrac{kE}{I}$ **49.** $r = \dfrac{k}{F}$ **51.** $R = \dfrac{E - Ir}{I}$ **53.** $r = \dfrac{S - a}{S}$ **55.** $y = \dfrac{xz}{x + z}$

57. $z = \dfrac{3y}{5 - 9xy}$ **59.** $r_1 = \dfrac{Rr_2}{r_2 - R}$ **61.** 3, $-1/3$ **63.** $1/2$, -6 **65** 0, -4

67. $r = \dfrac{d}{10}$ **69.** $\dfrac{1}{x}$

Supplementary Exercises on Rational Expressions (page 236)

1. $\dfrac{8}{m}$ **3.** $\dfrac{3}{4x^2(x + 3)}$ **5.** $\dfrac{r + 4}{r + 1}$ **7.** -43 **9.** $\dfrac{17}{3(y - 1)}$ **11.** 2, $1/5$ **13.** $\dfrac{-19}{18z}$

15. $\dfrac{3m + 5}{(m + 3)(m + 2)(m + 1)}$

Section 5.7 (page 242)

1. 9   **3.** $12/17$ **5.** 6, 9   **7.** $2/3$ or -3 **9.** Father's, \$120; son's, \$48   **11.** \$32,000
13. 2 miles per hour   **15.** 150 miles each way   **17.** $6/5$ hours or 1 hour, 12 minutes
19. $84/19$ hours (about 4.4 hours)   **21.** $28/5$ days or 5 $3/5$ days   **23.** $100/11$ minutes or 9.1 minutes
25. 27   **27.** 125   **29.** $4/9$ **31.** 69.08 centimeters   **33.** 20 pounds   **35.** 3 hours
37. $2/3$ or 1   **39.** $5/2$ **41.** $37/2$ miles per hour   **43.** 144 feet   **45.** (a) 1   (b) -23
47. (a) 7   (b) -17 **49.** (a) $4/7$ (b) $22/7$ **51.** (a) $-2/3$ (b) $4/3$

Chapter 5 Review Exercises (page 246)

1. 0   **3.** 4, -2 **5.** (a) $9/5$ (b) 1   **7.** (a) $24/7$ (b) 8   **9.** $3p$ **11.** $\dfrac{3x - 4}{2}$

13. $\dfrac{40}{3p^2}$ **15.** $\dfrac{7}{2}$ **17.** $\dfrac{2p - 1}{2p + 3}$ **19.** 90   **21.** $y(y + 2)(y + 4)$ **23.** $20/45$

25. $\dfrac{9m}{24m^3}$ **27.** $\dfrac{-10k}{15k + 75}$ **29.** $\dfrac{1}{r}$ **31.** $\dfrac{35 - 2k}{5k}$ **33.** $\dfrac{-y - 5}{(y + 1)(y - 1)}$

35. $\dfrac{8p - 30}{p(p - 2)(p - 3)}$ **37.** 4   **39.** $16/5$ **41.** 20   **43.** 5   **45.** No solution

47. 0, 3   **49.** $y = \dfrac{xz}{3 - 2x}$ **51.** $15/11$ **53.** 150 kilometers per hour   **55.** 2 days

57. $9/4$ **59.** $\dfrac{4p + 8q}{15}$ **61.** $\dfrac{zx + 1}{zx - 1}$ **63.** $\dfrac{16 - 3r}{2r^2}$ **65.** -4 **67.** -6

69. $6/5$ or $1/2$

Chapter 5 Test (page 248)

1. 3, 1 **2.** (a) 1/3 (b) Undefined **3.** $\dfrac{4}{3mp^3}$ **4.** $\dfrac{5y(y-1)}{2}$ **5.** a **6.** $-\dfrac{40}{27}$

7. $\dfrac{3m-2}{3m+2}$ **8.** $\dfrac{a+3}{a+4}$ **9.** $150p^5$ **10.** $(2r+3)(r+2)(r-5)$ **11.** $\dfrac{77r}{49r^2}$

12. $\dfrac{15m}{24m^2-48m}$ **13.** $-\dfrac{1}{x}$ **14.** $\dfrac{-13}{6(a+1)}$ **15.** $\dfrac{m^2-m-1}{m-3}$ or $\dfrac{1+m-m^2}{3-m}$

16. $\dfrac{-2k^2+4k+3}{(2k-1)(k+2)(k+1)}$ **17.** $\dfrac{2k}{3p}$ **18.** $\dfrac{-2p-7}{2p+9}$ **19.** 6 **20.** No solution

21. 1 **22.** 1/4 or 1/2 **23.** 20 miles per hour **24.** 20/9 hours or 2 2/9 hours **25.** 56/3

CHAPTER 6

Section 6.1 (page 255)
1. Yes **3.** Yes **5.** No **7.** Yes **9.** Yes **11.** No **13.** 11 **15.** 29
17. -4 **19.** 3 **21.** (3, 7), (0, 1), (-1, -1) **23.** (2, 2), (0, 8), (-3, 17) **25.** (0, 9),
(3, 3), (12, -15) **27.** (-4, 6), (-4, 2), (-4, -3) **29.** (-9, 8), (-9, 3), (-9, 0)
31. (-2, 6), (0, 0), (2, -6) **33.** (2, 5) **35.** (-5, 5) **37.** (7, 3)

39–49. **51.** I **53.** II **55.** III **57.** II **59.** IV **61.** None

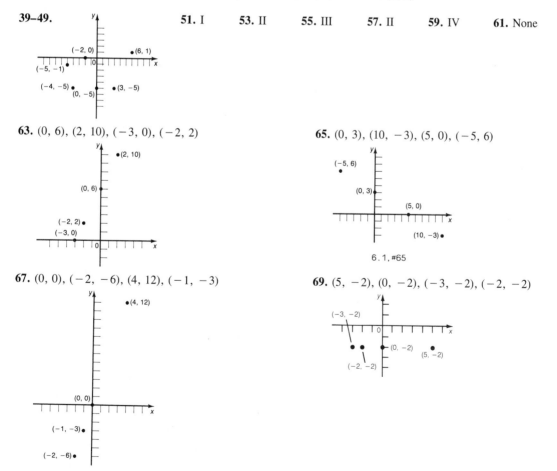

63. (0, 6), (2, 10), (-3, 0), (-2, 2) **65.** (0, 3), (10, -3), (5, 0), (-5, 6)

6.1, #65

67. (0, 0), (-2, -6), (4, 12), (-1, -3) **69.** (5, -2), (0, -2), (-3, -2), (-2, -2)

71.

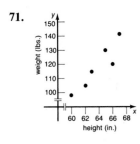

73.

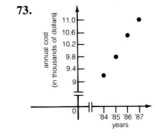

75. 8/3   **77.** −9   **79.** 1   **81.** 15

Section 6.2 (page 265)
1. (0, 5), (5, 0), (2, 3)

3. (0, 4), (−4, 0), (−2, 2)

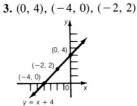

5. (0, −6), (2, 0), (3, 3)

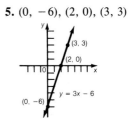

7. (0, 4), (10, 0), (5, 2)

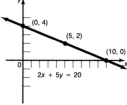

9. (−5, 2), (−5, 0), (−5, −3)

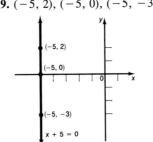

11. *x*-intercept: 3; *y*-intercept: 2   **13.** *x*-intercept: 3; *y*-intercept: −9/5   **15.** *x*-intercept: 0; *y*-intercept: 0
17. *x*-intercept: −4; *y*-intercept: none
19.

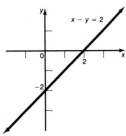

21.

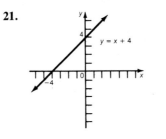

23.

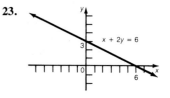
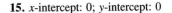

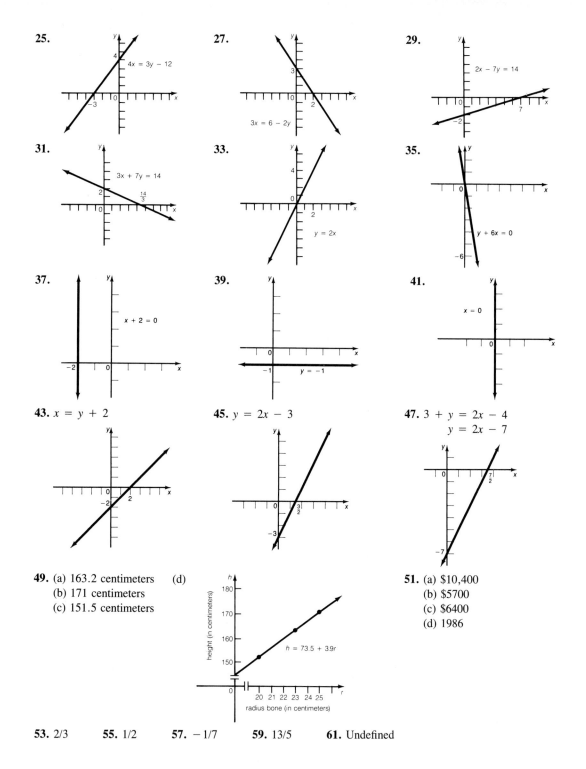

25. $4x = 3y - 12$

27. $3x = 6 - 2y$

29. $2x - 7y = 14$

31. $3x + 7y = 14$

33. $y = 2x$

35. $y + 6x = 0$

37. $x + 2 = 0$

39. $y = -1$

41. $x = 0$

43. $x = y + 2$

45. $y = 2x - 3$

47. $3 + y = 2x - 4$
$y = 2x - 7$

49. (a) 163.2 centimeters   (d)
(b) 171 centimeters
(c) 151.5 centimeters

$h = 73.5 + 3.9r$

height (in centimeters)

radius bone (in centimeters)

51. (a) \$10,400
(b) \$5700
(c) \$6400
(d) 1986

53. 2/3   **55.** 1/2   **57.** $-1/7$ **59.** 13/5   **61.** Undefined

Section 6.3 (page 273)
1. $-1/2$ **3.** $2/7$ **5.** 0 **7.** $7/6$ **9.** $9/2$ **11.** $-5/8$ **13.** 0 **15.** Undefined
17. -1.020 **19.** -1 **21.** -5 **23.** -2 **25.** $3/2$ **27.** $-2/5$ **29.** 0
31. Parallel **33.** Parallel **35.** Perpendicular **37.** Neither **39.** Neither **41.** Neither
43. $8/15$ **45.** -5 **47.** $3/18$ or $1/6$ **49.** $\dfrac{120}{140}$ or $\dfrac{6}{7}$ **51.** $y = 2x - 11$
53. $y = -x - 5$ **55.** $y = -3x - 1$ **57.** $y = -\dfrac{x}{2} - \dfrac{21}{10}$ or $y = -\dfrac{1}{2}x - \dfrac{21}{10}$

Section 6.4 (page 279)
1. $y = 3x + 5$ **3.** $y = -x - 6$ **5.** $y = \dfrac{2}{5}x - \dfrac{1}{4}$ **7.** $y = -5$ **9.** $y = 4.61x - 2.38$

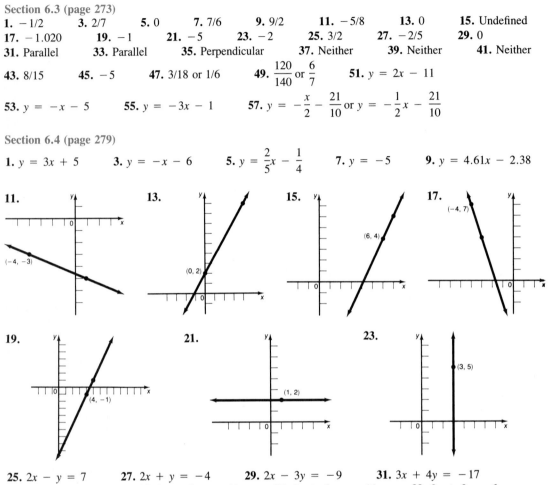

11. $(-4, -3)$ **13.** $(0, 2)$ **15.** $(6, 4)$ **17.** $(-4, 7)$
19. $(4, -1)$ **21.** $(1, 2)$ **23.** $(3, 5)$

25. $2x - y = 7$ **27.** $2x + y = -4$ **29.** $2x - 3y = -9$ **31.** $3x + 4y = -17$
33. $8x + 11y = 48$ **35.** $3x - 5y = -11$ **37.** $3x + 5y = -11$ **39.** $2x + 3y = 6$
41. $2x + y = -1$ **43.** $x - 3y = -4$ **45.** $88x - 72y = -133$ **47.** $y = 12x + 100$
49. (a) \$700 (b) \$1600 **51.** (a) $(1, 24)$; $(5, 48)$ (b) $y = 6x + 18$ **53.** $m < 7$
55. $p \le 1/2$ **57.** $p > -3$ **59.** $p < -11$

Section 6.5 (page 285)
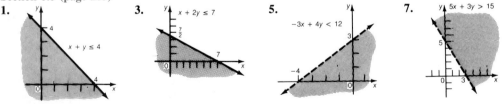

1. $x + y \le 4$ **3.** $x + 2y \le 7$ **5.** $-3x + 4y < 12$ **7.** $5x + 3y > 15$

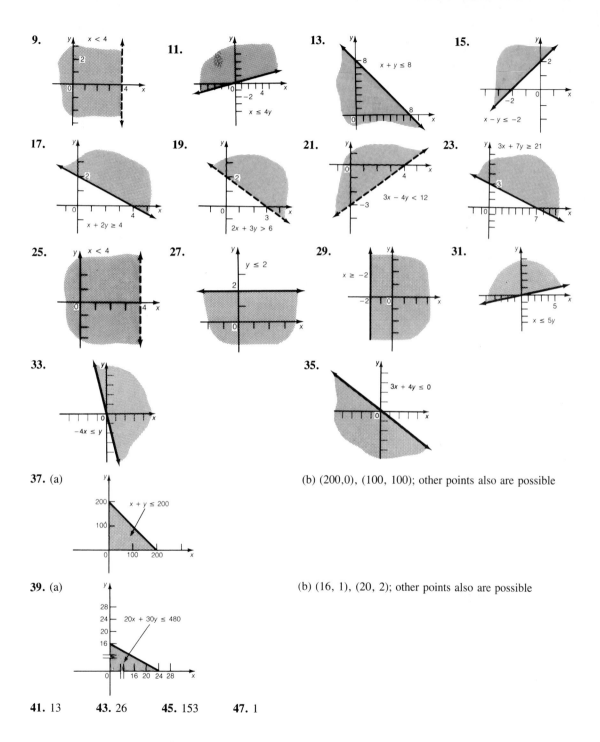

9. $x < 4$

11. $x \le 4y$

13. $x + y \le 8$

15. $x - y \le -2$

17. $x + 2y \ge 4$

19. $2x + 3y > 6$

21. $3x - 4y < 12$

23. $3x + 7y \ge 21$

25. $x < 4$

27. $y \le 2$

29. $x \ge -2$

31. $x \le 5y$

33. $-4x \le y$

35. $3x + 4y \le 0$

37. (a) $x + y \le 200$ (b) (200,0), (100, 100); other points also are possible

39. (a) $20x + 30y \le 480$ (b) (16, 1), (20, 2); other points also are possible

41. 13   **43.** 26   **45.** 153   **47.** 1

Section 6.6 (page 292)
1. No **3.** Yes **5.** No **7.** Yes **9.** No **11.** No **13.** Yes **15.** Yes
17. Yes **19.** No **21.** No **23.** Yes **25.** Both domain and range are set of all real
numbers **27.** Both domain and range are set of all real numbers **29.** Domain is set of all real
numbers; range is set of all $y \geq -3$ **31.** Domain is set of all real numbers; range is set of all $y \leq 6$
33. (a) 8 (b) 2 (c) -7 **35.** (a) 2 (b) 4 (c) 7 **37.** (a) -12 (b) -4 (c) 8
39. (a) 6 (b) 2 (c) 11 **41.** (a) 1 (b) 9 (c) 36 **43.** (a) -4 (b) -2 (c) -1
45. (a) 0 (b) 3 (c) 8 **47.** (a) 9 (b) 24 (c) -11 **49.** (a) 1 (b) -62 (c) -7 **51.** Yes
53. No **55.** Yes **57.** No **59.** $4x$ **61.** $2b$

Chapter 6 Review Exercises (page 295)
1. Yes **3.** No **5.** Yes **7.** $(-1, -5), (0, -2), (7/3, 5)$ **9.** $(-4, -3), (-4, 0), (-4, 5)$

11. **13.** **15.**

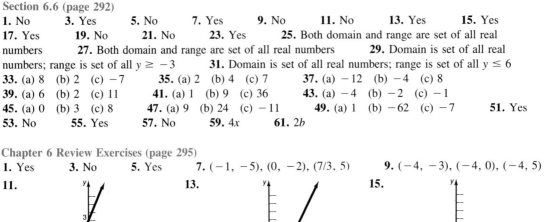

$y = 2x + 3$ $2x - y = 5$ $y + 3 = 0$

17. $x = 5/2; y = -5$ **19.** $x = 0; y = 0$ **21.** 2 **23.** 3 **25.** Undefined
27. Parallel **29.** Perpendicular **31.** $3x - y = 2$ **33.** $2x - 3y = -15$
35. $2x - 3y = -14$ **37.** $x + 4y = 6$

39. **41.**

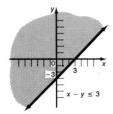

$x - y \leq 3$

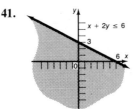

$x + 2y \leq 6$

43. No **45.** No **47.** Yes **49.** No **51.** Domain: all real numbers; range: $y \geq 1$
53. (a) 8 (b) -1 **55.** (a) 5 (b) 2 **57.** $1/3; x = 7; y = -7/3$ **59.** Yes **61.** $(0, 3),$
$(9/4, 0), (-2, 17/3)$ **63.** $x = 5$ **65.** $3x + y = 30$ **67.** Domain: all real numbers;
range: all real numbers

69. **71.**

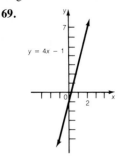

$y = 4x - 1$

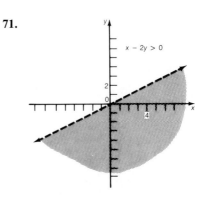

$x - 2y > 0$

Chapter 6 Test (page 297)
1. $(0, -6)$, $(-2, -16)$, $(4, 14)$ **2.** $(0, 3)$, $(21/2, 0)$, $(3, 15/7)$, $(7/2, 2)$ **3.** $(0, 0)$, $(6, 2)$, $(8, 8/3)$, $(-12, -4)$ **4.** $(5, 2)$, $(4, 2)$, $(0, 2)$, $(-3, 2)$

5. 4; 4 **6.** 3; 6 **7.** 6; $-\dfrac{9}{2}$ **8.** 0; 0

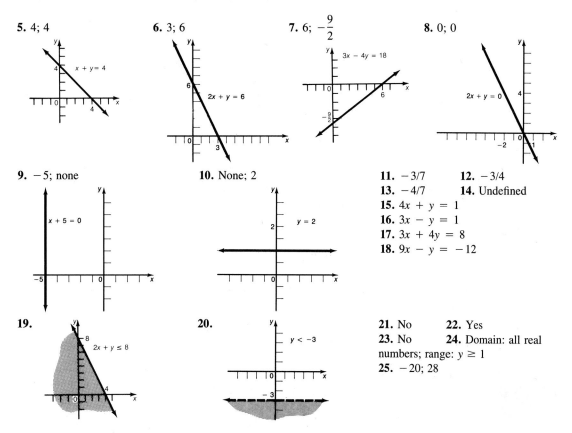

9. -5; none **10.** None; 2 **11.** $-3/7$ **12.** $-3/4$
13. $-4/7$ **14.** Undefined
15. $4x + y = 1$
16. $3x - y = 1$
17. $3x + 4y = 8$
18. $9x - y = -12$

19. **20.** **21.** No **22.** Yes
23. No **24.** Domain: all real
numbers; range: $y \geq 1$
25. -20; 28

CHAPTER 7

Section 7.1 (page 304)
1. Yes **3.** No **5.** No **7.** Yes **9.** No **11.** Yes **13.** $(4, 2)$ **15.** $(0, 4)$
17. $(2, -1)$ **19.** $(-4, 1)$ **21.** $(-4, 3)$ **23.** $(5, 0)$ **25.** $(6, -3)$ **27.** $(1, 0)$
29. $(10, 0)$ **31.** $(-1, 3)$ **33.** $(5, 6)$ **35.** No solution **37.** No solution
39. Infinite number of solutions **41.** $(0, 5)$ **43.** (a) Neither (b) Intersecting (c) One solution
45. (a) Dependent (b) One line (c) Infinite number of solutions **47.** (a) Neither (b) Intersecting
(c) One solution **49.** (a) Neither (b) Intersecting (c) One solution **51.** (a) Dependent
(b) One line (c) Infinite number of solutions **53.** $10p + 7q$ **55.** $8r$ **57.** $2a$ **59.** 0

Section 7.2 (page 312)
1. $(1, -2)$ **3.** $(2, 0)$ **5.** $(6, 2)$ **7.** $(-2, 6)$ **9.** $(1/2, 2)$ **11.** $(3/2, -2)$
13. $(3, 0)$ **15.** $(2, -3)$ **17.** $(4, 4)$ **19.** $(2, -5)$ **21.** $(4, 9)$ **23.** $(-4, 0)$
25. $(4, -3)$ **27.** $(-9, -11)$ **29.** $(6, 3)$ **31.** $(6, -5)$ **33.** $(-6, 0)$ **35.** $(1/2, 3)$

37. No solution   **39.** Infinite number of solutions   **41.** No solution   **43.** (2, −5)
45. (3/8, 5/6)   **47.** (2/3, −1/5)   **49.** (4, 9)   **51.** (−3, 2)   **53.** (0, 0)
55. (2.283, .071), rounded to nearest thousandth   **57.** (−7, −1)   **59.** (0, 0)   **61.** 4
63. 9   **65.** −7   **67.** 14/13

Section 7.3 (page 317)
1. (2, 4)   **3.** (6, 4)   **5.** (8, −1)   **7.** No solution   **9.** (2, −4)   **11.** (5, 1)
13. Infinite number of solutions   **15.** (4, 1)   **17.** (−2, 5)   **19.** (−6, 2)   **21.** (−5, 3)
23. (3, 2)   **25.** (7, 0)   **27.** Infinite number of solutions   **29.** (6, 5)   **31.** (4, −6)
33. (18, −12)   **35.** (3, 2)   **37.** (3, −1)   **39.** (12, 6)   **41.** 21   **43.** 8 feet
45. 13 twenties

Section 7.4 (page 323)
1. 43, 9   **3.** 72, 24   **5.** 55°, 35°   **7.** 22 tens, 63 twenties   **9.** 74 at 16¢, 96 at 20¢
11. 80 fives, 44 tens   **13.** 21 at $7, 18 at $4   **15.** 4 liters of 90%, 16 liters of 75%
17. 4 liters of 25%, 8 liters of 55%   **19.** 20 barrels at $60 per barrel, 30 barrels at $40 per barrel
21. boat, 10 miles per hour; current, 2 miles per hour   **23.** plane, 470 miles per hour; wind, 70 miles per
hour   **25.** $7500 at 10%, $2500 at 14%   **27.** 10 inches by 20 inches   **29.** 30 pounds of $6,
60 pounds of $3   **31.** John, 3 1/4 miles per hour; Harriet, 2 3/4 miles per hour   **33.** 4 girls, 3 boys

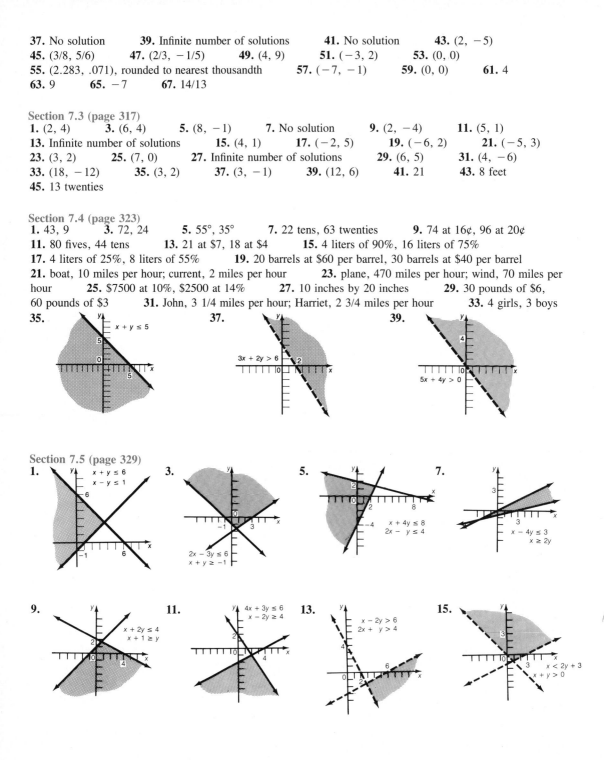

Section 7.5 (page 329)

17. 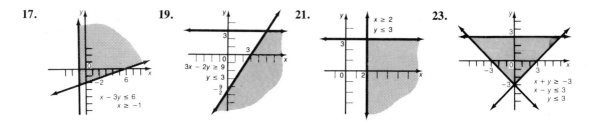 **19.**   **21.**   **23.**

25. 25   **27.** 121   **29.** 216   **31.** 16

Chapter 7 Review Exercises (page 330)

1. No   **3.** No   **5.** Yes   **7.** (2, 1)   **9.** (−1, 1)   **11.** (−1, −2)   **13.** (2, 4)
15. (3, −2)   **17.** Infinite number of solutions; dependent   **19.** (4, −1)   **21.** Infinite number of
solutions; dependent   **23.** (4, 2)   **25.** (5, 3)   **27.** (4, −1)   **29.** 8 meters by 12 meters
31. $7000 at 12%, $11,000 at 16%   **33.** 40 liters of 40%, 20 liters of 70%
35.   **37.**   **39.** (9, 2)   **41.** (−1, 2)   **43.**

45. (3, 1)   **47.** 28, 14   **49.** (− 6, 1)

Chapter 7 Test (page 332)

1. (4, −3)   **2.** (6, −5)   **3.** (4, 1)   **4.** (6, 5)   **5.** (4, 3)   **6.** Infinite number of
solutions   **7.** (3/2, −2)   **8.** No solution   **9.** (− 11/14, − 8/7)   **10.** (3, −5)
11. (−6, 8)   **12.** Infinite number of solutions   **13.** (−6, 4)   **14.** 12, 27   **15.** 2 at $5, 4 at
$7.50   **16.** 75 liters of 40%, 25 liters of 60%   **17.** 45 miles per hour, 60 miles per hour
18.   **19.**   **20.**

CHAPTER 8

Section 8.1 (page 339)

1. 3, −3   **3.** 11, −11   **5.** 20/9, −20/9   **7.** 25, −25   **9.** 39, −39   **11.** 63, −63
13. 2   **15.** 5   **17.** − 8   **19.** 13/7   **21.** − 7/3   **23.** Does not exist   **25.** Rational;
4   **27.** Irrational; 3.873   **29.** Irrational; 6.856   **31.** Irrational; 8.246   **33.** Rational; − 11
35. Irrational; 10.488   **37.** Rational; 20   **39.** Irrational; − 14.142   **41.** 23.896   **43.** 28.249

45. 1.985   **47.** .095   **49.** $c = 10$ **51.** $b = 15$ **53.** $c = \sqrt{164}$ or 12.806
55. $c = \sqrt{95}$ or 9.747   **57.** 10   **59.** 5   **61.** -2 **63.** -2 **65.** 1   **67.** Does not
exist   **69.** -3 **71.** 1   **73.** -2.378 **75.** 2.280   **77.** 1.092   **79.** 4.036
81. -4.1 **83.** 5.1   **85.** .07   **87.** .11   **89.** 8.485 inches   **91.** 1, 2, 3, 4, 6, 8, 12, 24
93. 1, 2, 3, 4, 5, 6, 10, 12, 15, 20, 25, 30, 50, 60, 75, 100, 150, 300   **95.** 1, 2, 3, 4, 6, 8, 9, 12, 18,
24, 36, 72   **97.** 1, 2, 3, 5, 6, 10, 15, 25, 30, 50, 75, 150

Section 8.2 (page 343)
1. 4   **3.** 6   **5.** 21   **7.** $\sqrt{21}$ **9.** $3\sqrt{3}$ **11.** $3\sqrt{2}$ **13.** $4\sqrt{3}$ **15.** $5\sqrt{6}$
17. $30\sqrt{3}$ **19.** $4\sqrt{5}$ **21.** 36   **23.** 60   **25.** $20\sqrt{3}$ **27.** $10\sqrt{10}$ **29.** 10/3
31. $\sqrt{30/7}$ **33.** 2/5   **35.** 4/25   **37.** 5   **39.** 4   **41.** $3\sqrt{5}$ **43.** $5\sqrt{10}$ **45.** y
47. $\sqrt{xz}$ **49.** x **51.** x^2 **53.** xy^2 **55.** $x\sqrt{x}$ **57.** $4/x$ **59.** $\sqrt{11}/r^2$
61. $2m\sqrt{7m}$ **63.** $5x\sqrt{5y}$ **65.** $2\sqrt[3]{5}$ **67.** $3\sqrt[3]{2}$ **69.** $4\sqrt[3]{2}$ **71.** $2\sqrt[4]{5}$ **73.** 2/3
75. 10/3   **77.** 2   **79.** 2   **81.** $2x\sqrt[3]{4}$ **83.** $\sqrt[3]{3m}/(2n)$ **85.** 6 centimeters
87. 27/39   **89.** 10/15   **91.** $6/(3x)$ **93.** $10/(2\sqrt{3})$

Section 8.3 (page 347)
1. $7\sqrt{3}$ **3.** $-5\sqrt{7}$ **5.** $2\sqrt{6}$ **7.** $3\sqrt{17}$ **9.** $4\sqrt{7}$ **11.** $10\sqrt{2}$ **13.** $3\sqrt{3}$
15. $20\sqrt{2}$ **17.** $19\sqrt{7}$ **19.** $12\sqrt{6} + 6\sqrt{5}$ **21.** $-2\sqrt{2} - 12\sqrt{3}$ **23.** $2\sqrt{2}$
25. $3\sqrt{3} - 2\sqrt{5}$ **27.** $3\sqrt{21}$ **29.** $5\sqrt{3}$ **31.** $-\sqrt[3]{2}$ **33.** $24\sqrt[3]{3}$ **35.** $10\sqrt[4]{2} + 4\sqrt[4]{8}$
37. $6\sqrt{x}$ **39.** $13\sqrt{a}$ **41.** $15x\sqrt{3}$ **43.** $2x\sqrt{2}$ **45.** $13p\sqrt{3}$ **47.** $42x\sqrt{5z}$
49. $6k^2h\sqrt{6} + 27hk\sqrt{6k}$ **51.** $6\sqrt[3]{p^2}$ **53.** $14m$ **55.** $36k\sqrt[3]{2k}$ **57.** $4m\sqrt[3]{4m}$
59. $27\sqrt{3}$ meters ≈ 46.765 meters   **61.** $p^2 + 5p - 14$ **63.** $20z^2 - 22z - 16$
65. $4y^2 + 20y + 25$ **67.** $4n^2 - 9$

Section 8.4 (page 350)
1. $6\sqrt{5}/5$ **3.** $\sqrt{5}$ **5.** $3\sqrt{7}/7$ **7.** $8\sqrt{15}/5$ **9.** $\sqrt{30}/2$ **11.** $8\sqrt{3}/9$ **13.** $3\sqrt{2}/10$
15. $\sqrt{2}$ **17.** $\sqrt{2}$ **19.** $2\sqrt{30}/3$ **21.** $\sqrt{2}/2$ **23.** $3\sqrt{5}/5$ **25.** $\sqrt{15}/10$
27. $3\sqrt{14}/4$ **29.** $\sqrt{3}/5$ **31.** $4\sqrt{3}/27$ **33.** $\sqrt{6p}/p$ **35.** $\sqrt{3y}/y$ **37.** $4\sqrt{m}/m$
39. $p\sqrt{3q}/q$ **41.** $x\sqrt{7xy}/y$ **43.** $p\sqrt{2pm}/m$ **45.** $x\sqrt{y}/(2y)$ **47.** $3a\sqrt{5r}/5$ **49.** $\sqrt[3]{4}/2$
51. $\sqrt[3]{2}/4$ **53.** $\sqrt[3]{121}/11$ **55.** $\sqrt[3]{50}/5$ **57.** $\sqrt[3]{6y}/(2y)$ **59.** $\sqrt[3]{42mn^2}/(6n)$
61. $9\sqrt{2}/4$ seconds   **63.** $4x$ **65.** $-2m - 3$ **67.** $3y + 8z$

Section 8.5 (page 356)
1. $27\sqrt{5}$ **3.** $21\sqrt{2}$ **5.** -4 **7.** $\sqrt{15} + \sqrt{35}$ **9.** $2\sqrt{10} + 10$ **11.** $-4\sqrt{7}$
13. $21 - \sqrt{6}$ **15.** $87 + 9\sqrt{21}$ **17.** $34 + 24\sqrt{2}$ **19.** $37 - 12\sqrt{7}$ **21.** $5 + 2\sqrt{6}$
23. 7   **25.** -4 **27.** 1   **29.** $2 + 2\sqrt{2}$ **31.** $x\sqrt{30} + \sqrt{15x} + 6\sqrt{5x} + 3\sqrt{10}$
33. $6t - 3\sqrt{14t} + 2\sqrt{7t} - 7\sqrt{2}$ **35.** $m\sqrt{15} - \sqrt{15mn} + \sqrt{10mn} - n\sqrt{10}$
37. $\sqrt{7} - 2$ or $-2 + \sqrt{7}$ **39.** $\dfrac{\sqrt{3} + 5}{4}$ **41.** $\dfrac{6 - \sqrt{10}}{2}$ **43.** $\dfrac{2 + \sqrt{2}}{3}$
45. $-10 + 5\sqrt{5}$ **47.** $-2 - \sqrt{11}$ **49.** $3 - \sqrt{3}$ **51.** $\dfrac{-3 + 5\sqrt{3}}{11}$
53. $\dfrac{\sqrt{6} + \sqrt{2} + 3\sqrt{3} + 3}{2}$ **55.** $\dfrac{-6\sqrt{2} + 12 + \sqrt{10} - 2\sqrt{5}}{2}$
57. $\dfrac{-4\sqrt{3} - \sqrt{2} + 10\sqrt{6} + 5}{23}$ **59.** $\sqrt{21} + \sqrt{14} + \sqrt{6} + 2$ **61.** $\dfrac{3\sqrt{2} + \sqrt{6}}{2}$
63. $\sqrt{3} + 1$ **65.** $\dfrac{4 + \sqrt{6}}{2}$ **67.** $2 - 3\sqrt[3]{4}$ **69.** $12 + 10\sqrt[4]{8}$ **71.** $-1 + 3\sqrt[3]{2} - \sqrt[3]{4}$
73. 1   **75.** $-6 + 2\sqrt{13}$ centimeters   **77.** 1, 3   **79.** 5/2, -3 **81.** 1, 2

Section 8.6 (page 363)

1. 4   **3.** 1   **5.** 7   **7.** −7   **9.** −5   **11.** No solution   **13.** 49   **15.** No solution   **17.** 9   **19.** 16   **21.** 6   **23.** 7   **25.** No solution   **27.** 6   **29.** 12   **31.** 5   **33.** 0, −1   **35.** 5   **37.** 3   **39.** 9   **41.** 12   **43.** 9   **45.** 21   **47.** 8   **49.** (a) 90 miles per hour (b) 120 miles per hour (c) 60 miles per hour (d) 60 miles per hour   **51.** $1/(3^5)$ **53.** $1/(2x^3)$ **55.** $1/(2^2y^8)$

Section 8.7 (page 367)

1. 4   **3.** 5   **5.** 2   **7.** 4   **9.** 2   **11.** 2   **13.** 8   **15.** 9   **17.** 8   **19.** 4   **21.** 4   **23.** −4   **25.** 2^3 **27.** $1/(6^{1/2})$ **29.** $1/(15^{1/2})$ **31.** $11^{1/7}$ **33.** 8^3 **35.** $6^{1/2}$ **37.** $3^{3/4}$ **39.** 125/8   **41.** 4/9   **43.** $1/(2^{8/5})$ **45.** $6^{2/9}$ **47.** p^3 **49.** z **51.** $m^2n^{1/6}$ **53.** $x^{3/8}y^{1/2}$ **55.** $(a^{2/3})/(b^{4/9})$ **57.** $1/(c^{4/3})$ **59.** $1/(k^{7/2})$ **61.** 2   **63.** 2   **65.** $\sqrt{a}$ **67.** $\sqrt[3]{k^2}$ **69.** (a) $122\sqrt{2}$ miles (b) $122\sqrt{3}$ miles   **71.** 5, −5   **73.** $\sqrt{14}$, $-\sqrt{14}$ **75.** $3\sqrt{2}$, $-3\sqrt{2}$ **77.** $4\sqrt{5}$, $-4\sqrt{5}$

Chapter 8 Review Exercises (page 369)

1. 7, −7   **3.** 15, −15   **5.** 4   **7.** Does not exist   **9.** 2   **11.** Irrational; 3.873   **13.** Rational; −13   **15.** $5\sqrt{3}$ **17.** 70   **19.** $\sqrt{10}/13$ **21.** p **23.** $x\sqrt{x}$ **25.** $\sqrt[3]{6}/5$ **27.** $8\sqrt{2}$ **29.** $2\sqrt{3} - 3\sqrt{10}$ **31.** $5\sqrt[3]{2}$ **33.** 0   **35.** $11k^2\sqrt{2n}$ **37.** $3\sqrt{10}/5$ **39.** $\sqrt{30}/15$ **41.** $\sqrt[3]{10}/2$ **43.** $r\sqrt{5rt}/5w$ **45.** $3\sqrt{6} + 12$ **47.** $179 + 20\sqrt{7}$ **49.** $\dfrac{-\sqrt{3} + 3}{2}$ **51.** $\dfrac{2\sqrt{3} + 2 + 3\sqrt{2} + \sqrt{6}}{2}$ **53.** 99   **55.** −2   **57.** 9   **59.** $7^{7/3}$ **61.** $x^{3/4}$ **63.** $2\sqrt[3]{2}$ **65.** $6/p$ **67.** $\sqrt{15}/5$ **69.** 5   **71.** $3m\sqrt{2pm}/2p$ **73.** $21\sqrt{3}$ **75.** −13   **77.** $4\sqrt{3}$ **79.** 2/15

Chapter 8 Test (page 371)

1. 10   **2.** 8.775   **3.** −13.784   **4.** −3   **5.** 5   **6.** $3\sqrt{3}$ **7.** $8\sqrt{2}$ **8.** $-2\sqrt[3]{4}$ **9.** Cannot be simplified   **10.** $9\sqrt{7}$ **11.** $-m\sqrt{5}$ **12.** $-5\sqrt{3x}$ **13.** $2y\sqrt[3]{4x^2}$ **14.** $2np\sqrt[4]{2m^3p^2}$ **15.** 31   **16.** $6\sqrt{2} + 2 - 3\sqrt{14} - \sqrt{7}$ **17.** $11 + 2\sqrt{30}$ **18.** $\sqrt{3}$ **19.** $4p\sqrt{k}/k$ **20.** $\sqrt[3]{15}/3$ **21.** $\dfrac{-12 - 3\sqrt{3}}{13}$ **22.** $\dfrac{3\sqrt{2} + \sqrt{14} + 3 + \sqrt{7}}{2}$ **23.** No solution   **24.** 4   **25.** 1, 25   **26.** 1   **27.** 16   **28.** $5^{5/2}$ **29.** 1/3   **30.** $m^{1/6}$

CHAPTER 9

Section 9.1 (page 375)

1. 5, −5   **3.** 8, −8   **5.** $\sqrt{13}$, $-\sqrt{13}$ **7.** No real number solution   **9.** $\sqrt{2}$, $-\sqrt{2}$ **11.** $\sqrt{5}/2$, $-\sqrt{5}/2$ **13.** $2\sqrt{6}$, $-2\sqrt{6}$ **15.** $2\sqrt{5}/5$, $-2\sqrt{5}/5$ **17.** 1.6, −1.6   **19.** 8.8, −8.8   **21.** 6, −2   **23.** $-4 + \sqrt{10}$, $-4 - \sqrt{10}$ **25.** No real number solution   **27.** $1 + 4\sqrt{2}$, $1 - 4\sqrt{2}$ **29.** 2, −1   **31.** 13/6, −3/2   **33.** $\dfrac{5 + \sqrt{30}}{2}$, $\dfrac{5 - \sqrt{30}}{2}$ **35.** $\dfrac{1 + 3\sqrt{2}}{3}$, $\dfrac{1 - 3\sqrt{2}}{3}$ **37.** $\dfrac{5 + 7\sqrt{2}}{2}$, $\dfrac{5 - 7\sqrt{2}}{2}$ **39.** $\dfrac{-4 + 2\sqrt{2}}{3}$, $\dfrac{-4 - 2\sqrt{2}}{3}$

41. $-.15, -3.09$ **43.** About 1/2 second   **45.** .1 or 10%   **47.** $\dfrac{3 + 3\sqrt{3}}{2}$ **49.** $\dfrac{18 + \sqrt{6}}{3}$

51. $(x + 4)^2$ **53.** $\left(p - \dfrac{5}{2}\right)^2$

Section 9.2 (page 381)

1. 1   **3.** 81   **5.** 81/4   **7.** 25/4   **9.** $-1, -3$ **11.** $-1 + \sqrt{6}, -1 - \sqrt{6}$

13. $-2, -4$ **15.** $3 + 2\sqrt{2}, 3 - 2\sqrt{2}$ **17.** $\dfrac{-3 + \sqrt{17}}{2}, \dfrac{-3 - \sqrt{17}}{2}$ **19.** No real number

solution   **21.** No real number solution   **23.** $3 + \sqrt{5}, 3 - \sqrt{5}$ **25.** $1, -1/3$
27. $5 + \sqrt{17}, 5 - \sqrt{17}$ **29.** $2.633, -.633$ **31.** $-.268, -3.732$ **33.** 16 feet and -8 feet;
only 16 feet is a reasonable answer   **35.** 8.703 hours (rounded to nearest thousandth)   **37.** $1 + \sqrt{3}$
39. $\dfrac{2 + \sqrt{7}}{4}$ **41.** $\dfrac{4 + 3\sqrt{3}}{2}$ **43.** $\dfrac{2 + \sqrt{5}}{4}$

Section 9.3 (page 386)

1. $3, 4, -8$ **3.** $-8, -2, -3$ **5.** $2, -3, 2$ **7.** $1, 0, -2$ **9.** $3, -8, 0$
11. $1, 1, -12$ **13.** $9, 9, -26$ **15.** -3 **17.** -2 **19.** $1, -13$ **21.** 3/2
23. $1/5, -1$ **25.** $5/2, -1$ **27.** $-1 + \sqrt{3}, -1 - \sqrt{3}$ **29.** $\dfrac{-6 + \sqrt{26}}{2}, \dfrac{-6 - \sqrt{26}}{2}$
31. $0, -1$ **33.** $0, 5/3$ **35.** $2\sqrt{5}, -2\sqrt{5}$ **37.** $4/3, -4/3$ **39.** $0, -2$
41. No real number solutions   **43.** No real number solutions   **45.** $1.618, -.618$
47. $.681, -.881$ **49.** $\dfrac{-1 + \sqrt{73}}{6}, \dfrac{-1 - \sqrt{73}}{6}$ **51.** $\dfrac{2 + \sqrt{22}}{6}, \dfrac{2 - \sqrt{22}}{6}$
53. $1 + \sqrt{2}, 1 - \sqrt{2}$ **55.** $\dfrac{1 + \sqrt{3}}{3}, \dfrac{1 - \sqrt{3}}{3}$ **57.** No real number solutions   **59.** No real
number solutions   **61.** 4 seconds   **63.** 1 hour   **65.** $4\sqrt{3}$ **67.** $2\sqrt{11}$ **69.** $12\sqrt{2}$
71. $3\sqrt{30}$

Supplementary Quadratic Equations (page 387)

1. $\dfrac{-3 + \sqrt{5}}{2}, \dfrac{-3 - \sqrt{5}}{2}$ **3.** $-1/2, 1$ **5.** $-5/4, 3/2$ **7.** $\dfrac{1 + \sqrt{10}}{2}, \dfrac{1 - \sqrt{10}}{2}$
9. $2/5, 4$ **11.** $\dfrac{2 + \sqrt{7}}{3}, \dfrac{2 - \sqrt{7}}{3}$ **13.** $\dfrac{-9 + 4\sqrt{3}}{2}, \dfrac{-9 - 4\sqrt{3}}{2}$ **15.** $-6/5, -8/3$
17. $\dfrac{-5 + \sqrt{5}}{2}, \dfrac{-5 - \sqrt{5}}{2}$ **19.** $-2/3, 2$ **21.** $\dfrac{5 + \sqrt{17}}{4}, \dfrac{5 - \sqrt{17}}{4}$
23. No real number solutions   **25.** $\dfrac{-5 + \sqrt{13}}{6}, \dfrac{-5 - \sqrt{13}}{6}$ **27.** $5, -17$ **29.** $2, -3/2$

Section 9.4 (page 393)

1. $3i$ **3.** $2i\sqrt{5}$ **5.** $6i$ **7.** $5i\sqrt{5}$ **9.** $5 + 3i$ **11.** $6 - 9i$ **13.** $6 - 7i$
15. $14 + 5i$ **17.** $7 - 22i$ **19.** 45   **21.** $-\dfrac{3}{2} - \dfrac{1}{2}i$ **23.** $-\dfrac{1}{4} - \dfrac{7}{4}i$ **25.** $-3 - 4i$
27. $-1 + 2i, -1 - 2i$ **29.** $3 + \sqrt{5}i, 3 - \sqrt{5}i$ **31.** $-\dfrac{2}{3} + \sqrt{2}i, -\dfrac{2}{3} - \sqrt{2}i$

33. $1 + i,\ 1 - i$ **35.** $-\dfrac{3}{4} + \dfrac{\sqrt{31}}{4}i,\ -\dfrac{3}{4} - \dfrac{\sqrt{31}}{4}i$ **37.** $\dfrac{3}{2} + \dfrac{\sqrt{7}}{2}i,\ \dfrac{3}{2} - \dfrac{\sqrt{7}}{2}i$

39. $\dfrac{1}{5} + \dfrac{\sqrt{14}}{5}i,\ \dfrac{1}{5} - \dfrac{\sqrt{14}}{5}i$ **41.** $-\dfrac{1}{2} + \dfrac{\sqrt{13}}{2}i,\ -\dfrac{1}{2} - \dfrac{\sqrt{13}}{2}i$ **43.** $1 + \dfrac{\sqrt{6}}{2}i,\ 1 - \dfrac{\sqrt{6}}{2}i$

45.   **47.**   **49.**

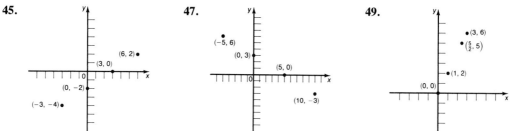

Section 9.5 (page 400)

1.   **3.**   **5.**   **7.**

9.   **11.**   **13.**   **15.**

17.   **19.**   **21.**

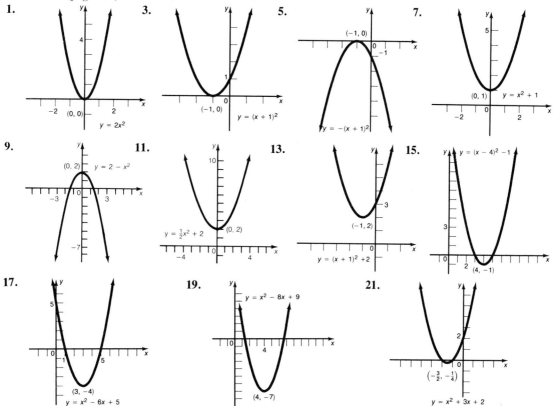

23. One solution: 2 numbers; range: $y \geq 0$ **25.** Two solutions: -2 and 2 **27.** No solution   **29.** Domain: all real numbers; range: $y \geq 1$ **31.** Domain: all real numbers; range: $y \leq 4$ **33.** Domain: all real numbers;   **35.** 3   **37.** 21

Chapter 9 Review Exercises (page 402)

1. $7, -7$ **3.** $4\sqrt{3}, -4\sqrt{3}$ **5.** $3 + \sqrt{7}, 3 - \sqrt{7}$ **7.** $\dfrac{-2 + 2\sqrt{3}}{3}, \dfrac{-2 - 2\sqrt{3}}{3}$

9. $-5, -1$ **11.** $-1 + \sqrt{6}, -1 - \sqrt{6}$ **13.** $1, -2/5$ **15.** $1 + \sqrt{5}, 1 - \sqrt{5}$ **17.** No

real number solution   **19.** $\dfrac{2 + \sqrt{10}}{2}, \dfrac{2 - \sqrt{10}}{2}$ **21.** $\dfrac{-3 + \sqrt{41}}{2}, \dfrac{-3 - \sqrt{41}}{2}$ **23.** $5 - i$

25. $-3 + 2i$ **27.** 20 **29.** 29 **31.** i **33.** $\dfrac{1}{5} - \dfrac{7}{5}i$ **35.** $-2 + \sqrt{3}i, -2 - \sqrt{3}i$

37. $\dfrac{2}{3} + \dfrac{2\sqrt{2}}{3}i, \dfrac{2}{3} - \dfrac{2\sqrt{2}}{3}i$ **39.** $\dfrac{1}{3} + \dfrac{\sqrt{2}}{3}i, \dfrac{1}{3} - \dfrac{\sqrt{2}}{3}i$ **41.** $\dfrac{3}{8} + \dfrac{\sqrt{23}}{8}i, \dfrac{3}{8} - \dfrac{\sqrt{23}}{8}i$

43.   **45.**   **47.**

49.   **51.**   **53.**

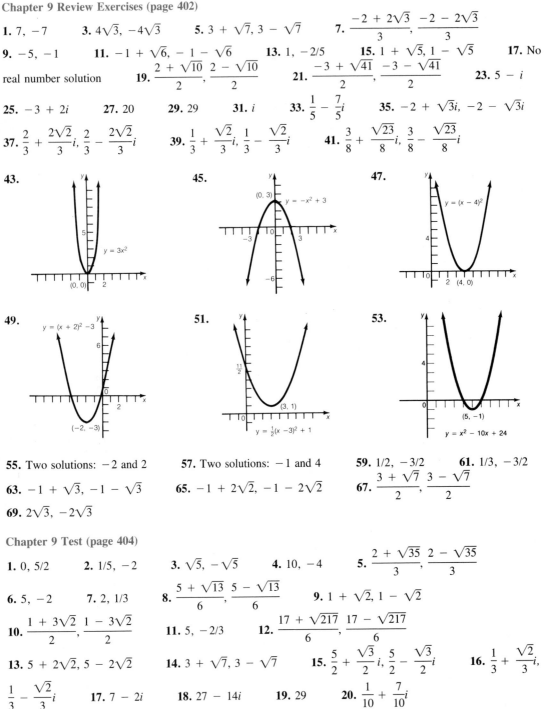

55. Two solutions: -2 and 2 **57.** Two solutions: -1 and 4 **59.** $1/2, -3/2$ **61.** $1/3, -3/2$

63. $-1 + \sqrt{3}, -1 - \sqrt{3}$ **65.** $-1 + 2\sqrt{2}, -1 - 2\sqrt{2}$ **67.** $\dfrac{3 + \sqrt{7}}{2}, \dfrac{3 - \sqrt{7}}{2}$

69. $2\sqrt{3}, -2\sqrt{3}$

Chapter 9 Test (page 404)

1. $0, 5/2$ **2.** $1/5, -2$ **3.** $\sqrt{5}, -\sqrt{5}$ **4.** $10, -4$ **5.** $\dfrac{2 + \sqrt{35}}{3}, \dfrac{2 - \sqrt{35}}{3}$

6. $5, -2$ **7.** $2, 1/3$ **8.** $\dfrac{5 + \sqrt{13}}{6}, \dfrac{5 - \sqrt{13}}{6}$ **9.** $1 + \sqrt{2}, 1 - \sqrt{2}$

10. $\dfrac{1 + 3\sqrt{2}}{2}, \dfrac{1 - 3\sqrt{2}}{2}$ **11.** $5, -2/3$ **12.** $\dfrac{17 + \sqrt{217}}{6}, \dfrac{17 - \sqrt{217}}{6}$

13. $5 + 2\sqrt{2}, 5 - 2\sqrt{2}$ **14.** $3 + \sqrt{7}, 3 - \sqrt{7}$ **15.** $\dfrac{5}{2} + \dfrac{\sqrt{3}}{2}i, \dfrac{5}{2} - \dfrac{\sqrt{3}}{2}i$ **16.** $\dfrac{1}{3} + \dfrac{\sqrt{2}}{3}i,$

$\dfrac{1}{3} - \dfrac{\sqrt{2}}{3}i$ **17.** $7 - 2i$ **18.** $27 - 14i$ **19.** 29 **20.** $\dfrac{1}{10} + \dfrac{7}{10}i$

21.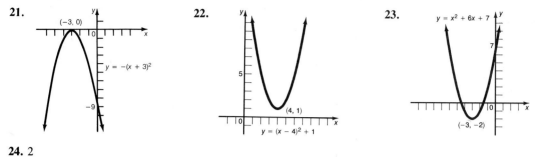

22.

23.

24. 2

APPENDICES

Appendix A (page 407)
1. 117.385   **3.** 13.21   **5.** 150.49   **7.** 96.101   **9.** 4.849   **11.** 166.32
13. 164.19   **15.** 1.344   **17.** 4.14   **19.** 2.23   **21.** 4800   **23.** .53   **25.** 1.29
27. .96   **29.** .009   **31.** 80%   **33.** .7%   **35.** 67%   **37.** 12.5%   **39.** 109.2
41. 238.92   **43.** 5%   **45.** 110%   **47.** 25%   **49.** 148.44   **51.** 7839.26
53. 7.39%   **55.** $2760   **57.** 805 miles   **59.** 63 miles   **61.** $2400   **63.** 6%
65. 11.2%   **67.** $205.45 (rounded to the nearest cent)

Appendix B (page 410)
1. {1, 2, 3, 4, 5, 6, 7}   **3.** {winter, spring, summer, fall}   **5.** $\emptyset$ **7.** True   **9.** True
11. True   **13.** False   **15.** True   **17.** True   **19.** True   **21.** {a, c, e} = B
23. {a}   **25.** {a, c, d, e}   **27.** $\emptyset$

Index

Absolute value, 26
Addition
  of complex numbers, 390
  of fractions, 4
  of polynomials, 138
  of radicals, 345
  of rational expressions, 218
  of real numbers, 28
Addition method for linear
  systems, 306
Addition property of equality,
  63
Addition property of inequality,
  106
Additive inverse, 25
Algebraic expression, 16
Associative property, 47
Axes, 253

Base, of an exponential
  expression, 11, 120
Binomial, 137
  square of, 148
Binomial products, 146
Braces, set, 18

Centimeter, 411
Coefficient, 135
  numerical, 59
Combining like terms, 60
Common denominator, 4, 214
Common factor, 3, 163
Commutative property, 47
Completing the square, 377
Complex fractions, 225
Complex numbers, 389
Conjugates, 355, 391

Consecutive integers, 81, 192
Consistent system, 302
Conversion table, metric to
  English, 412
Coordinate system, 253
Cross multiplication, 92
Cross products, 92
Cube root, 338
Cubes
  difference of, 179
  sum of, 180

Decimal, 405
Degree, of a polynomial, 137
Demand, 266
Denominator, 1
  common, 4, 214
  rationalizing, 348
Dependent equations, 302
Descending powers, 136
Difference, 5, 33
Difference of two cubes, 179
Difference of two squares, 177
Direct variation, 240
Disjoint sets, 410
Distributive property, 49
Division
  of complex numbers, 391
  of fractions, 4
  of polynomials, 151, 154
  of radicals, 342
  of rational expressions, 211
  of real numbers, 42
  by zero, 43
Domain, 19, 287
Double negative rule, 25

Elements, 409
Empty set, 409

English system, 411–12
Equal sets, 410
Equation, 18
  dependent, 302
  independent, 302
  linear, 63, 250
  quadratic, 183
  radical, 358
  with rational expressions, 229
  in two variables, 250
Equation of a line, 275
  general form, 279
  point-slope form, 277
  slope-intercept form, 276
Exponent, 11, 20
  fractional, 364
  negative, 127
  power rules, 122–23
  product rule, 121
  quotient rule, 128
  rules for, 129
  variable, 130
  zero, 126
Exponential expression, 120
Expression
  algebraic, 16
  exponential, 120
  rational, 204

Factor, 2, 44, 162
  common factor, 3, 163
  greatest common factor, 2,
   163
Factored form, 3, 167
  prime, 2, 163
Factoring, 167
  by grouping, 165
  perfect square trinomial, 178
Factoring out, 164
Factoring trinomials, 167

Table 1 Prime Factors of the Numbers 2 through 100

$2 = 2$	$26 = 2 \cdot 13$	$51 = 3 \cdot 17$	$76 = 2^2 \cdot 19$
$3 = 3$	$27 = 3^3$	$52 = 2^2 \cdot 13$	$77 = 7 \cdot 11$
$4 = 2^2$	$28 = 2^2 \cdot 7$	$53 = 53$	$78 = 2 \cdot 3 \cdot 13$
$5 = 5$	$29 = 29$	$54 = 2 \cdot 3^3$	$79 = 79$
$6 = 2 \cdot 3$	$30 = 2 \cdot 3 \cdot 5$	$55 = 5 \cdot 11$	$80 = 2^4 \cdot 5$
$7 = 7$	$31 = 31$	$56 = 2^3 \cdot 7$	$81 = 3^4$
$8 = 2^3$	$32 = 2^5$	$57 = 3 \cdot 19$	$82 = 2 \cdot 41$
$9 = 3^2$	$33 = 3 \cdot 11$	$58 = 2 \cdot 29$	$83 = 83$
$10 = 2 \cdot 5$	$34 = 2 \cdot 17$	$59 = 59$	$84 = 2^2 \cdot 3 \cdot 7$
$11 = 11$	$35 = 5 \cdot 7$	$60 = 2^2 \cdot 3 \cdot 5$	$85 = 5 \cdot 17$
$12 = 2^2 \cdot 3$	$36 = 2^2 \cdot 3^2$	$61 = 61$	$86 = 2 \cdot 43$
$13 = 13$	$37 = 37$	$62 = 2 \cdot 31$	$87 = 3 \cdot 29$
$14 = 2 \cdot 7$	$38 = 2 \cdot 19$	$63 = 3^2 \cdot 7$	$88 = 2^3 \cdot 11$
$15 = 3 \cdot 5$	$29 = 3 \cdot 13$	$64 = 2^6$	$89 = 89$
$16 = 2^4$	$40 = 2^3 \cdot 5$	$65 = 5 \cdot 13$	$90 = 2 \cdot 3^2 \cdot 5$
$17 = 17$	$41 = 41$	$66 = 2 \cdot 3 \cdot 11$	$91 = 7 \cdot 13$
$18 = 2 \cdot 3^2$	$42 = 2 \cdot 3 \cdot 7$	$67 = 67$	$92 = 2^2 \cdot 23$
$19 = 19$	$43 = 43$	$68 = 2^2 \cdot 17$	$93 = 3 \cdot 31$
$20 = 2^2 \cdot 5$	$44 = 2^2 \cdot 11$	$69 = 3 \cdot 23$	$94 = 2 \cdot 47$
$21 = 3 \cdot 7$	$45 = 3^2 \cdot 5$	$70 = 2 \cdot 5 \cdot 7$	$95 = 5 \cdot 19$
$22 = 2 \cdot 11$	$46 = 2 \cdot 23$	$71 = 71$	$96 = 2^5 \cdot 3$
$23 = 23$	$47 = 47$	$72 = 2^3 \cdot 3^2$	$97 = 97$
$24 = 2^3 \cdot 3$	$48 = 2^4 \cdot 3$	$73 = 73$	$98 = 2 \cdot 7^2$
$25 = 5^2$	$49 = 7^2$	$74 = 2 \cdot 37$	$99 = 3^2 \cdot 11$
	$50 = 2 \cdot 5^2$	$75 = 3 \cdot 5^2$	$100 = 2^2 \cdot 5^2$

Table 2 Selected Powers of Numbers

n	n^2	n^3	n^4	n^5	n^6
2	4	8	16	32	64
3	9	27	81	243	729
4	16	64	256	1024	4096
5	25	125	625	3125	
6	36	216	1296		
7	49	343	2401		
8	64	512	4096		
9	81	729	6561		
10	100	1000	10,000		